GLOBAL CLIMATE GOVERNANCE BEYOND 2012

Architecture, Agency and Adaptation

This book provides a cutting-edge assessment of policy options for future global climate governance, written by a team of 30 leading experts from the European Union and developing countries.

Global climate governance is at a crossroads. The 1997 Kyoto Protocol was merely a first step, and its core commitments expire in 2012. This book addresses three questions that will stand at the centre of any new climate agreement:

- What is the most effective overall legal and institutional architecture for successful and equitable climate politics?
- What role should non-state actors play, including multinational corporations, non-governmental organizations, public–private partnerships and market mechanisms in general?
- How can we deal with the growing challenge of the necessity to adapt our existing institutions to a substantially warmer world?

The 19 chapters integrate a variety of approaches that range from quantitative research and formal modelling to qualitative and legal analysis.

The book will be attractive both to practitioners and academics from a variety of backgrounds. Practitioners active in mitigation and adaptation policy development, analysis and negotiation will refer to this book for in-depth qualitative and quantitative assessments of the costs and benefits of a range of novel policy options. Academics from a number of disciplines – including international relations, international law, environmental studies, economics, geography and development studies, and ranging from final-year undergraduates to researchers – will benefit from description of innovative approaches of their disciplines towards international climate negotiations, and will also learn about the possibilities and advantages of integrating insights across disciplines.

Three other books arise from the ADAM Project, all published by Cambridge University Press, and, together with this volume, derive from research funded by DG-RTD as part of the Sixth Framework Programme of the European Commission:

Making Climate Change Work for Us: European Perspectives on Adaptation and Mitigation Strategies
Edited by Mike Hulme and Henry Neufeldt

Climate Change Policy in the European Union: Confronting the Dilemmas of Mitigation and Adaptation?
Edited by Andrew Jordan, Dave Huitema, Harro van Asselt, Tim Rayner and Frans Berkhout

Mainstreaming Climate Change in Development Cooperation: Theory, Practice and Implications for the European Union
Edited by Joyeeta Gupta and Nicolien van der Grijp

Frank Biermann is professor and head of the Department of Environmental Policy Analysis in the Institute for Environmental Studies at the Vrije Universiteit Amsterdam, the Netherlands. He is also general director of the Netherlands Research School for the Socio-Economic and Natural Sciences of the Environment (SENSE); director of the Global Governance Project, a network of 12 European research institutions (glogov.org); and chair of the Earth System Governance Project, a ten-year global research programme under the auspices of the International Human Dimensions Programme on Global Environmental Change. His most recent publications are *Managers of Global Change: The Influence of International Environmental Bureaucracies* (edited with B. Siebenhüner, MIT Press, 2009) and *International Organizations in Global Environmental Governance* (edited with B. Siebenhüner and A. Schreyögg, Routledge, 2009).

Philipp Pattberg is an assistant professor of international relations, Department of Environmental Policy Analysis, Institute for Environmental Studies, Vrije Universiteit Amsterdam, the Netherlands. He is also the management committee chair of the European COST Action 'The Transformation of Global Environmental Governance: Risks and Opportunities' and deputy director and research coordinator of the Global Governance Project, a network of 12 leading European institutions in the field of global environmental governance. In addition, Philipp Pattberg is the academic director of a series of PhD training schools on the Human Dimensions of Global Environmental Change, supported by the EU Marie Curie Programme. His most recent publication is *Private Institutions and Global Governance: The New Politics of Environmental Sustainability* (Edward Elgar, 2007).

Fariborz Zelli is a research fellow at the German Development Institute in Bonn, Germany. He is also a visiting fellow at the Tyndall Centre for Climate Change Research, United Kingdom, where he was a senior research associate from 2006 to early 2009. Since 2004, he has been a research fellow of the Global Governance Project, where he co-coordinates the research group Multiple Options, Solutions and Approaches: Institutional Interplay and Conflict (MOSAIC).

GLOBAL CLIMATE GOVERNANCE BEYOND 2012

Architecture, Agency and Adaptation

Edited by

FRANK BIERMANN
PHILIPP PATTBERG
and
FARIBORZ ZELLI

CAMBRIDGE UNIVERSITY PRESS
Cambridge, New York, Melbourne, Madrid, Cape Town, Singapore,
São Paulo, Delhi, Dubai, Tokyo, Mexico City

Cambridge University Press
The Edinburgh Building, Cambridge CB2 8RU, UK

Published in the United States of America by Cambridge University Press, New York

www.cambridge.org
Information on this title: www.cambridge.org/9780521180924

First published 2010
First paperback edition 2010

A catalogue record for this publication is available from the British Library

[Insert Library of Congress data if available from input material]

ISBN 978-0-521-19011-4 Hardback
ISBN 978-0-521-18092-4 Paperback

Contents

Contributors

Mozaharul Alam is a research fellow with the Bangladesh Centre for Advanced Studies. He also holds an international fellowship as part of the Capacity Strengthening of the Least Developed Countries in Adaptation to Climate Change programme. He specializes in impacts, adaptation and vulnerability assessment to climate change, integrated environmental assessment and natural resource management. He has been a contributing author to the Fourth Assessment Report of the Intergovernmental Panel on Climate Change. He holds an MSc degree in geography from Jahangirnagar University, Bangladesh.

Knut H. Alfsen is research director at the Centre for International Climate and Environmental Research (CICERO), Oslo, Norway. Previous positions include research director in the Research Department of Statistics Norway with a focus on natural resources and environmental economics, research director at the Institute for Energy Research (a technological research institute) and director at CICERO. He holds a PhD in theoretical physics from the University of Oslo, but has worked as an environmental economist for several decades.

Harro van Asselt is a researcher with the Department of Environmental Policy Analysis at the Institute for Environmental Studies (IVM) of the Vrije Universiteit Amsterdam, the Netherlands. He is also a research fellow with the EU-based Global Governance Project (glogov.org). His work focuses on international climate policy and law, trade and environment, environmental law, international environmental governance, and European environmental policy. He is working on his doctoral thesis on institutional interactions in global climate governance. He was a visiting researcher at the Department of Value and Decision Science at the Tokyo Institute of Technology, Japan (2007) and at the Dean Rusk Centre of the University of Georgia School of Law, United States (2008). He holds an LLM (International Law) degree from Vrije Universiteit Amsterdam.

Jessica Ayers is a PhD researcher in the Development Studies Institute, London School of Economics and Political Science, and a research consultant for the International Institute for Environment and Development, both in the United Kingdom. Her work focuses on the relationship between community-based adaptation and wider climate change and development governance structures at the national and international levels, as well as on exploring the interface between science and policy in environmental governance in developing countries more broadly. She is also an independent consultant for development, environment and climate change organizations in the area of climate change adaptation, and has published numerous articles and chapters in peer-reviewed journals, books and popular publications on these issues.

Lavinia Baumstark is a PhD student at the Potsdam Institute for Climate Impact Research, Germany, where she is part of the Research Domain 'Sustainable Solutions'. She works on the hybrid model REMIND-R which has been developed at the Potsdam Institute. Her research interests are endogenous growth, trade and technological spillover and the analysis of the impact of climate policy on these economic variables. She holds a degree in mathematics from the University of Potsdam.

Frank Biermann is professor and head of the Department of Environmental Policy Analysis of the Institute for Environmental Studies at the Vrije Universiteit Amsterdam, the Netherlands. He is also general director of the Netherlands Research School for the Socio-Economic and Natural Sciences of the Environment (SENSE); director of the EU-based Global Governance Project; and chair of the Earth System Governance Project, a ten-year global research programme under the auspices of the International Human Dimensions Programme on Global Environmental Change.

Ingrid Boas is a political scientist with the Department of Environmental Policy Analysis of the Institute for Environmental Studies at the Vrije Universiteit Amsterdam, the Netherlands. She is also the coordinator of the Climate Refugee Policy Forum of the EU-based Global Governance Project, and coordinator of the Amsterdam-based research activities of the Earth System Governance Project.

Kelly de Bruin is a PhD researcher with the Social Sciences Group at Wageningen University, the Netherlands. She is completing her doctoral dissertation on the trade-offs between adaptation and mitigation, coalition formation and stability in international climate policy. She is involved in several projects on the modelling of adaptation in climate change integrated assessment models.

Rob Dellink is an economist with the Organization for Economic Co-operation and Development in Paris. He is also an assistant professor at the Environmental

Economics and Natural Resources Group at Wageningen University and Research Centre, the Netherlands. He is an expert in the field of environmental economic modelling and involved in several projects on the modelling of climate change. He published a thesis on dynamic computable general equilibrium modelling for the Netherlands with special attention to pollution and abatement. His current research interests include trade-offs between adaptation and mitigation, coalition formation and stability in international climate policy, economic consequences of the European Water Framework Directive, and sustainable national income.

Ottmar Edenhofer is professor of the economics of climate change at the Technical University Berlin, Germany, and co-chair of Working Group III of the Intergovernmental Panel on Climate Change. He is also deputy director and chief economist at the Potsdam Institute for Climate Impact Research, Germany, where he leads the research domain 'Sustainable Solutions', which focuses on the economics of atmospheric stabilization.

Michel den Elzen is a senior climate policy analyst with the Netherlands Environmental Assessment Agency (PBL). His main research area is integrated environmental assessment models and climate policy, including climate attribution, post-2012 regimes for differentiation of future commitments, development of interactive policy-decision support tools (FAIR model), long-term stabilization scenarios, mitigation and adaptation costs, carbon markets and Kyoto mechanisms. He has been a contributing author for the 1990, 2001 and 2007 Working Group III reports of the Intergovernmental Panel on Climate Change. He has published about 50 articles in peer-reviewed journals and more than ten book chapters in the field of global change, as well as many policy reports. He holds a degree in mathematics from the Vrije Universiteit Amsterdam; his PhD research focuses on the modelling of global environmental change and the integrated environmental assessment model IMAGE.

Gunnar S. Eskeland is climate professor at the Norges Handelshøyskole, Bergen, Norway. His research focuses on international themes attached to economic politics and sustainable development. Eskeland has co-authored the World Bank *World Development Report 2003*. His other research areas are technology development and climate politics, and political questions connected to the collaboration between nations. Eskeland worked previously as a senior researcher at the Centre for International Climate and Environmental Research (CICERO), Oslo, and has additional work experience from the World Bank, SNF, Klosters and Statoil. He is a civil economist with a doctorate degree from the Norges Handelshøyskole, Bergen, Norway.

Christian Flachsland is a PhD student with the research domain 'Sustainable Solutions' at the Potsdam Institute for Climate Impact Research, Germany. He studied sociology, economics and philosophy at the universities of Frankfurt am Main, Stockholm and Potsdam. His work focuses on international emissions trading and linking of regional cap-and-trade systems.

Nitu Goel is a researcher with The Energy and Resources Institute (TERI) in New Delhi, India. She specializes in development studies, climate change policy, low carbon growth, technology transfer and renewable energy. She has authored or co-authored chapters and articles on climate policy and actively participated in international policy circles, among others as expert advisor to the Climate Change Working Group of a consortium led by McKinsey. She holds a MBA degree from the Indian Institute of Forest Management, Bhopal, India.

Andries Hof is a policy researcher with the Netherlands Environmental Assessment Agency (PBL). His expertise is in integrated assessment models, climate policy, discounting, and evaluating the costs and benefits of climate policy mitigation and adaptation. He holds a master's degree in environmental economics from the University of Groningen, the Netherlands.

Saleemul Huq is a senior fellow of the International Institute for Environment and Development in London, United Kingdom, where he was previously head of climate change. His interests are in the interlinkages between climate change (mitigation and adaptation) and sustainable development from the perspective of the least developed countries. He has published numerous scientific and popular books and articles on these issues, and served as lead author of the Intergovernmental Panel on Climate Change's Third Assessment Report (chapter on adaptation and sustainable development) and Fourth Assessment Report (chapter on adaptation and mitigation).

Anne Jerneck is a senior lecturer in economic history at Lund University Centre for Sustainability Studies (LUCSUS), Sweden. Her research interests include the role of the state in Vietnam's transition from plan to market, gender inequalities, (un)sustainable livelihoods in coastal Vietnam and subsistence farming in Kenya.

Marian Leimbach is a climate economist and senior researcher with the Potsdam Institute for Climate Impact Research, Germany. He has been involved in the development and application of the integrated assessment model REMIND-R. His research focuses on macroeconomic modelling, endogenous economic growth and international trade.

Kristin Linnerud is a researcher with the Centre for International Climate and Environmental Research (CICERO), Oslo, Norway, where she is leading a major research project on climate change and the power sector, funded by the Research Council of Norway. She has a varied work experience, including membership in the board of directors in the industry and the energy sector. She holds a PhD in economics from the Norwegian School of Economics and Business Administration and an MSc in finance from the London School of Economics.

Eva Lövbrand is an associate professor at the Centre for Climate Science and Policy Research at Linköping University in Sweden. Her research revolves around the role of science and expertise in global environmental politics and the marketization of climate governance. Her work has been published in journals such as *Review of International Studies*, *Global Environmental Politics*, *Environmental Science and Policy*, *Global Environmental Change* and *Climatic Change*.

Paul Lucas is an environmental scientist and integrated assessment modeller with the Climate Change and Sustainable Development team at the Netherlands Environmental Assessment Agency (PBL). He has worked for several years as climate policy analyst, studying post-2012 climate regimes, climate mitigation strategies and mitigation costs estimates. Currently, his work addresses the Millennium Development Goals in a long-term perspective and the role of the environment in sustainable development.

Robert Marschinski is a researcher at the Potsdam Institute for Climate Impact Research (PIK), Germany, where he specializes in economic and policy issues related to climate change. He holds an MSc degree in physics from the University of Bologna, Italy, and is finalizing a doctoral dissertation on climate change economics for submission to the Technical University of Berlin. He has been a visiting researcher at The Energy and Resource Institute (TERI) in New Delhi, and a short-term consultant with the Development Economics Research Group of the World Bank.

Lennart Olsson is professor of physical geography and founding director of the Lund University Centre for Sustainability Studies (LUCSUS), Sweden. His research focus is human–nature interaction in the context of land degradation, climate change and food security. He has held research positions in Australia, the United States and Hong Kong. He has contributed as lead author in the Assessment Reports of the Intergovernmental Panel on Climate Change and the UNEP Global Environmental Outlook.

Philipp Pattberg is an assistant professor of international relations, Department of Environmental Policy Analysis, Institute for Environmental Studies, Vrije Universiteit Amsterdam, the Netherlands. He is also the deputy director and research coordinator of the Global Governance Project, a network of 12 leading European institutions in the field of global environmental governance.

Manish Kumar Shrivastava is a research associate with The Energy and Resources Institute (TERI), New Delhi, India. He specializes in technological change, evolution of markets, institutional economics, co-evolutionary economics, science and law, and ethics and climate change. He has published on the relocation of communities from protected areas in India, education policy in India, gender and trade, technology transfer, and the Clean Development Mechanism. He holds an MPhil in science policy from Jawaharlal Nehru University, New Delhi, India.

Johannes Stripple is an assistant professor at the Department of Political Science, Lund University, Sweden. His research interests lie at the intersection of international relations theory and global environmental politics, particularly climate change. His work has appeared in *Global Governance*, *Review of International Studies*, *Global Environmental Change* and *International Environmental Agreements*, among others.

Jasper van Vliet is a climate policy analyst with the Netherlands Environmental Assessment Agency (PBL), specializing on the integrated assessment framework IMAGE, using the FAIR-SiMCaP model to investigate emission pathways to reach long-term climate targets and to analyse post-2012 climate policy strategies, and the TIMER model to analyse the implications of these strategies for global energy use. He has been involved in several projects including the evaluation of post-2012 climate regimes, the EMF-22 study, the OECD Environmental Outlook and the assessment of global bio-energy potentials.

Detlef van Vuuren is a senior researcher at the Netherlands Environmental Assessment Agency (PBL). His work concentrates on the integrated assessment of global environmental change and more specifically on long-term projection of climate change. He was involved as Coordinating Lead Author and Lead Author in several international assessments including the Millennium Ecosystem Assessment and the Fourth Assessment Report of the Intergovernmental Panel on Climate Change. He has published over 50 articles in peer-reviewed journals. He is also involved in activities of the Stanford University-based Energy Modelling Forum.

Harald Winkler is an associate professor at the Energy Research Centre at the University of Cape Town, South Africa, where he also heads the Energy, Environment and Climate Change programme. His research interests focus on energy and environment, in particular climate change and the economics of mitigation. Recent work has addressed the post-2012 climate policy at international, national and local levels. He was a lead author for the Intergovernmental Panel on Climate Change's Working Group III on mitigation and sustainable development, and is a member of the South African delegation to negotiations under the United Nations Framework Convention on Climate Change.

Fariborz Zelli has been a research fellow at the German Development Institute in Bonn, Germany, since February 2009. He is also a visiting fellow at the Tyndall

Centre for Climate Change Research, United Kingdom, where he was a senior research associate from 2006 to early 2009. Since 2004, he has been a research fellow of the Global Governance Project, where he co-coordinates the research group Multiple Options, Solutions and Approaches: Institutional Interplay and Conflict (MOSAIC). In 2001–2003, he was a research assistant at the Centre for International Relations, Peace and Conflict Studies at the University of Tübingen, Germany.

Preface
www.adamproject.eu

This book is the result of a collective effort of more than two dozen scientists, all sharing an interest in finding effective solutions to the imminent crisis of global warming and large-scale alterations of the Earth system. Our common goal was to develop new ideas and insights that may assist negotiations of new global agreements on global climate governance for the period after 2012, when the current commitment period under the 1997 Kyoto Protocol to the 1992 United Nations Framework Convention on Climate Change expires.

Many have described the creation of a stable long-term architecture for global climate governance as one of the largest political challenges of our time, with tremendous implications for most areas of human life. These implications range from far-reaching reforms in the richer industrialized countries with high per capita emissions of greenhouse gases to the parallel quest of the many poorer societies in the developing world to lift the living standards and eradicate poverty while limiting growth in greenhouse gas emissions to the extent possible. While mitigation of global warming must have centre stage in current policies to prevent further build-up of greenhouse gases in the atmosphere, it is also vital to prepare for a world that may be substantially warmer than today due to failed or belated climate policies in the past. This book thus addresses both governance for mitigation and governance for adaptation, and, in particular, possible synergies and conflicts between both policy objectives.

The research documented in this volume has been part of a larger research programme on Adaptation and Mitigation Strategies: Supporting European Climate Policy (the ADAM Project). The ADAM Project lasted from 2006 to 2009 and was funded as an 'integrated project' by a major grant from the European Commission under its sixth framework research programme (Global Change and Ecosystem Priority, contract No. 018476). In total, more than 100 researchers from 26 institutes in Europe, India and China were part of the ADAM Project at one stage. This book brings together the results of one sub-project within the ADAM

Project, the working group 'Post-2012 Options in Global Climate Governance'. While funding as well as most researchers for this project came from the European Union, the analysis presented here is motivated by a global perspective that seeks to advance stable global governance without protecting or preserving the parochial interest of any region. The book thus draws from insights of scientists from countries like Bangladesh, India and South Africa, and many more developing country experts have contributed their views in a series of workshops that were held under the auspices of this project.

The core research for this book has been carried out by seven institutions in six countries: the Institute for Environmental Studies, Vrije Universiteit Amsterdam, the Netherlands, which also coordinated this effort; the Centre for International Climate and Environmental Research (CICERO), Oslo, Norway; the Lund University Centre for Sustainability Studies (LUCSUS) and the Department of Political Science at Lund University, Sweden; the Netherlands Environmental Assessment Agency (PBL) in the Netherlands; the Potsdam Institute for Climate Impact Research, Germany; The Energy and Resources Institute (TERI), New Delhi, India; and the Tyndall Centre for Climate Change Research, United Kingdom, which also provided the overall coordination of the ADAM Project.

Several researchers from other research institutions participated in this effort, including from the Bangladesh Centre for Advanced Studies, Bangladesh; the Energy Research Centre at the University of Cape Town, South Africa; the German Development Institute, Bonn, Germany; the International Institute for Environment and Development, United Kingdom; the Centre for Climate Science and Policy Research at Linköping University, Sweden; and the London School of Economics and Political Science, United Kingdom.

In addition, the design of this research programme has benefited substantially from the Science and Implementation Plan of the Earth System Governance Project, a ten-year global research effort under the auspices of the International Human Dimensions Programme on Global Environmental Change (www.earthsystem governance.org).

This book draws on the work and input of many colleagues and stakeholders. First of all, we wish to thank for their various contributions and insights all our team members who have contributed chapters to this book, as well as Karin Bäckstrand, Alex Haxeltine, Richard Klein, Eric Massey and Åsa Persson, who have all provided valuable additional input. In addition, we are grateful to the members of the ADAM Contact Group who provided an invaluable 'reality check' for the proposals that have been developed in this research programme: Marcel Berk, Daniel Bodansky, Chandrashekhar Dasgupta, Dagmar Droogsma, Bo Kjellén, Benito Müller, Lars Müller, Willem Thomas van Ierland and Michael Wriglesworth. In addition, Sebastian Oberthür, Youba Sokona, Simon Tay and Oran R. Young provided important comments at conference presentations. Last but not least, this book benefited substantially from the useful suggestions and critique from Steinar Andresen, Mike Hulme, Norichika Kanie and Henry Neufeldt.

Putting 19 chapters from 30 authors into one single coherent volume is no easy task, and we are highly grateful for the efforts of the project's student assistant in Amsterdam, Jonathan Berghuis, for implementing our house-style and making the many parts a whole. Moreover, we would like to give our special thanks to the staff at Cambridge University Press – in particular Laura Clark, Anna Hodson, Abigail Jones and Matt Lloyd – for their professionalism, assistance and understanding throughout the preparation of this volume.

This research programme benefited also from the lively discussions and pleasant interactions at numerous 'ADAM weeks' with the more than 100 researchers involved in the many other work packages in the project's four core domains A-ADAPTATION, M-MITIGATION, P-POLICY and S-SCENARIOS. We wish to thank here for all their efforts in bringing this large consortium together – and keeping it together! – in particular Mike Hulme, director of the ADAM Project, and Henry Neufeldt, the project manager.

Abbreviations

ADAM	Adaptation and Mitigation Strategies: Supporting European Climate Policy
AD-FAIR	Adaptation in the Framework to Assess International Regimes for differentiation of commitments
AD-RICE	Adaptation in the Dynamic Integrated model of Climate and the Economy
CCP	Cities for Climate Protection
CDM	Clean Development Mechanism
CDP	Carbon Disclosure Project
CSD	Commission on Sustainable Development
DICE	Dynamic Integrated model of Climate and the Economy
EU	European Union
FAIR	Framework to Assess International Regimes for differentiation of commitments
FAO	United Nations Food and Agriculture Organization
FUND	climate Framework for Uncertainty, Negotiation and Distribution
GAMS	General Algebraic Modelling System
GATS	General Agreement on Trade in Services
GATT	General Agreement on Tariffs and Trade
GD	gross or potential Damages
GDP	gross domestic product
GEF	Global Environment Facility
IATAL	International Air Travel Adaptation Levy
ICLEI	Local Governments for Sustainability
IEA	International Energy Agency
IPCC	Intergovernmental Panel on Climate Change
MAC	marginal abatement costs
MAGICC	Model for the Assessment of Greenhouse-gas Induced Climate Change

NGO(s)	non-governmental organization(s)
NPV	net present value
OECD	Organization for Economic Co-operation and Development
OPEC	Organization of the Petroleum Exporting Countries
PC	adaptation cost
ppm	parts per million
RD	residual damage
REDD	Reducing Emissions from Deforestation and Forest Degradation in Developing Countries
RICE	Regional Integrated model of Climate and the Economy
TIST	International Small Group and Tree Planting Programme
TRIPS	Trade-Related Aspects of Intellectual Property Rights
UN	United Nations
UNDP	United Nations Development Programme
UNEP	United Nations Environment Programme
UNFCCC	United Nations Framework Convention on Climate Change
WSSD	World Summit on Sustainable Development
WTO	World Trade Organization

1

Global climate governance beyond 2012: an introduction

FRANK BIERMANN, PHILIPP PATTBERG AND FARIBORZ ZELLI

Future historians might remember the period 2009–2012 as a turning point in the political response to global warming and climate change. The 1980s were a time of agenda-setting in which climate change became accepted as a political problem; the 1990s saw the first institutionalization through adoption of the United Nations Framework Convention on Climate Change in 1992 and its Kyoto Protocol in 1997. The 2000s marked the period of ratification of the protocol and further institutionalization of its means of implementation. Yet the Kyoto Protocol was merely a first step, and its core commitments expire in 2012. Even full compliance with the Kyoto agreement will not prevent 'dangerous anthropogenic interference with the climate system' – the overall objective of the climate convention. Concentrations of greenhouse gases in the atmosphere are rising, while drastic reductions of emissions are needed according to current scientific consensus (IPCC 2007).

These years are thus a crucial moment for human societies to change current economic, social and political development paths and to embark on a transition to new ways of production and consumption that emit less carbon – or to adapt to a world that is substantially warmer and hence different from the world that human and natural systems have been adapted to so far. At the planetary level, this is the quest for long-term, stable and effective 'global governance'. The term governance derives from the Greek word for navigating, and this challenge of turning around the wheel and charting a new course is indeed what is at stake in current negotiations on climate change.

Yet what is this new course that societies should navigate? What is the direction to follow? What systems of governance will promise to deliver the steering mechanisms needed to achieve drastic cuts in greenhouse gas emissions? These are the main questions and motivations underlying this book, which seeks to chart new directions for global climate governance beyond 2012, when the commitments of the Kyoto Protocol expire and new agreements are needed.

Global Climate Governance Beyond 2012: Architecture, Agency and Adaptation, eds. F. Biermann, P. Pattberg and F. Zelli. Published by Cambridge University Press.

Within the large array of important issues that need to be addressed in shaping global climate governance beyond 2012,[1] the contributions in this book focus on three issues that are, we believe, pivotal for any new governance structure.

First, we look into different options for the overarching architecture of global climate governance beyond 2012. While the term governance 'architecture' is found in policy debates in a variety of meanings, we use it here for a specific empirical phenomenon: the overall system of often overlapping, not always coordinated and at times conflicting institutions, norms and decision-making procedures in the area of climate governance (see Biermann *et al.*, this volume, Chapter 2). We are interested in particular in assessing whether higher or lower degrees of fragmentation of this overall architecture promise to be more effective in steering societies towards lower emission levels of greenhouse gases. We also look into whether more centralized governance – organized around and steered by a central treaty, for example a new comprehensive protocol to the climate convention – will increase overall effectiveness, or whether pluricentric, diverse and possibly redundant systems of governance would bring better outcomes. This question, which is at the centre of the first part of this book, has major policy implications, since governments are constantly confronted with different demands and must define strategies that may point towards centralization and integration, or rather towards fragmentation, diversity and pluricentric governance.

Second, this book looks into the role of a particular type of actors, and of a particular type of institutions, that have taken a more visible and possibly more relevant place in global climate governance in recent years: non-state actors and, more generally, governance beyond the state (see the conceptualization by Stripple and Pattberg, this volume, Chapter 9). These agents include a vast array of purely private actors, such as environmentalist groups, business associations or scientific networks. They also include public actors beyond central governments, for example cities, provinces or intergovernmental bureaucracies. Importantly, the question of agency beyond the state embraces new types of institutions and networks that assume governance functions with no or only marginal involvement of central governments, for instance the many 'partnerships for sustainable development' that have been agreed around the 2002 World Summit on Sustainable Development. Last but not least, governance beyond the state is governance beyond traditional means of public policy and intergovernmental rule-making. Especially in climate governance, market-based approaches have become prominent elements of many programmes and strategic proposals, from the hybrid Clean Development Mechanism that combines public and private steering to regional emissions trading

[1] For overviews of the many proposals on future global climate governance see Baumert *et al.* 2002; Bodansky *et al.* 2004; Aldy and Stavins 2007; Kuik *et al.* 2008.

and the many voluntary schemes for 'offsetting' emissions. How can we better understand the role and relevance of this increasing trend towards privatized and market-based governance mechanisms for climate change mitigation and the host of private actors that surround these new mechanisms? To what extent, and under what conditions, do private or public–private transnational governance mechanisms produce policy outcomes that are comparable, or even superior, to (traditional) forms of intergovernmental cooperation? To what extent should policies beyond 2012 include, or rely on, private and market-based governance? These questions are analysed in more depth in the second part of this book.

The third part of this volume studies global adaptation governance, building on the assumption that despite all mitigation efforts, some degree of global warming cannot be prevented. Adaptation to climate variation has always been a factor in human development, and adaptation to global warming is today part of discourses and decisions in many nations, in both North and South (see for example Jordan *et al.* 2010 on European policies in this field). Yet what is still uncertain is the global response to the myriad local problems of adaptation. The expected climate changes are likely to affect many core areas of global society, from the world economic system to global health, food security, trade, the provision of water, energy and other basic services, up to major humanitarian crises through climate-related migration or violent conflicts (see the conceptualization by Biermann and Boas, this volume, Chapter 14). Global adaptation governance must thus be part of climate governance beyond 2012. For many countries, notably the low-lying or semi-arid developing countries, global governance in support of local adaptation may evolve as one of the most crucial issues in this century. But what are promising policy options for the adaptation of regions, countries and international institutions to the impacts of climate change? To what extent do effective adaptation policies require global regulatory mechanisms, as opposed to local policy-making? To what extent does effective adaptation governance require the integration of adaptation policies in the overall climate governance architecture, and/or in other policy domains? These questions are the focus of the third part of this book.

These three core research themes – the *architecture* of global climate governance; *agency* in climate governance that goes beyond the central nation state; and *adaptation* to climate change at the level of global institutions and organizations – are not mutually exclusive. For instance, questions of architecture are also relevant when developing institutions for future adaptation governance, and non-state actors are important for adaptation as well. Instead of providing a clear-cut taxonomy, the three themes rather provide diverse lenses to approach the complexity of global climate governance and to direct attention to key issues and trends.

The selection of the three themes has been informed by current political processes, but also by broader debates in international relations and international law, for example on globalization, transnationalization, fragmentation and legitimacy (Ruggie 2001; Rosenau 2003; Hafner 2004; Börzel and Risse 2005). Linking research on climate governance to these broader theoretical and conceptual discourses in the social sciences (such as the role and relevance of the state versus non-state actors) increases understanding of contemporary climate governance while also contributing to theory consolidation within and across disciplines.

In addition, this book is part of two larger research programmes. First, our research is an integral contribution to the European research programme 'Adaptation and Mitigation Strategies: Supporting European Climate Policy' (the 'ADAM Project'), funded by a major grant from the European Commission. The ADAM Project is characterized by a high degree of interdisciplinarity that brought together more than 100 experts in disciplines as diverse as economics, engineering, political science or climate modelling. The project is also innovative in its focus on combining research on mitigation and adaptation in one integrated research design. The research in this volume presents the core results of one work package within this larger project, concentrating on 'Post-2012 Options in Global Climate Governance' (see Hulme and Neufeldt 2010 on the overall results of the ADAM Project; as well as Jordan *et al.* 2010 on European climate governance and Gupta and van der Grijp, 2010 on the relationship between climate governance and development cooperation).

Second, this volume is one of the first publications that respond to the science and implementation plan of the Earth System Governance Project, a new long-term research programme on governance and institutions under the auspices of the International Human Dimensions Programme on Global Environmental Change, which will last from 2009 to 2018 (Biermann *et al.* 2009). This science and implementation plan of the Earth System Governance Project identifies five core analytical problems, three of which are studied in this volume: different architectures of global and local governance, the role and relevance of different types of agents and agency in earth system governance, and the adaptiveness of governance systems.

As part of both the ADAM Project and the Earth System Governance Project, this book draws on the systematic and comprehensive integration of different disciplinary bodies of knowledge and of different methodological tools and approaches, from international law, political science and global governance studies to place-based development research and computer-based scenarios and modelling exercises. In particular, the three research themes of architectures, agency and global adaptation have been analysed from the perspective of three methodological approaches, each contributing to a comprehensive examination.

First, we analysed each theme by means of policy analysis. These studies advanced understanding of opportunities and barriers for policy-making at different stages of the policy process, as well as of institutional interlinkages and barriers to rule-making. We covered criteria of inclusiveness and legitimacy (regarding the participation of different types of actors), social acceptability and political feasibility. These methods helped determine the viability and the legal and political effectiveness of policy strategies, that is, their chances to materialize as concrete legal provisions (for example new rules under a future climate regime) and to change the compliance incentives of actors. Theoretical approaches applied in our research include institutional theory and global governance research, bargaining and game theory, international law analysis and economic analysis.

Second, the use of modelling tools helped to create a structured and quantitative framework for analysis. These methods focus less on political or legal implications but rather on criteria of long-term effectiveness and efficiency of policy options. They assist in determining the structural effects of selected strategies on both the global climate and social systems, for example regarding long-term emission reductions or effects on national incomes. Methods applied in this research include the FAIR meta-model, developed by the Netherlands Environmental Assessment Agency (Hof *et al.*, this volume, Chapter 4). FAIR is a stylized multi-region formal model that integrates modelling of the climate system (the relation between greenhouse gas emissions, concentrations and temperature) with the social–economic system (costs of mitigation, emissions trading and effects of climate change on national income). A second model employed is REMIND, developed by the Potsdam Institute for Climate Impact Research. REMIND is a hybrid model designed to integrate macroeconomic, energy system and climate modules. It is a multi-region endogenous economic growth model that can focus on regional interactions such as trade flows, foreign investments or technological spill-over.

Third, many contributions to this volume draw on participatory assessment approaches. Such tools give voice to stakeholders' perspectives. They allow for a critical examination of policy recommendations against the interests and concerns of key stakeholders, and can assist in refining recommendations into feasible and socially robust strategies. Participatory assessments hence complement the examination of political feasibility criteria provided by policy analysis. Participatory methods applied here include a series of structured international workshops with experts and policy-makers; regular consultations with an advisory group of senior experts and policy-makers; and a major survey of Southern policy-makers, academics and representatives of non-governmental organizations. The participatory appraisal exercises were held in New Delhi, India, on developing country perspectives; in Geneva, Switzerland, jointly with the Economics and Trade Branch of the UN Environment Programme, on climate and trade policies; in Lund, Sweden, on the reform of the Clean Development Mechanism; in Brussels, Belgium, on adaptation funding; in Brussels, Belgium, jointly with the Centre for European Policy Studies, on the overall research results; and finally a dialogue-event at the thirteenth conference of the parties of the climate convention in Bali.

The research in this volume has been policy-relevant in orientation while remaining academic in nature. Most efforts were directed at scoping or developing policy options that could provide a basis for future climate governance, and at appraising these options through multi-disciplinary assessment methodologies. While many of these policy options are derived from current debates, their appraisal took a much broader, long-term perspective, in a search for solutions that may be relevant and viable long after the current negotiations have ended. Also, while core elements of this research drew on local facts and findings – for example in studies on vulnerabilities of the poorest of the poor – our focus remained at the global level and at the most important elements of an overarching governance architecture for mitigating, and adapting to, global climate change.

This book is structured along the three research themes of architecture (Part I, Chapters 2–8), agency (Part II, Chapters 9–13) and adaptation (Part III, Chapters 14–18) (Table 1.1). We conclude with a summary of our research results and a number of concrete policy recommendations for global climate governance beyond 2012 (Biermann *et al.*, this volume, Chapter 19).

Part I addresses the problem of governance architecture. In this part, Biermann and colleagues (this volume, Chapter 2) first conceptualize global governance architectures as the overarching system of public and private institutions – that is, organizations, regimes and other forms of principles, norms, regulations and decision-making procedures – that are valid or active in a given area of world politics.

Based on this understanding and a comprehensive qualitative policy analysis, Zelli and colleagues (this volume, Chapter 3) appraise the consequences of different degrees of fragmentation for climate governance. They argue that different types of fragmentation are likely to have different degrees of performance. While synergistic fragmentation may bring both costs and benefits, there are hardly any convincing arguments in favour of more conflictive fragmentation.

Hof and colleagues (this volume, Chapter 4) complement this qualitative analysis with a quantitative assessment of different types of global climate architectures. They apply the FAIR meta-model to different governance scenarios and review quantitative studies about the costs and environmental effectiveness of universal and fragmented regimes. With this overview, they close a crucial research gap: in recent years, scholars have devised numerous proposals on universal and fragmented regimes. Yet while many of these proposals have been quantitatively assessed, no attempt has yet been made to compare cost estimates of these studies for specific regions under different architectures.

Flachsland and colleagues (this volume, Chapter 5) similarly provide a quantitative account of different fragmentation scenarios, focusing on emissions trading. Based on the REMIND model, they analyse different integration scenarios for

Table 1.1 *Research themes and methodologies*

	Architecture	Agency beyond the state	Adaptation
Policy analysis	Institutional fragmentation *(institutional theory, bargaining theory, international law)* UN climate regime and world trade regime *(institutional theory, bargaining theory, international law)* Equity-based architecture for North–South cooperation *(qualitative policy analysis)*	Transnational climate governance *(institutional theory)* CDM reform *(institutional theory)* Research and development, and technological change *(economic analysis)*	Climate refugees *(institutional theory, international law)* Food insecurity *(institutional theory)* Adaptation funding *(qualitative economic analysis)* Interests and perspectives of developing countries *(institutional theory, international law)* Vulnerability of the poorest of the poor *(socio-economic analysis)*
Modelling	Institutional fragmentation *(FAIR meta-model)* Linking of emission trading systems *(REMIND model)*	Sectoral mitigation *(FAIR meta-model)*	Cost–benefit interlinkages between adaptation and mitigation *(FAIR meta-model)*
Participatory approaches	Institutional fragmentation *(side-events at conferences of the parties, UNEP workshop, policy workshop in Brussels, developing country conference in Delhi, interviews, survey)* UN climate regime and world trade regime *(UNEP workshop, policy workshop in Brussels, interviews)* Southern perspectives *(developing country conference in Delhi)*	Transnational climate governance *(interviews, survey)* Reform of Clean Development Mechanism *(policy workshop in Lund, policy workshop in Brussels)* Market-based mechanisms and developing countries *(developing country conference in Delhi, survey)*	Climate refugees *(side-events at conferences of the parties, policy workshop in Brussels, interviews)* Food insecurity *(side-events at conferences of the parties, developing country conference in Delhi, policy workshop in Brussels, interviews)* Adaptation in developing countries *(developing country conference in Delhi)* Adaptation funding *(policy workshop in Brussels, interviews)*

carbon markets – for top–down trading schemes, like the one established by the Kyoto Protocol, but also for linking bottom–up schemes with decentralized decision-making systems that are emerging in the United States, Canada and other countries. In examining these different scenarios, they quantify likely changes in global and regional mitigation costs.

In Chapter 6, Zelli and van Asselt concentrate on a related aspect of the fragmentation of the global climate governance architecture, namely overlaps between the climate convention and the World Trade Organization. Both arenas address similar topics such as emissions trading or the transfer of climate-friendly goods, services and technologies. This duplication of debates and the associated lack of legal clarity may imply detrimental ramifications for the climate regime. For instance, parties to the climate convention that fear incompatibility with WTO rules might refrain from implementing ambitious domestic climate policies. Based on the results of a joint workshop with the United Nations Environment Programme, Zelli and van Asselt explore a range of policy options to tackle such negative implications.

While these chapters focus on particular components of the global climate governance architecture and its fragmentation, Winkler (this volume, Chapter 7) explores options for a future global architecture from the perspective of North–South cooperation and the principles of equity and common but differentiated responsibilities. In his qualitative analysis, he assesses different approaches and concludes that there is not just one option for compromise, but a number of feasible ways to strike a balance between developing and industrialized countries. He therefore suggests conceptualizing options for future climate governance architecture as a continuum of feasible scenarios and negotiation packages. Winkler explores two of these compromises in further detail.

Shrivastava and Goel (this volume, Chapter 8) also review the options for an effective and equitable architecture of global climate governance from a perspective of the developing countries, yet from a different angle that focuses on technological capability and financial support from industrialized countries to developing countries. They see it as critical that the future global governance architecture is guided by national requirements of developing countries. To this end, they suggest as the best policy option a two-tier architecture with two distinct but integrated components: a set of institutions, policies and programmes at the national level to identify the direction of technological development within the country; and a network of global institutions, financial mechanisms and technological programmes to support the institutions, policies and programmes in developing countries.

Part II of this volume deals with non-state agency in global climate governance. Stripple and Pattberg (this volume, Chapter 9) first map this multifarious and complex emerging transnational arena of global climate governance, in which agency is constructed, maintained and challenged not only by central governments but also by a host of other actors such as non-governmental organizations, business actors, scientists and sub-national governments.

In more detail, Pattberg (this volume, Chapter 10) then discusses different forms of networked climate governance and evaluates their impacts with regard to problem-solving capacity, the democratic legitimacy of global environmental governance and the nature of their linkage to the international climate regime. Empirical illustrations include global city networks, public–private partnerships concluded within the context of the World Summit on Sustainable Development and disclosure-based corporate social responsibility schemes.

Stripple and Lövbrand (this volume, Chapter 11) complement this study by offering a detailed case study on the creation and transformation of carbon markets. Rather than asking which entities govern carbon markets, they address the question of how and by which procedures carbon markets are rendered thinkable and operational in the first place. To that end, they study baseline-and-credit markets in particular, where a complex measurement of counterfactuals (current emissions vis-à-vis a business-as-usual scenario) enables reductions of carbon dioxide-equivalents to be assigned a market value and be transformed into various 'offset currencies'.

Den Elzen and colleagues (this volume, Chapter 12) take a different perspective by focusing on the Triptych approach that differentiates allocation of emissions reductions based on sectoral targets that involve non-state actors through economic sectors in greenhouse gas mitigation. They argue that decomposing targets according to sectors provides for a more direct involvement of non-state actors in emission reduction targets. The framework also allows for discussions on sectors that compete worldwide. The disadvantage is that it requires projections of sectoral growth rates for each country.

In Chapter 13, Alfsen and colleagues address the question of innovation by looking at research-and-development policies and the role of agency therein. They argue that international agreements are best suited to boost research and development on climate-friendly technologies, and that research-and-development agreements and cap-and-trade agreements are mutually supportive because research and development reduces future abatement costs and thus allows politicians to agree on tighter caps. Cap-and-trade strengthens a research and development programme because the latter becomes more efficient when a price on emissions stimulates innovation. Research and development and cap-and-trade should thus not be seen as alternatives or substitutes, but as mutually supportive elements in an effort to tackle climate change.

Part III of this volume then presents the core findings of this research programme on global adaptation governance.

First, Biermann and Boas (this volume, Chapter 14) map the challenge of global adaptation governance. They emphasize the difficulties in designing effective research designs that appraise the performance of governance options for future climate change impacts that are merely predicted, but in essence unknowable in type and degree of harm. Also, they sketch the core areas of global governance that are likely to be negatively affected by the impacts of global warming in the twenty-first century. Chapter 14 thus serves as an outline of a major research effort on global adaptation governance. Chapters 15–18 present first findings from the ADAM Project in this field, all of which require further research and refinement.

Hof and colleagues (this volume, Chapter 15) present a state-of-the-art study on adaptation in integrated assessment models. They start with the observation that the explicit consideration of adaptation is still in its infancy in integrated assessment models that aim at supporting climate policy by analysing economic and environmental consequences and by formulating efficient responses. Hof and colleagues try to fill this gap in integrated assessment models by integrating adaptation and residual damage functions from the AD-RICE model with the FAIR model. This version of the FAIR model (called AD-FAIR) allows for an analysis of the interactions between mitigation, emission trading, adaptation and residual damages on a global as well as regional scale. Adaptation is modelled here explicitly as a policy variable, which provides insights in the economic consequences of adaptation.

The question of the costs of climate change is addressed from a different perspective also by Biermann and Boas in Chapter 16. This chapter presents a policy analysis of possible governance systems to recognize, protect and resettle millions of climate refugees that may have to give up their homes over the course of this century due to sea-level rise or water scarcity. Biermann and Boas study a number of existing governance mechanisms and conclude that new approaches and institutions are needed. In particular, they sketch a proposal for a new intergovernmental agreement on the recognition, protection and resettlement of climate refugees that could be adopted as a protocol or otherwise integral part of future climate agreements.

Chapter 17 reviews the debate on global adaptation governance – similar to Chapters 7 and 8 – from the perspective of developing countries. Ayers, Alam and Huq (this volume, Chapter 17) argue that past policies resulted in a framing of adaptation that is inappropriate for addressing the myriad developing country concerns in this field. What is needed, according to Ayers, Alam and Huq, is thus a reframing of the adaptation agenda to ensure that developing country priorities can be met comprehensively and consistently. In particular, they argue that adaptation must be taken as seriously as mitigation and that a more comprehensive and operational approach to adaptation must be taken, including substantial and mandatory financial commitments and a legal framework for adaptation. Eventually, they suggest that adaptation concerns may better be achieved under a different type of international architecture, outside mitigation governance. This could be an independent ‘adaptation protocol’, with a more flexible definition of adaptation and operationizable targets and guidance on adaptation funding and action, which is in line with proposals by Biermann and Boas (this volume, Chapter 16) for a climate refugee agreement.

Jerneck and Olsson (this volume, Chapter 18) complement these reflections on the Southern perspective by adding a particular emphasis on the poorest of the poor, described as the 'bottom billion' of human society. As they analyse in detail, it is these people who are likely to be the most affected by the impacts of climate change, who are the least responsible for the causation of the problem, and who have the least means to respond to the emerging crisis. Jerneck and Olsson thus focus on the South–North conflict in climate governance but add another perspective of those people in the developing world that are likely to suffer most. They suggest that policies intended for poverty eradication may have unintended consequences in marginalizing some groups, strengthening social stratification of the poor and contributing to the reproduction of 'the poorest of the poor'. Policy-making for adaptation must thus seek to avoid such marginalization. Jerneck and Olsson eventually call for a rethinking of development from a sustainability perspective rather than mainstreaming climate change and adaptation into the narrower paradigm of development.

Finally, in Chapter 19 the editors provide an extensive summary and review of the overall results of this three-year research effort that involved more than 30 researchers. There is no single answer but a patchwork of findings. The overall findings of this programme, as summarized in Chapter 19, reflect the diversity of the problem by offering a diversity of elements that can point to more effective and equitable governance beyond 2012. The findings emphasize the benefits of more integrated governance architectures as opposed to more fragmented architectures; emphasize problems of privatization in terms of possibly lower performance, legitimacy and equity, while acknowledging also some benefits of the privatization of parts of global climate governance; and emphasize the need for better integrated, focused and financed systems of global adaptation governance. Last but not least, this book shows the urgent need not only for new governance, but also for new and additional governance research. We hope this volume will make a contribution to this important debate on developing policy options for effective, equitable and legitimate climate governance beyond 2012.

References

Aldy, J. and R. B. Stavins (eds.) 2007. *Architectures for Agreement: Addressing Global Climate Change in the Post-Kyoto World*. Cambridge, UK: Cambridge University Press.

Baumert, K. A., O. Blanchard, S. Llosa and J. F. Perkaus (eds.) 2002. *Building on the Kyoto Protocol: Options for Protecting the Climate*. Washington, DC: World Resources Institute.

Biermann, F., M. M. Betsill, J. Gupta, N. Kanie, L. Lebel, D. Liverman, H. Schroeder and B. Siebenhüner 2009. *Earth System Governance: People, Places and the Planet – Science and Implementation Plan of the Earth System Governance Project*. Bonn: Earth System Governance Project of the International Human Dimensions Programme on Global Environmental Change.

Bodansky, D., S. Chou and C. Jorge-Tresolini 2004. *International Climate Efforts beyond 2012: A Survey of Approaches*. Arlington, VA: Pew Center on Global Climate Change.

Börzel, T. A. and T. Risse 2005. 'Public–private partnerships: effective and legitimate tools of international governance', in E. Grande and L. W. Pauly (eds.), *Reconstructing Political Authority: Complex Sovereignty and the Foundations of Global Governance*. Toronto: University of Toronto Press, pp. 195–216.

Gupta, J. and N. van der Grijp (eds.) 2010. *Mainstreaming Climate Change in Development Cooperation: Theory, Practice and Implications for the European Union*. Cambridge, UK: Cambridge University Press.

Hafner, G. 2004. 'Pros and cons ensuing from fragmentation of international law', *Michigan Journal of International Law* **25**: 849–863.

Hulme, M. and H. Neufeldt (eds.) 2010. *Making Climate Change Work for Us: European Perspectives on Adaptation and Mitigation Strategies*. Cambridge, UK: Cambridge University Press.

IPCC 2007. *Climate Change 2007: Mitigation of Climate Change. Contribution of Working Group III to the Fourth Assessment Report of the Intergovernmental Panel on Climate Change*. Geneva: IPCC.

Jordan, A., D. Huitema, H. van Asselt, F. Berkhout and T. Rayner (eds.) 2010. *Climate Change Policy in the European Union: Confronting the Dilemmas of Mitigation and Adaptation*. Cambridge, UK: Cambridge University Press.

Kuik, O., J. Aerts, F. Berkhout, F. Biermann, J. Bruggink, J. Gupta and R. S. J. Tol 2008. 'Post-2012 climate change policy dilemmas: how do current proposals deal with them?', *Climate Policy* **8**: 317–336.

Rosenau, J. N. 2003. *Distant Proximities: Dynamics beyond Globalization*. Princeton, NJ: Princeton University Press.

Ruggie, J. G. 2001. 'Global_governance.net: the global compact as learning network', *Global Governance* **7**: 371–378.

Part I

Architecture

2

The architecture of global climate governance: setting the stage

FRANK BIERMANN, FARIBORZ ZELLI, PHILIPP PATTBERG AND HARRO VAN ASSELT

2.1 Introduction

This chapter introduces the first main part of this volume, on the overarching 'architecture' of global climate governance beyond 2012.[1] In particular, the central question that guides all chapters in this part is about the causes and consequences of fragmentation versus integration of governance architectures. We ask which type of governance architectures promises a higher degree of institutional performance in terms of social and environmental effectiveness, and in particular whether a well-integrated governance architecture is likely to be more effective than a fragmented governance architecture. This question of increasing fragmentation of systems of global governance and of its relative benefits and problems has become a major source of concern for observers and policy-makers alike. Yet there is little consensus in the academic literature on this issue: in different strands of academic research, we find different predictions that range from a positive, affirmative assessment of fragmentation to a rather negative one (Zelli *et al.*, this volume, Chapter 3).

A key example is global climate governance, where advantages and disadvantages of a fragmented governance architecture have become important elements in proposals and strategies for future institutional development. Several proposals for a future climate governance architecture have been put forward that explicitly assert the value of fragmentation or diversity, or at least implicitly accept it. Others, however, remain supportive of a more integrated overall architecture. And yet, political science lacks a conceptual framework for the comparative study of different types and degrees of fragmentation of global governance architectures.

In this chapter, we attempt to help resolve this problem, thereby guiding the analyses in the following chapters under the 'architecture' theme. We first conceptualize the notion of global governance architectures and of different types and

[1] This chapter draws on Biermann *et al.* (2009).

Global Climate Governance Beyond 2012: Architecture, Agency and Adaptation, eds. F. Biermann, P. Pattberg and F. Zelli. Published by Cambridge University Press.

degrees of their fragmentation (Section 2.2) and then illustrate these concepts in global governance in response to climate change (Section 2.3). The following chapters substantiate our discussion by detailed assessments of specific questions within this larger analytical framework.

2.2 Conceptualization

There is no commonly agreed definition of the term 'global governance architecture'. We define the term in this book as the overarching system of public and private institutions – that is, organizations, regimes and other forms of principles, norms, regulations and decision-making procedures[2] – that are valid or active in a given issue area of world politics. Architecture can thus be described as the meta-level of governance.

Through its focus on a particular issue area – such as climate policy – the concept of governance architecture is narrower than the notion of order. Both concepts share a focus on the overarching governance structures that reach beyond single regimes. Yet while international order reflects the organization of the entire system of international relations (Bull 1977), architecture is a more appropriate concept for distinct issue areas of global governance. Moreover, the concept of international order often implies an optimistic bias regarding the coherence and internal coordination of the international system. Architecture, on its part, is more neutral and accounts for dysfunctional and non-intended effects too. Architecture, in this book, does not presuppose order in a normatively loaded understanding.

Instead, a degree of fragmentation is a frequent characteristic of global governance architectures. Conceptualizing governance architectures in different issue areas allows for the comparative analysis of different degrees and types of fragmentation. We advance the notion of global governance architecture in particular for this reason: because it allows for the analysis of (the many) policy domains in international relations that are not regulated, and often not even dominated, by a single international regime in the traditional understanding. Many policy domains are instead marked by a patchwork of international institutions that are different in their character (organizations, regimes and implicit norms), their constituencies (public and private), spatial scopes (from bilateral to global) and subject matters (from specific policy fields to universal concerns). These situations we understand as *fragmented* global governance architectures. As discussed in the following chapters in more detail, climate governance is a prime example of such situations.

[2] International regimes are usually defined as 'sets of implicit or explicit principles, norms, rules, and decision-making procedures around which actors' expectations converge in a given area of international relations' (Krasner 1983: 2). International institutions, as the more generic term, comprise international regimes, international organizations and implicit norms and principles (Keohane 1989: 3–4).

The notion of fragmentation is widely employed in international legal literature (for example Hafner 2000, 2004; Koskenniemi and Leino 2002; International Law Commission 2006). Some see fragmentation here as a sign of the expansion of international law to previously unregulated fields, such as international commerce, human rights or the environment (Lindroos and Mehling 2005). Increasingly, also scholars in international relations and international economics refer to the 'fragmentation' of arrangements, especially regarding environmental governance (for example Andresen 2001; Bernstein and Ivanova 2007; Kanie 2007). Similar phenomena are captured at times under different terminology, including 'multiplicity' of global environmental governance (Ivanova and Roy 2007), 'division of labour' among international norms and institutions (Siebert 2003; Haas 2004: 8), or, with a more negative connotation, 'treaty congestion' (Brown Weiss 1993: 697).

Regarding the conceptualization of fragmentation in this book, we emphasize three points. First, we use the term fragmentation as a relative concept: all global governance architectures are fragmented to some degree; that is, they consist of distinct parts that are hardly ever fully interlinked and integrated. Non-fragmented, 'universal' architectures are theoretically conceivable as opposites of fragmentation; an architecture would be universal if all countries relevant in an issue area are subject to the same regulatory framework; participate in the same decision-making procedures; and agree on a core set of common commitments. Empirically, however, such a situation is difficult to trace in current world politics. For instance, even one of the most widely supported international treaties, the Convention on the Rights of the Child, has been ratified by 193 parties yet not by the United States and Somalia, and its optional protocols on children in armed conflicts and on child pornography and prostitution lack ratification by all nations. Fragmentation, in other words, is ubiquitous. Yet the degree of fragmentation varies from case to case. The concept of architecture allows for the comparative analysis of issue areas and policy domains and for the study of overarching phenomena that the more restricted concept of regimes could not capture.

Second, we use the concepts of both architecture and fragmentation value-free. We assume neither an *a priori* existing state of universal order nor a universal trend towards order. In most empirical cases, architectures are likely to result from incremental processes of institutionalization in international affairs that are decentralized and hardly planned. In other words, the concept of architecture does not assume the existence of an architect.

Third, empirical research on fragmentation of global governance architectures depends on the perceived scale of the problem. The larger the perceived scale of the problem, the higher the degree of fragmentation is likely to be. Fragmentation is evident in more narrowly defined global governance architectures, that is, between parallel policies and regimes in the same issue area, such as climate governance or

Table 2.1 *Typology of fragmentation of governance architectures*

	Synergistic	Cooperative	Conflictive
Institutional nesting	One core institution, with other institutions being closely integrated	Core institutions with other institutions that are loosely integrated	Different, largely unrelated institutions
Norm conflicts	Core norms of institutions are integrated	Core norms are not conflicting	Core norms conflict
Actor constellations	All relevant actors support the same institutions	Some actors remain outside main institutions, but maintain cooperation	Major actors support different institutions

the governance of plant genetic resources (Raustiala and Victor 2004; McGee and Taplin 2006; van Asselt 2007). It is here where the concept of architecture and the comparative analysis of different degrees and types of fragmentation are likely to be most fruitful. Yet fragmentation is likely to be more significant the broader the issues areas that are defined, for example with a view to the entire domain of global environmental governance or economic governance.

To assess degrees of fragmentation, we employ the following three criteria to differentiate between degrees of fragmentation: degree of institutional nesting and of overlaps between decision-making systems; existence and degree of norm conflicts; and type of actor constellations. Based on these criteria, we distinguish three types of fragmentation (Table 2.1): (1) synergistic fragmentation; (2) cooperative fragmentation; and (3) conflictive fragmentation. In empirical research, boundaries between these three types will not be clear-cut; the criteria and types are meant as a conceptual tool to determine and compare degrees of fragmentation of different issue areas in comparative research. Likewise, long-term analyses might find that an architecture has shifted from one type of fragmentation to another.

(1) We speak of a situation of *synergistic fragmentation* when the core institution (a) includes (almost) all countries and (b) provides for effective and detailed general principles that regulate the policies in distinct yet substantially integrated institutional arrangements. An example is the 1985 Vienna Convention and its 1987 Montreal Protocol on Substances that Deplete the Ozone Layer and its amendments from London (1990), Copenhagen (1992), Montreal (1997) and Beijing (1999) (United Nations Environment Programme 2007). Each amendment to the protocol adds new substances to the regulative system, including decision-making procedures on further policies on these substances. The governance architecture on ozone depletion comes close to a system of five concentric circles, with the 1987 Montreal Protocol having the most parties, and each of the four amendments a more restrictive reach. However, the overarching Vienna Convention and Montreal Protocol govern all amendments in every

important aspect, serving as an integrative umbrella and authority in linking the different amendments and political processes. No significant institutions exist on this issue outside the framework of the Vienna Convention and the Montreal Protocol, which shows a high degree of integration within this governance architecture.

(2) We speak of a situation of *cooperative fragmentation* when an issue area is marked by: (a) different institutions and decision-making procedures that are loosely integrated; (b) when the core institution does not comprise all countries that are important in the issue area; and/or (c) when the relationship between norms and principles of different institutions is ambiguous. Policies in the same area are then defined, decided and monitored through different institutions, or through core institutions, on the one hand, and individual countries that are not part of this institution on the other. However, overall integration within the governance architecture in the issue area is sufficient to prevent open conflicts between different institutions. One example is the relationship between the United Nations Framework Convention on Climate Change ('climate convention') and its Kyoto Protocol, which we discuss in Section 2.3 in more detail (see also Zelli *et al.*, this volume, Chapter 3).

(3) We speak of a situation of *conflictive fragmentation* when an issue area is marked by different institutions that: (a) are hardly connected and/or have different, unrelated decision-making procedures; (b) have conflicting sets of principles, norms and rules; and (c) have different memberships and/or are driven by actor coalitions that accept, or even advance, these conflicts. One prominent example is the regulation of access and benefit sharing of plant genetic resources. Here, two regimes attempt to regulate this issue, the Convention on Biological Diversity and the Agreement on Trade-Related Aspects of Intellectual Property Rights (TRIPS) under the World Trade Organization (WTO). The latter seeks to strengthen and harmonize systems of intellectual property rights, whereas the former reaffirms sovereign rights of states over biological resources. The negotiations of both regimes, which partly took place in parallel, were marked by intense conflicts between developing and industrialized countries. Consequently, the relevant rules of the biodiversity convention remain rather abstract and imprecise, and the United States did not ratify the convention. As Rosendal (2006: 94) suggests, a virtual 'arms race' has taken place through additional agreements that try to flesh out the regulations of both regimes.

In empirical research, the boundaries between these three ideal types of fragmentation in global governance architectures may remain difficult to ascertain in specific cases. In addition, the three types are not mutually exclusive, but may coexist within the same architecture. The three types are thus meant to serve as a conceptual tool for comparative empirical analysis in order to advance understanding of the causes and consequences of fragmentation in global governance architectures. Based on the conceptualization of these three ideal-types of governance fragmentation, comparative empirical research can shed light on the core question of the relative costs and benefits of different types and degrees of fragmentation. In addition, it becomes possible to analyse in much more detail possible political, legal and institutional solutions to problems of fragmentation, which may depend on the types and degrees of fragmentation at hand.

2.3 The case of global climate governance

How can this typology be applied on the case of global climate governance? We now show that the governance architecture in this area has elements of all three types of fragmentation – synergistic, cooperative and conflictive – but that the overall situation is best described as a case of cooperative fragmentation.

(1) First, the core of the climate governance architecture has elements of *synergistic fragmentation*. The institutional core of the architecture is the climate convention, ratified by almost all nations. The convention lays down a number of fundamental principles. These include the 'ultimate objective' of climate governance to prevent 'dangerous anthropogenic interference with the climate system' (article 2), the principle of common but differentiated responsibilities and respective capabilities and a precautionary approach (article 3). In addition, the convention provides for a sizeable international bureaucracy for administrative support, data collection and policy development, as the organizational nodal point of the governance architecture in this area (Busch 2009). The 1997 Kyoto Protocol is part of the larger climate convention and shares its basic principles.

(2) Yet in addition, the climate governance architecture has strong elements of *cooperative fragmentation*, which is the most fitting overall description. The protocol provides for quantified emissions limitation and reduction obligations only for industrialized countries. Moreover, one of the world's largest greenhouse gas emitters, the United States, is party only to the convention and not to the protocol, which creates a higher degree of fragmentation within the regime. This fragmentation has become obvious in the negotiations on future climate governance, which occur in separate negotiating tracks for the convention and the protocol (Clémençon 2008). The 2007 and 2008 conferences of the parties showed the increased complexity, with dozens of agenda items discussed in numerous contact groups and informal negotiations, and many items postponed to later sessions of subsidiary bodies.

In addition to the UN climate regime, there are an increasing number of additional institutional governance arrangements at different levels. Some arrangements, such as the Methane to Markets partnership, are public–private partnerships registered with the UN Commission on Sustainable Development after the 2002 World Summit on Sustainable Development. Other initiatives, such as the Carbon Sequestration Leadership Forum and the International Partnership for a Hydrogen Economy, are not registered with the Commission on Sustainable Development, even though their form is similar. Other initiatives are high-level ministerial dialogues, such as the

Dialogue on Climate Change, Clean Energy and Sustainable Development, initiated by the meeting of the Group of Eight in Gleneagles, Scotland, in July 2005. The start of the European emissions trading scheme in 2005 marked the launch of another UN-independent initiative. Although based on the Kyoto Protocol, the trading scheme's start did not depend on the protocol's entry into force (Flachsland *et al.*, this volume, Chapter 5). Finally, there are sub-national initiatives such as California's Global Warming Solution Act and the Regional Greenhouse Gas Initiative in the United States, as well as private institutions that attempt to regulate issue areas relevant for climate governance, such as the Carbon Disclosure Project (Pattberg and Stripple 2008; Stripple and Pattberg, this volume, Chapter 9).

In sum, some arrangements explicitly relate to the institutional core, such as the EU emissions trading scheme (which in 2008 connected to the transaction log of the climate convention) (Flachsland *et al.*, this volume, Chapter 5; van Asselt 2010) or public–private partnerships to implement the climate convention. Other initiatives are connected to the UN regime mainly through the participation of key actors in various forums. Most initiatives acknowledge the UN process, even though many do not provide for a coordination mechanism that could ensure mutual compatibility.

(3) In addition, the climate governance architecture shows indications of *conflictive fragmentation*. Notably, the 2005 Asia–Pacific Partnership on Clean Development and Climate departs from key features of the UN climate regime, including the consideration of climate change impacts and differentiation between industrialized and developing countries (McGee and Taplin 2006; van Asselt 2007). At the same time, while not comparable to the UN regime in terms of financial endowment or membership, the partnership still provides an alternative to international climate action that may reduce incentives for complying with, or signing up to, international legally binding commitments. A similar initiative is the Major Economies Process on Energy Security and Climate Change launched by the United States in 2007. This Process includes 17 of the world's largest economies and aims at a long-term greenhouse gas emissions reduction goal (White House 2007); its relation to the UN climate regime is ambiguous and partially conflictive. For example, during the 2007 conference of the parties to the climate convention, the delegation of the European Union threatened to boycott the next session of the US-initiated Major Economies Process. Representatives from the Group of 77 and China, too, argued that the UN climate regime should remain the central platform for addressing action on climate change (International Institute for Sustainable Development 2007).

Importantly, these instances of fragmentation in climate governance are intentional (on the problems of intentional interplay see Young 2008). The Asia–Pacific Partnership and similar proposals – backed by the United States – were created not out of ignorance of the climate regime but *because* of it, at a time when the climate convention and the Kyoto Protocol were well established and in force. In addition, the emergence of numerous initiatives outside the climate regime indicates that the global climate governance architecture may become more fragmented over time. Many new initiatives include the United States, which has rejected the Kyoto Protocol; most are not or only loosely linked to the UN climate regime; and the compatibility of some norms and principles with those of the core institution is often ambiguous at best.

Nonetheless, the overall architecture of climate governance, at present, can be best characterized as an example of cooperative fragmentation.

2.4 Conclusions

In this chapter, we have introduced the term architecture as the overarching system of public and private institutions in a given issue area. We have conceptualized governance fragmentation as a situation of multiple international institutions in an issue area that differ in character, constituencies, spatial scope and subject matter. We have argued that fragmentation is an inherent structural characteristic of international policy – in other words: when comparing the international architectures of different issue areas, fragmentation is a matter of degree. Based on a threefold typology of fragmentation, we characterized the global climate governance architecture as a case of cooperative fragmentation: apart from one core institution, which does not comprise all relevant countries, an increasing number of other organizations, regimes and arenas are addressing climate change, while the relationship among these different institutions remains often ambiguous, but by and large cooperative.

Having identified this advanced degree of fragmentation, the following chapters address the implications of fragmentation of global climate governance beyond 2012. While assessing different scenarios of a future climate governance architecture, they attend to our core appraisal question for this first part of this volume: is an almost universal, strongly integrated governance architecture likely to be more effective than a heavily fragmented, heterogeneous governance architecture? Moreover, the following chapters explore policy options to address the increasing fragmentation of global climate governance. They approach this phenomenon from different disciplinary backgrounds, namely: qualitative policy assessment in Chapters 3, 6, 7 and 8, quantitative analysis and modelling in Chapters 4 and 5 and participatory assessment in Chapters 6 and 8.

References

Andresen, S. 2001. 'Global environmental governance: UN fragmentation and co-ordination', in O. Schram Stokke and Ø.B. Thommessen (eds.), *Yearbook of International Co-operation on Environment and Development 2001/2002*. London: Earthscan, pp. 19–26.

Asselt, H. van 2007. 'From UN-ity to diversity? The UNFCCC, the Asia-Pacific Partnership and the future of international law on climate change', *Carbon and Climate Law Review* **1**: 17–28.

Asselt, H. van 2010 (in press). 'Emissions trading: the enthusiastic adoption of an alien instrument?', in A. Jordan, D. Huitema, H. van Asselt, F. Berkhout and T. Rayner (eds.), *Climate Change Policy in the European Union: Confronting the Dilemmas of Mitigation and Adaptation*? Cambridge, UK: Cambridge University Press.

Bernstein, S. and M. Ivanova 2007. 'Institutional fragmentation and normative compromise in global environmental governance: what prospects for re-embedding?', in S. Bernstein and L. W. Pauly (eds.), *Global Liberalism and Political Order: Towards a New Grand Compromise*? Albany, NY: State University of New York Press, pp. 161–185.

Biermann, F., P. Pattberg, H. van Asselt and F. Zelli 2009. 'The fragmentation of global governance architectures: a framework of analysis', *Global Environmental Politics* **9**(4): 14–40.

Brown Weiss, E. 1993. 'International environmental law: contemporary issues and the emergence of a new order', *Georgetown Law Journal* **81**: 675–710.

Bull, H. 1977. *The Anarchical Society: A Study of Order in World Politics*. New York: Columbia University Press.

Busch, P. O. 2009. 'The climate secretariat: making a living in a straitjacket', in F. Biermann and B. Siebenhüner (eds.), *Managers of Global Change: The Influence of International Environmental Bureaucracies*. Cambridge, MA: MIT Press, pp. 245–264.

Clémençon, R. 2008. 'The Bali road map: a first step on the difficult journey to a post-Kyoto Protocol agreement', *Journal of Environment and Development* **17**: 70–94.

Haas, P. M. 2004. 'Addressing the global governance deficit', *Global Environmental Politics* **4**: 1–15.

Hafner, G. 2000. 'Risks ensuing from fragmentation of international law', in *Official Records of the General Assembly*, Fifty-fifth session, Supplement No. 10 (A/55/10, 2000), Annex, pp. 326–354.

Hafner, G. 2004. 'Pros and cons ensuing from fragmentation of international law', *Michigan Journal of International Law* **25**: 849–863.

International Institute for Sustainable Development 2007. *Earth Negotiations Bulletin* **12**, No. 342. www.iisd.ca/vol12.

International Law Commission 2006. *Fragmentation of International Law: Difficulties Arising from the Diversification and Expansion of International Law*, Report of the Study Group of the International Law Commission. UN Doc. A/CN.4/L.682. Geneva: International Law Commission.

Ivanova, M. and J. Roy 2007. 'The architecture of global environmental governance: pros and cons of multiplicity', in L. Swart and E. Perry (eds.), *Global Environmental Governance: Perspectives on the Current Debate*. New York: Center for UN Reform, pp. 48–66.

Kanie, N. 2007. 'Governance with multilateral environmental agreements: a healthy or ill-equipped fragmentation', in L. Swart and E. Perry (eds.), *Global Environmental Governance: Perspectives on the Current Debate*. New York: Center for UN Reform, pp. 69–86.

Keohane, R. O. 1989. *International Institutions and State Power: Essays in International Relations Theory*. Boulder, CO: Westview Press.

Koskenniemi, M. and P. Leino 2002. 'Fragmentation of international law? Postmodern anxieties', *Leiden Journal of International Law* **15**: 553–579.

Krasner, S. D. 1983. 'Structural causes and regime consequences: regimes as intervening variables', in S. D. Krasner (ed.), *International Regimes*. Ithaca, NY: Cornell University Press, pp. 1–21.

Lindroos, A. and M. Mehling 2005. 'Dispelling the chimera of "self-contained regimes": international law and the WTO', *European Journal of International Law* **16**: 857–877.

McGee, J. and R. Taplin 2006. 'The Asia-Pacific Partnership on Clean Development and Climate: a competitor or complement to the Kyoto Protocol', *Global Change, Peace and Security* **18**: 173–192.

Pattberg, P. and J. Stripple 2008. 'Beyond the public and private divide: remapping transnational climate governance in the 21st century', *International Environmental Agreements: Politics, Law and Economics* **8**: 367–388.

Raustiala, K. and D. Victor 2004. 'The regime complex for plant genetic resources', *International Organization* **58**: 277–309.

Rosendal, G. K. 2006. 'The Convention on Biological Diversity: tensions with the WTO TRIPS Agreement over access to genetic resources and the sharing of benefits', in S. Oberthür and T. Gehring (eds.), *Institutional Interaction in Global Environmental Governance: Synergy and Conflict among International and EU Policies*. Cambridge, MA: MIT Press, pp. 79–102.

Siebert, H. 2003. 'On the fears of the international division of labor: eight points in the debate with anti-globalizationers', in H. Siebert (ed.), *Global Governance: An Architecture for the World Economy*. Berlin: Springer, pp. 3–23.

United Nations Environment Programme 2007. *Evolution of the Montreal Protocol*. http://ozone.unep.org/Ratification_status/evolution_of_mp.shtml.

White House 2007. *Fact Sheet: Major Economies Meeting on Energy Security and Climate Change*. http://georgewbush-whitehouse.archives.gov/news/releases/2007/09/20070927.html.

Young, O. R. 2008. 'Deriving insights from the case of the WTO and the Cartagena Protocol', in O. R. Young, W. B. Chambers, J. A. Kim and C. ten Have (eds.), *Institutional Interplay: Biosafety and Trade*. Tokyo: United Nations University Press, pp. 131–158.

3

The consequences of a fragmented climate governance architecture: a policy appraisal

FARIBORZ ZELLI, FRANK BIERMANN, PHILIPP PATTBERG AND HARRO VAN ASSELT

3.1 Introduction

This chapter complements the analysis in Chapter 2 by a policy-oriented inquiry of how different degrees of fragmentation of governance architectures are likely to affect the environmental effectiveness of policies. Our study relates here to an area of widespread contestation in academic and policy writing. It is often maintained, as we describe further below, that a more integrated climate governance architecture would promise higher effectiveness. This claim, however, is also contested, and several authors emphasize the potential benefits of a multitude of agreements, institutions and approaches within an overall fragmented architecture. Claims in favour and against stronger or lesser fragmentation are found in a variety of literatures, ranging from international relations and international law to the comparative study of environmental policy. We review these claims here,[1] organized along the questions of: (1) the relative speed of reaching agreements; (2) the level of regulatory ambition that can be realized; (3) the level of potential participation of actors and sectors; and (4) the equity concerns involved.

The four aspects of speed, ambition, participation and equity are interrelated and eventually will have a bearing on overall governance performance.[2] Based on our typology in the previous chapter (Biermann *et al.*, this volume, Chapter 2), we view the propositions as a continuum of different claims as to the relative positive or negative consequences of higher (conflictive) or lower (synergistic) degrees of fragmentation.

3.2 Methodology

For this qualitative assessment, we reviewed and discussed the state of the art in the scholarly literature regarding the promises or perils of fragmentation of global

[1] This chapter draws on Biermann *et al.* (2009).

[2] While we use these four aspects here to structure arguments on the *consequences* of fragmentation, the criteria presented in the previous chapter (Biermann *et al.*, this volume, Chapter 2) in Table 2.1 help assess the *degree* of fragmentation.

Global Climate Governance Beyond 2012: Architecture, Agency and Adaptation, eds. F. Biermann, P. Pattberg and F. Zelli. Published by Cambridge University Press.

governance architectures. We analysed different bodies of literature, comprising writings on international law, international relations and cooperation theory in general as well as more specific writing on global environmental governance and institutional interlinkages. We contrasted these bodies of literature with evidence from current climate negotiations. As a further 'reality check', we discussed the pros and cons of fragmentation repeatedly with international experts of the Contact Group of the ADAM Project and other experts of the project.

3.3 Analysis

3.3.1 Speed

Proponents of fragmentation in governance architectures emphasize, first, that agreements that encompass merely a few yet important countries may on average be faster to negotiate and to enter into force. Fragmentation, in its cooperative form with different memberships, loosely integrated institutions and common core norms, could thus be a positive quality of governance architectures, or at least not a reason for concern. Concerning climate governance, Victor for instance favours a 'club' approach that involves few nations that would negotiate and review climate policy packages (Victor 2007). Others have suggested that the United States should conclude alternative, regional agreements with like-minded countries, for example in Latin America or with China and, possibly, other key developing countries (Stewart and Wiener 2003). Bodansky, for instance, argued for an 'institutional hedging strategy' with the United States becoming the creator of 'a more diversified, robust portfolio of international climate change policies in the long term' (Bodansky 2002: 1). In terms of the criterion of actor constellation, such regional or small-party agreements could cover only the world's largest greenhouse gas emitters and allow for experimentation of alternative international climate regulatory frameworks. For some, such an approach would allow to negotiate only with the more 'moderate' developing countries, while disabling 'the hard-line developing countries … to prevent more moderate developing states from joining' (Bodansky 2002: 6). Likewise, Barrett (2007) argues for a 'multi-track climate treaty system, with protocols for research and development into mitigation technologies; the development and diffusion of these technologies; funding for adaptation; and geo-engineering'. Similarly, Sugiyama and Sinton (2005) suggest an 'orchestra of treaties' that would have many elements described here as cooperative fragmentation. This orchestra of treaties would complement the climate convention with a focus on mitigation and adaptation technologies, clean development in developing countries and carbon markets. Countries could apply a pick-and-choose strategy and sign only those treaties that promote their interests.

However, it is doubtful whether the speed of reaching small-n initial agreements may indeed improve the overall governance performance. An architecture with a cooperative or conflictive degree of fragmentation may produce solutions that fit the interests only of the few participating countries. There is no guarantee that other countries will join. A quick success in negotiating small-n agreements might run counter to the long-term success, when important structural regime elements have not sufficiently been resolved. A certain degree of instant problem solving through a small-n agreement might provide disincentives for third countries to engage in climate action and could further disintegrate the overall negotiation system. For example, McGee and Taplin argue that specific features of the Asia–Pacific Partnership reduced compliance incentives for parties to the Kyoto Protocol or may even motivate countries to leave the protocol based on utilitarian calculations (McGee and Taplin 2006).

The 1987 Montreal Protocol illustrates many of these problems: even though the protocol was negotiated relatively quickly within the OECD group, major developing countries did not accept it. Two years after adoption of the protocol, only 10 had ratified the treaty, and of the 13 developing countries whose chlorofluorocarbon consumption appeared to rise in 1987 most sharply, only Mexico, Nigeria and Venezuela had joined (Kohler *et al.* 1987). The architecture of ozone governance was thus, in the beginning, rather fragmented. In August 1989, a UN working group[3] hence warned that 'for the Protocol to be fully effective … all countries must become Parties'. Both China and India agreed to ratify the treaty only after substantial changes to its basic structure had been made. In the ozone regime, the Southern contribution to the problem was small, yet threatened to grow. In climate governance, the Southern role is much larger from the outset. Regional agreements of a few like-minded actors, in the hope that others will later follow, do not promise to bring the long-term trust and regime stability that is needed in the climate domain. An 'institutional hedging strategy' (Bodansky 2002) with different policies and regimes scattered around the globe might hence eventually move towards a more conflictive degree of fragmentation with conflicting norms and different actors supporting different institutions. This however might cause havoc to the larger goal of building long-term stable climate governance (Müller *et al.* 2003; Biermann 2005).

3.3.2 Ambition

Some strands of cooperation theory suggest that small-n agreements within a fragmented architecture might prove more progressive and far-reaching. While a

[3] Informal Working Group of Experts on Financial Mechanisms for the Implementation of the Montreal Protocol (1989: para. 8).

universal architecture might include all nations and ideally even reach full compliance, its eventual norms and standards could be rather low and modest. 'Narrow-but-deep' agreements that achieve substantial policy goals with relatively little participation may be superior to a situation of a less demanding regime even if it has full participation and compliance ('broad-but-shallow') (Aldy *et al.* 2003). A fragmented architecture could also increase opportunities for side-payments. Bilateral agreements among countries may allow for concessions that governments would find unacceptable to grant to a larger group of states. Such concessions could include bilateral trade concessions, the bilateral exchange of technology or support for enhanced political influence in international organizations.

Some strands in the literature on environmental policy analysis also suggest that fragmentation and regulatory diversity increase innovation and thus overall governance performance (Jänicke and Jacob 2006). In federal political systems, for instance, regulatory competition may allow for the development of different solutions in different regulatory contexts, of which the most effective will 'survive' and be diffused to other regulatory contexts. Fragmentation may enhance innovation at the level of the firm or public agency and increase innovation in the entire system. A key tenet is the notion of diffusion of innovation, including innovations of policies, technologies, procedures and ideas. This is also central to the claim of environmentally beneficial consequences of trade, which would reduce artificial barriers to the free transfer of technologies and products and thus increase efficiency and innovation (Tews *et al.* 2003). One example of this line of thought is Stewart and Wiener, who proposed that the United States should initially stay outside the Kyoto framework and rather seek a new framework with China and, possibly, other key developing countries. This would address the world's two largest greenhouse gas emitters and allow for experimentation of alternative international climate regulatory frameworks (Stewart and Wiener 2003).

However, it is doubtful whether short-term benefits through small-n agreements will increase the long-term performance of the governance system. A quick success in negotiating small-n agreements might run counter to long-term success, when important structural regime elements (for example inclusion of the principle of common but differentiated responsibilities) have not sufficiently been resolved (Biermann 2005). At a later stage, when interest-constellations change and new situations arise, it might be difficult to reach agreement within the international community without an existing overall agreement that includes those structural elements. In addition, smaller agreements only with few like-minded countries will decrease the opportunity for creating package deals, which will minimize the overall policy acceptance and effectiveness (also Zelli and van Asselt, this volume, Chapter 6).

Economic modelling projects that compared different hypothetical universal and fragmented climate regimes – based on criteria of environmental effectiveness, cost effectiveness and cost distribution – also concluded that the more fragmented a regime is, the higher the costs are to stabilize greenhouse gas concentrations at low levels, because more ambitious reduction targets need to be achieved by a smaller number of countries (Hof *et al.*, this volume, Chapter 4). As Aldy *et al.* (2003: 378) concur, '[c]urrent understanding of the benefit and cost functions characterizing climate change suggest that the latter type of policy [broad-but-shallow] is more likely to satisfy the dynamic efficiency criterion. Since marginal emissions control costs increase steeply, a broad-but-shallow policy would result in lower overall costs.'

Similarly, economic model calculations show that emission trading brings both higher environmental effectiveness and cost effectiveness if based on a universal architecture. If one compares the relative costs of four possible architectures for emissions trading – global trading based on the Kyoto Protocol, formal linking of regional emission trading systems, indirect linkages of regional emissions trading through common acceptance of credits, and a mixed approach that combines elements of these three scenarios – then one finds that an environmentally ambitious global trading approach is best for controlling global emissions. Formal linking of emission trading systems can be a fallback option. A more fragmented architecture, for example through indirect linking, may enhance the efficiency of reduction efforts but will not lead to a comprehensive and effective response (Flachsland *et al.*, this volume, Chapter 5).

In addition, regulatory fragmentation in combination with free trade and economic competition might result in the general decline of environmental standards – a 'race to the bottom'. This hypothesis has only limited empirical support regarding current environmental policies. However, the increasing future needs of more stringent environmental policies, notably in climate governance, will also increase costs of regulation, which will then make regulatory differentials in some sectors more relevant for a 'race to the bottom' scenario. This problem is central to domestic complaints by energy-intensive industries in many countries (van Asselt and Biermann 2007). Related is the concern of a general regulatory 'chaos' in environmental policy, but also in associated areas such as energy, transport or agriculture (Massey 2008). For example, investors in the Kyoto Protocol's Clean Development Mechanism have emphasized the importance of clear signals of a long-term commitment of all actors to one stable process (Stripple and Lövbrand, this volume, Chapter 11). In sum, in particular governance architectures with conflictive types of fragmentation – that is, that do not unite all major actors in one coherent and consistent regulatory framework and that include conflicting norms and principles – are likely to send confusing messages to all, thus reducing the overall performance of the system.

3.3.3 Participation

Some suggest that a higher degree of fragmentation might reduce entry costs for actors, including private entities such as industry and business. The role of private actors and new forms of governance beyond the state are a key concern in recent institutional scholarship on the environment (Part II, this volume, Chapters 9–13; also Falkner 2003; Jagers and Stripple 2003; Pattberg 2005). A loose network of various institutions, many of which might be public–private, could make it easier for business actors to engage in rule-making and thus help creating regulatory systems that are easy to implement and affordable from a business perspective. In addition, a fragmented governance architecture might make it easier to broaden the coverage of relevant sectors. A positive understanding of fragmentation, in particular in its cooperative and synergistic variations, could circumvent negotiation stalemates among countries that may have been caused by the attempt of finding universal agreement. For example, the Kyoto Protocol does not yet require emission reductions from aviation and international maritime transport, whereas the European Commission took up aviation in the EU emissions trading scheme. Thus, higher degrees of cooperative fragmentation where key norms are not in conflict may allow for more and different policy approaches, which could allow for the inclusion of more relevant actors and areas than would be feasible through a more integrated but static architecture.

Yet again, serious problems may outweigh benefits. First, conflictive fragmentation, where different actors pull in different directions, may complicate linkages with other policy areas. There may be strong economic implications – in terms of international competitiveness – if one coalition of states adopts a stringent policy (for example binding emission caps), while other coalitions opt for a less rigorous way of reducing emissions (for example voluntary pledges). This, in turn, could have severe ramifications for the world trade regime that unites both coalitions under one uniform umbrella. A less fragmented architecture, on the other hand, could allow for systematic and stable agreements between the institutional frameworks of the world trade regime and climate institutions. Since a fragmented architecture may decrease entry-costs for private actors, it is also conceivable that business actors use regulatory fragmentation to choose among different levels of obligation, thereby starting a race to the bottom within and across industry sectors (Vormedal 2008).

3.3.4 Equity

A fragmented architecture might offer solutions that are specifically tailored for specific regions and thus increase equity by better accounting for special circumstances. Reinstein, for example, proposed a bottom–up process in which countries – similar to trade negotiations – would put on the table acceptable climate policies and measures in line with national circumstances (Reinstein 2004). Some lawyers also argue that increased fragmentation in international law is a way of accommodating different interests of states. As a result, specialized regimes may better serve the interests of governments and have higher compliance rates. On this account, Hafner (2004: 859) argues that a 'less-than-global approach seems particularly necessary when different States clearly hold different beliefs about what basic values should be preserved by international regulation'.

Yet, fragmented architectures also raise serious concerns of equity and fairness (Winkler, this volume, Chapter 7; Shrivastava and Goel, this volume, Chapter 8). Cooperation theory assumes that bilateral and small-n agreements grant more bargaining power to larger and more influential countries, while large-n agreements allow smaller countries to enter into coalitions, such as the Group of 77 and China, that protect their collective interests from the interest of the larger countries (Biermann 1998). In the end, perceptions of inequity and unfairness are linked to policy effectiveness through its legitimacy – a governance system that is not seen as fair by all parts of the international community is likely to lack in overall effectiveness. As stressed by Benvenisti and Downs (2007: 626), 'powerful states have increasingly turned to fragmentation to maintain their control'. Fragmentation allows powerful states to opt for a mechanism that best serves their interests, in the form of forum shopping (Hafner 2000), or to create new agreements if the old ones no longer fit their interest.

In the same vein, many climate-related initiatives like the Asia–Pacific Partnership include leading industrialized and developing countries while excluding least-developed countries (Ott 2007). The investment agendas of these initiatives hence do not reflect the immediate interests of many of those countries that are most affected by climate change. The bulk of developing countries thus continue to support the multilateral approach in climate policy, similar to other policy domains (Shrivastava and Goel, this volume, Chapter 8). Less fragmented and more integrated architectures allow the South to count on its numbers in diplomatic conferences and gain bargaining power from a uniform negotiation position. They allow for side-payments across negotiation clusters within a policy domain and across different policies and they minimize the risk for developing countries to be coerced into bilateral agreements with powerful nations that might offer them suboptimal negotiation outcomes (Abrego *et al.* 2003). For the many smaller and medium-sized developing countries, unity is strength, and multilateralism may seem its core guarantee. Since the emergence of the climate issue, the South has therefore sought to bring all negotiations under the UN framework and to frame global warming as an overarching political problem with implications far beyond mere environmental policy.

3.4 Conclusions and policy recommendations

This chapter has discussed the potential consequences of different degrees of fragmentation of global governance architectures. We found that different types of fragmentation are likely to have different degrees of performance. While cooperative forms of fragmentation may entail both significant costs and benefits, we did not find convincing arguments in favour of a high, or conflictive, degree of fragmentation. On balance, conflictive fragmentation of global governance architectures puts a burden on the overall performance of the system. On the other hand, what we described as 'synergistic fragmentation' might often be a realistic second-best option in a world of diversity and difference in which purely universal governance architectures are more a theoretical postulate than a real-life possibility.

This raises the policy question of how to minimize extreme cases of conflictive fragmentation and how to address some of the rather negative effects of cooperative fragmentation. This policy question is particularly important for the area of climate governance. To increase synergies within UN climate governance, it seems crucial to better integrate processes under the climate convention and the Kyoto Protocol and to reduce duplication, for instance in the current parallel negotiations on technology transfer in different arenas (Zelli and van Asselt, this volume, Chapter 6). Negotiations leading to future agreements ought to address key topics – such as deforestation, technology transfer or capacity-building – in only one forum. Regarding the cooperative and partially conflictive fragmentation between UN climate governance and climate arrangements outside this umbrella, it is imperative to open these institutions to additional members. For example, the Asia–Pacific Partnership could be broadened to also include least developed countries and small-island developing states, and to ensure through formal declarations or clauses better integration with the overall UN processes. Furthermore, formal coordination between these arrangements and the UN negotiations could ensure that they work towards common objectives. The UN climate regime also needs to be better coordinated with non-environmental institutions in order to minimize conflictive fragmentation, most importantly with regard to the WTO.

Russia's ratification of the Kyoto Protocol has evidenced that linking both arenas can create additional incentives for countries to support climate policies. Better integration can help identifying similar constellations of actors. For instance, like the climate regime, the WTO is hosting discussions on the transfer of climate-friendly goods and services in the special session of the WTO Committee on Trade and Environment (Zelli and van Asselt, this volume, Chapter 6). As long as this WTO-internal discussion is not linked to similar debates in the climate regime, a comprehensive solution is unlikely. Policy-makers have recognized this problem: in 2007, trade ministers, senior trade officials and the WTO Director-General met for the first time during a conference of the parties to the climate convention to discuss trade-related aspects of climate change. Yet also this meeting reflected the increasing fragmentation of the climate governance domain, with only few countries – and none from Africa – represented.

Our qualitative analysis also shows that – with regard to our main appraisal question for the 'architecture' part of this volume – major scholarly literatures offer conflicting statements on the relative advantages and disadvantages of fragmentation. This calls, we argue, for a continuation of this line of work through more in-depth studies of fragmentation. Such studies could also provide theory-driven explanations for the causes and consequences of fragmentation of given architectures, as well as for possible changes of the degree of fragmentation over time. This chapter offers a starting point on which further research can build.

Acknowledgements

We are indebted to the members of the ADAM Contact Group who provided an invaluable 'reality check' for the arguments in favour and against the institutional fragmentation of the global climate governance architecture: Marcel Berk, Daniel Bodansky, Chandrashekhar Dasgupta, Dagmar Droogsma, Bo Kjellén, Benito Müller, Lars Müller, Willem Thomas van Ierland and Michael Wriglesworth.

References

Abrego, L, C. Perroni, J. Whalley and R. M. Wigle 2003. 'Trade and environment: bargaining outcomes from linked negotiations', *Review of International Economics* **9**: 414–428.

Aldy, J. E., S. Barrett and R. N. Stavins 2003. 'Thirteen plus one: a comparison of global climate policy architectures', *Climate Policy* **3**: 373–397.

Asselt, H. van and F. Biermann 2007. 'European emissions trading and the international competitiveness of energy-intensive industries: a legal and political evaluation of possible supporting measures', *Energy Policy* **35**: 297–306.

Barrett, S. 2007. 'A multitrack climate treaty system', in J. E. Aldy and R. N. Stavins (eds.), *Architectures for Agreement: Addressing Global Climate Change in the Post-Kyoto World*. Cambridge, UK: Cambridge University Press, pp. 237–259.

Benvenisti, E. and G. W. Downs 2007. 'The empire's new clothes: political economy and the fragmentation of international law', *Stanford Law Review* **60**: 595–632.

Biermann, F. 1998. *Weltumweltpolitik zwischen Nord und Süd: Die neue Verhandlungsmacht der Entwicklungsländer [Global Environmental Policy between North and South: The New Bargaining Power of Developing Countries]*. Baden-Baden: Nomos.

Biermann, F. 2005. 'Between the United States and the South: strategic choices for European climate policy', *Climate Policy* **5**: 273–290.

Biermann, F., P. Pattberg, H. van Asselt and F. Zelli 2009. 'The fragmentation of global governance architectures: a framework of analysis[1]. *Global Environmental Politics* **9**(4): 14–40.

Bodansky, D. 2002. *U.S. Climate Policy after Kyoto: Elements for Success*. Washington, DC: Carnegie Endowment for International Peace. www.carnegieendowment.org/files/Policybrief15.

Echersley, R. 2004. 'The big chill: The WTO and multilateral environmental agreements', *Global Environmental Politics* **4**(2): 24–40.

Falkner, R. 2003. 'Private environmental governance and international relations: exploring the links', *Global Environmental Politics* **3**: 72–87.
Hafner, G. 2000. 'Risks ensuing from fragmentation of international law', in *Official Records of the General Assembly* Fifty-fifth session. Supplement No. 10 (A/55/10, 2000), Annex, pp. 326–354.
Hafner, G. 2004. 'Pros and cons ensuing from fragmentation of international law', *Michigan Journal of International Law* **25**: 849–863.
Informal Working Group of Experts on Financial Mechanisms for the Implementation of the Montreal Protocol on Substances that Deplete the Ozone Layer 1989. Report of the meeting in Geneva, 3–7 July 1989. UN Doc. UNEP/OzL.Pro.Mech.1/Inf.1 of 16 August 1989.
Jagers, S. C. and J. Stripple 2003. 'Climate governance beyond the state', *Global Governance* **9**: 385–399.
Jänicke, M. and K. Jacob 2006. 'Lead markets for environmental innovations: a new role for the nation state', in M. Jänicke and K. Jacob (eds.), *Environmental Governance in Global Perspective: New Approaches to Ecological Modernisation*. Berlin: Centre for Environmental Policy Analysis, pp. 30–50.
Kohler, D. F., J. Haaga and F. Camm 1987. *Projections of Consumption of Products Using Chlorofluorocarbons in Developing Countries*. Santa Monica, CA: RAND Corporation.
Massey, E. 2008. 'Global governance and adaptation to climate change for food security', in F. Zelli (ed.), *Integrated Analysis of Different Possible Portfolios of Policy Options for a Post-2012 Architecture*, ADAM Project report No. D-P3a.2b. Norwich, UK: Tyndall Centre for Climate Change Research, pp. 143–153.
McGee, J. and R. Taplin 2006. 'The Asia-Pacific Partnership on Clean Development and Climate: a competitor or complement to the Kyoto Protocol', *Global Change, Peace and Security* **18**: 173–192.
Müller, B., with contributions of J. Drexhage, M. Grubb, A. Michaelowa and A. Sharma 2003. *Framing Future Commitments: A Pilot Study on the Evolution of the UNFCCC Greenhouse Gas Mitigation Regime*. Oxford, UK: Oxford Institute for Energy Studies.
Ott, H. E. 2007. 'Climate policy post-2012 – a roadmap: the global governance of global change', a discussion paper for the 2007 Tällberg Forum. www.wupperinst.org/uploads/tx_wibeitrag/Ott_Taellberg_Post-2012.
Pattberg, P. 2005. 'The institutionalization of private governance: how business and non-profit organizations agree on transnational rules', *Governance* **18**: 589–610.
Reinstein, R. A. 2004. 'A possible way forward on climate change', *Mitigation and Adaptation Strategies for Global Change* **9**: 245–309.
Stewart, R. B. and J. B. Wiener 2003. *Reconstructing Climate Policy: Beyond Kyoto*. Washington, DC: AEI Press.
Sugiyama, T. and J. Sinton 2005. 'Orchestra of treaties: a future climate regime scenario with multiple treaties among like-minded countries', *International Environmental Agreements: Politics, Law and Economics* **5**: 65–88.
Tews, K., Busch, P. O. and H. Jörgens 2003. 'The diffusion of new environmental policy instruments', *European Journal of Political Research* **42**: 569–600.
Victor, D. G. 2007. 'Fragmented carbon markets and reluctant nations: implications for the design of effective architectures', in J. E. Aldy and R. N. Stavins (eds.), *Architectures for Agreement: Addressing Global Climate Change in the Post-Kyoto World*. Cambridge, UK: Cambridge University Press, pp. 123–163.
Vormedal, I. 2008. 'The influence of business and industry NGOs in the negotiation of the Kyoto mechanisms: the case of carbon capture and storage in the CDM', *Global Environmental Politics* **8**: 36–65.

4

Environmental effectiveness and economic consequences of fragmented versus universal regimes: what can we learn from model studies?

ANDRIES HOF, MICHEL DEN ELZEN AND DETLEF VAN VUUREN

4.1 Introduction

The thirteenth conference of the parties of the climate convention had launched a negotiation process to craft a new international climate change agreement by the end of 2009. This agreement would need to stipulate emission reduction commitments, specify essential actions to adapt to the impacts of climate change and mobilize the necessary funding and technological innovation. Given these enormous challenges, the structure and design of a future climate agreement are still unclear. Besides the negotiations within the UN climate regime, major greenhouse gas emitting countries are also leading ad hoc debates in other forums, for example in the context of the Group of Eight and the Asia–Pacific Partnership on Clean Development and Climate. Depending on the course of these processes, a new climate governance regime could develop in different directions; it could end somewhere between a universal, inclusive governance architecture and a strongly fragmented, heterogeneous governance architecture (Biermann *et al.*, this volume, Chapter 2).

In recent years, numerous universal and fragmented climate regimes have been proposed (for an overview, see Bodansky 2004; Blok *et al.* 2005; Philibert 2005; IPCC 2007: 770–773). Many of these regimes are quantitatively or qualitatively assessed, but no attempt has yet been made to compare the costs estimates of these studies for specific regions under different regimes. Nevertheless, the available material allows us to make an assessment of the regional costs of several universal and fragmented regimes, based on different models. This chapter presents a literature review concerning the economic effectiveness of a number of possible universal and fragmented regimes. We use only studies that quantitatively assess both emission reductions and costs. From a quantitative perspective, this chapter tries to answer the appraisal question of the 'architecture' domain of this book, namely whether a universal or a fragmented regime will be more effective to reduce greenhouse gas emissions.

Global Climate Governance Beyond 2012: Architecture, Agency and Adaptation, eds. F. Biermann, P. Pattberg and F. Zelli. Published by Cambridge University Press.

The chapter is structured as follows. Section 4.2 describes the methodology, including the criteria used for inclusion of the assessment, a typology of regimes, how we compared the studies and how we dealt with emissions trading (see Flachsland *et al.*, this volume, Chapter 5, for a detailed analysis on emissions trading). Section 4.3 analyses the universal and fragmented regimes based on our criteria. Finally, Section 4.4 concludes and briefly maps the policy options discussed.

4.2 Methodology

4.2.1 Criteria for inclusion in the assessment

While there are many criteria to evaluate climate regimes (den Elzen *et al.* 2003: 186; Höhne *et al.* 2003: 33–34; Bell *et al.* 2005: 33), the most important ones according to the IPCC are: environmental effectiveness, cost effectiveness, distributional effects and institutional feasibility (IPCC 2007: 751). Our focus is on the first three criteria, in order to bring our criteria in line with those of Flachsland *et al.* (this volume, Chapter 5). Environmental effectiveness relates to the reduction in greenhouse gas emissions that can be achieved by a regime. Other environmental effects, such as air quality, could also be included, but we focus only on greenhouse gas emissions. We address the economic consequences of climate regimes in terms of their cost effectiveness and distributional effects. Cost effectiveness relates to the extent to which the policy can achieve its objectives at a minimum cost to society (also see Section 4.2.5), while distributional effects relate to the distributional consequences of a policy, which includes dimensions such as equity or fairness (IPCC 2007: 751). These aspects are important because the probability of achieving an agreement will be reduced if the cost effectiveness is low or if the distributional effects – the differences in abatement costs between individual countries or groups of countries – are high.

So far, more than 50 climate regimes with different goals and/or actions have been proposed in the literature (IPCC 2007: 770–773). Our review includes only those studies that quantitatively assess regimes in terms of emission reductions and regional costs and which focus on the post-2012 period. The only exception is the assessment of a carbon tax, for which we use two different models. Direct abatement costs are projected using the Integrated Model to Assess the Global Environment (IMAGE 2.3) framework, which includes the energy model Targets Image Energy Regional (TIMER) 2.0 coupled to the Framework to Assess International Regimes for differentiation of commitments (FAIR) (van Vuuren *et al.* 2007). GDP losses are estimated by linking the IMAGE 2.3 model with the ENV-Linkages model (Bakkes and Bosch 2008). IMAGE is a dynamic integrated assessment modelling framework for global change, aimed at supporting decision-making by quantifying the relative importance of major processes and interactions in the society–biosphere–climate system.[1]

[1] For more information, see www.mnp.nl/image.

ENV-Linkages is a global macroeconomic general equilibrium model containing 26 sectors and 34 world regions and provides economic projections for multiple time periods.

4.2.2 Typology of regimes

The fundamental difference between a universal and fragmented regime is that the former involves a single comprehensive climate regime that applies to all countries (Biermann *et al.*, this volume, Chapter 2). This means that universal climate regimes require full participation of all countries, at least gradually, in the same international agreement, whereas fragmented climate regimes never achieve full participation in a single international agreement.

Figure 4.1 classifies the assessed regimes according to the number of participating countries. Regimes with more than one agreement or without full (gradual) participation are fragmented regimes. Since for our modelling analysis we have to rely on quantifiable criteria, our distinction of fragmented and universal regimes is not completely congruent with the distinction introduced by Biermann *et al.* (this volume, Chapter 2). These authors also refer to participation in terms of 'actor constellation', but in addition, they rely on qualitative criteria, namely 'institutional overlaps' and 'norm conflicts'. We, however, focus on the quantifiable criteria of participation and number of agreements. We regard a regime as more universal when more countries participate and when there are fewer different agreements involved. Therefore, it is possible to have a fragmented regime in which all countries contribute in some way to reduce emissions, although using different agreements. An example would be a regime in which the United States, Australia, India, China, Japan and South Korea continue to focus on cooperation on development and transfer of technology within the Asia–Pacific Partnership on Clean Development and Climate, while the rest of the world uses a system of absolute emission targets with the possibility of emission trading (Biermann *et al.*, this volume, Chapter 2). Universal regimes can have a higher or lower degree of participation as well. Regimes in which all countries participate immediately, based on one common rule, are perfectly universal. An example would be 'contraction and convergence', in which all countries participate according to the rule of converging per capita emissions. A number of universal regimes, however, have a gradual participation approach or staged system approach. In the latter approach, countries participate in a system with stages and stage-specific targets, where the transition between stages is a function of various indicators, such as per capita income thresholds (Gupta 1998; Berk and den Elzen 2001; Höhne *et al.* 2003). Another crucial factor that qualifies regimes is the type of target, where two broad distinctions can be made:

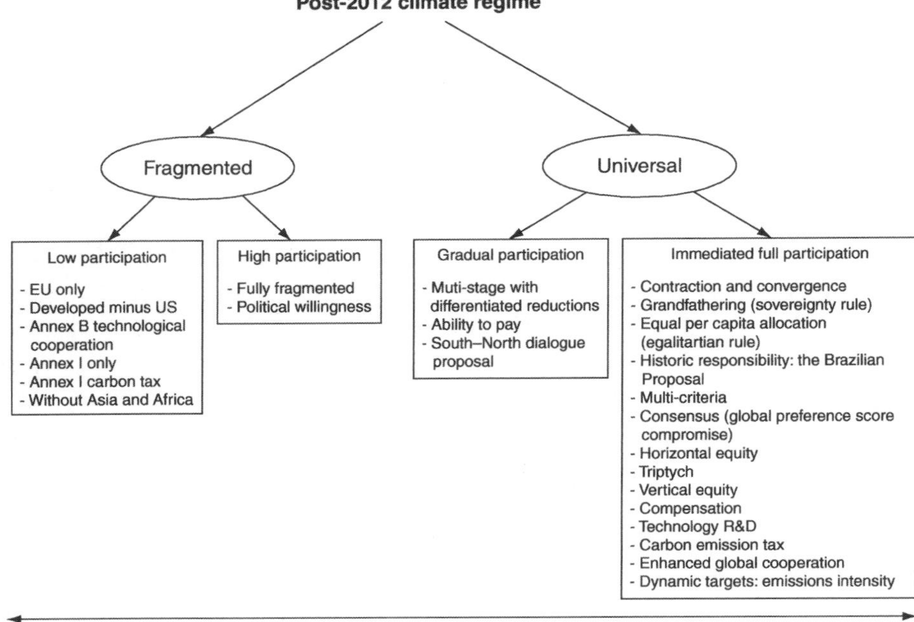

Figure 4.1 Quantitatively assessed post-2012 fragmented and universal climate regimes. Tables 4.1, 4.2 and 4.7 explain the different climate regimes mentioned in Figure 4.1.

(1) *Regimes with predefined emission targets* These regimes specify targets in terms of emission reductions. Often these emission targets are derived from a certain climate target, like the concentration stabilization level. Most universal regimes are of this type.
(2) *Regimes without predefined emission targets* These regimes do not include emission targets, but instead technological cooperation or a carbon emission tax, for instance. Fragmented and universal regimes can both be of this type.

Because these two types of regimes are difficult to compare, we analyse them separately.

4.2.3 Comparing quantitative assessments focusing on costs and reductions

Comparing studies that assess various climate regimes is not always a straightforward task. Studies use different assumptions and report different cost measurements. General equilibrium models usually measure costs in GDP or welfare loss, whereas in partial equilibrium models (energy system models) costs are measured in direct abatement costs. These two cost measurements are in absolute values not directly comparable. Therefore, we focus on the relative costs of different regions using both metrics, while Flachsland *et al.* (this volume, Chapter 5) consider direct abatement costs only.

For most regimes, the actual design ('regime parameters') is crucial for the outcome. Such regime parameters can include the overall concentration target (for example 450, 550 or 650 ppm carbon dioxide-equivalent), the baseline (high or low), discount rate (level and type) and marginal abatement costs. Other assumptions are regime-specific. For example, for the 'contraction and convergence' regime the convergence year can strongly determine the outcomes. Our analysis states for every assessment the major parameter settings, the cost measurement and the details of the regime.

4.2.4 Emission trading

Almost all regimes can be constructed so that emission trading is either allowed or not allowed. Many studies analyse the effect of emission trading or the clean development mechanism on the costs of emission abatement; all studies conclude that emission trading and the clean development mechanism decreases the cost of abatement (see, for example, Leimbach 2003: 1041; Böhringer and Welsch 2004: 32; Bollen *et al.* 2005: 16; Russ *et al.* 2005: 21; European Commission 2007: 48). This holds for all universal regimes (both full and gradual participation) and to a lesser extent for the fragmented climate regimes (depending on the size of coalitions). Emission trading not only reduces the overall costs of the regime, but also decreases the costs for every country participating in the coalition. To effectively compare the economic consequences of various climate regimes, we restrict our study to regimes that assume full emission trading between countries participating in the same coalition. However, in the case of fragmented regimes, none of the studies we consider assumes the possibility of trade between countries belonging to different coalitions. Flachsland *et al.* (this volume, Chapter 5) provide such an analysis of the environmental and cost effectiveness of fragmented carbon markets as compared to more integrated approaches for emission trading.

4.3 Analysis

4.3.1 Universal regimes

The majority of climate regime proposals presume global international climate negotiations with the goal of a single, comprehensive regime (IPCC 2007: 770–773). Most regimes predefine a global emission target and then apply allocation rules specifying the allocation of global emission reductions to countries or regions. In universal regime proposals, the global emission target can be set at any preferred level. In most cases, the authors looked at global short-term emission targets that are compatible with meeting long-term concentration stabilization targets in the range of

Table 4.1 *Universal regimes with a global emission target*

Name	Short description	Evaluated in terms of emission reductions and costs by
Full participation regimes		
Contraction and convergence	Distribute permits so per capita emissions converge in a certain year	Blanchard (2002), Böhringer and Helm (2008), Böhringer and Welsch (2004; 2006), Bollen *et al.* (2004, 2005), Criqui *et al.* (2003), den Elzen and Lucas (2005), den Elzen *et al.* (2005, 2008), Leimbach (2003), Manne and Richels (1995), Manne *et al.* (1995), Nakicenovic and Riahi (2003), Persson *et al.* (2006), Peterson and Klepper (2007)
Grandfathering (sovereignty rule)	Distribute permits in proportion to current emissions	Böhringer and Welsch (2006), Böhringer and Löschel (2005), Bollen *et al.* (2005), Buchner and Carraro (2003), den Elzen and Lucas (2005), Peterson and Klepper (2007), Rose *et al.* (1998)
Equal per capita allocation (egalitarian rule)	Distribute permits in proportion to population	Böhringer and Helm (2008), Böhringer and Welsch (2006), den Elzen and Lucas (2005), Persson *et al.* (2006), Rose *et al.* (1998), Wicke and Böhringer (2005)
Historic responsibility – the Brazilian Proposal	Distribute permits in proportion to the contribution of climate change over a certain period	Blanchard (2002), den Elzen and Lucas (2005), den Elzen *et al.* (2005), Rive *et al.* (2006)
Multi-criteria	Distribute permits based on a formula including several variables, such as population, GDP and others	den Elzen and Lucas (2005), Vaillancourt and Waaub (2006)
Consensus (global preference score compromise)	A combination of per capita allocation and grandfathering	den Elzen and Lucas (2005), Rose *et al.* (1998)
Horizontal equity	Distribute permits to equalize net welfare change as per cent of GDP	Rose *et al.* (1998), Vaillancourt *et al.* (2008)
Vertical equity	Progressively distribute permits proportions inversely correlated with per capita GDP	Rose *et al.* (1998), Vaillancourt *et al.* (2008)

Table 4.1 (*cont.*)

Name	Short description	Evaluated in terms of emission reductions and costs by
Triptych	National emission targets based on sectoral considerations	den Elzen and Lucas (2005)
Gradual participation regimes		
Multi-stage with differentiated reductions	Countries participate with different stages and stage-specific types of targets; countries transition between stages as a function of indicators	Boeters *et al.* (2007), Criqui *et al.* (2003), den Elzen and Lucas (2005), den Elzen *et al.* (2005, 2008)
Ability to pay	Permits are distributed in order to equalize abatement costs as per cent of GDP	Böhringer and Löschel (2005), den Elzen and Lucas (2005)
South–North dialogue proposal	Countries participate in the system with different stages and stage-specific types of targets	den Elzen *et al.* (2007a)

450–650 ppm carbon dioxide-equivalent. A relatively strong point of all universal regimes with predefined targets is that environmental effectiveness is secured. However, in reality this effectiveness is obviously a function of compliance.

Tables 4.1 and 4.2 show allocation proposals with and without predefined global emission targets that have been quantitatively assessed regarding their economic consequences for the post-2012 period. (On studies that evaluated proposals solely in terms of emission reduction targets but not in terms of costs see Höhne *et al.* 2003 and Blok *et al.* 2005.)

Universal regimes with a predefined emission target

'Contraction and convergence'

Of all regimes, the 'contraction and convergence' regime has been analysed most often. The most crucial reason is its simple formulation – which makes it a good reference for any form of allocation. The first step in the 'contraction and convergence' regime is to establish a long-term global emission profile. Then emission rights are allocated so that the per capita emissions converge from their current

Table 4.2 *Economic consequences of universal regimes without a global emission target*

Name	Short description	Evaluated in terms of costs by
Carbon emission tax	All countries agree to a common, international greenhouse gas emission tax	Bollen *et al.* (2005), Manne *et al.* (1995), Peterson and Klepper (2007), Vaillancourt *et al.* (2008)
Dynamic targets: emissions intensity	Targets are defined as a certain reduction of the ratio of carbon dioxide emissions to GDP	Blanchard (2002)
Technology research and development	A regime based on enhanced coordinated technology research and development	Buchner and Carraro (2004)
Enhanced global cooperation	Joint welfare maximization with an additional 10 per cent emission reduction	Buchner and Carraro (2003)

values to a global average in a specific target year. Table 4.3 shows some of the chief parameters used in the various assessments of the 'contraction and convergence' regime.

Comparing the results from these studies is challenging due to the large variations in cost measurements, convergence year and targets. In order to decouple results from specific cost indicators and baseline or reduction targets, we analyse the costs of certain key regions relative to the global average costs as share of GDP. Table 4.4 shows the results.[2] Every study finds substantial cost differences between regions. All studies, except Criqui *et al.* (2003), project high costs for the former Soviet Union and Eastern Europe due to (1) an unfavourable combination of high per capita emissions and low GDP and (2) reduced fossil fuel exports. For the same reasons, most studies also project high costs for the Middle East/North Africa. A few studies, however, expect net benefits for this region. Different projections of the abatement costs in this region are the major reason for these variations in cost projections. All studies agree that India and sub-Saharan Africa will profit from a 'contraction and convergence' regime with a 2050 convergence year, while they expect China to incur lower costs as compared to the global average. Most studies predict that the costs for Europe and the United States will be somewhat above the global average.

[2] The studies of Nakicenovic and Riahi (2003), Manne and Richels (1995) and Manne *et al.* (1995) are not included in this table because they only reported results from a few highly aggregated regions.

Table 4.3 *Parameters of 'contraction and convergence' regime cost assessments*

Study	Regions	Cost measurement	Convergence year	Target
Blanchard (2002)	17	Direct costs 2030	2050	9.4 GtC in 2030[a]
Böhringer and Welsch (2004, 2006), Böhringer and Helm (2008)	11	NPV[b] income	2050	25 per cent below 1990 in 2050[c]
Bollen *et al.* (2004)	6	GDP and income 2040	2050	S550e[d]
Bollen *et al.* (2005)	13	Income 2020	2024	S550e
Criqui *et al.* (2003)	11	Welfare 2025, direct 2025/ 2050	2050 and 2100	S550e and S650e
den Elzen and Lucas (2005)	18	Direct costs 2025 and 2050	2050	S550e and S650e
den Elzen *et al.* (2005)	11	Direct costs 2025 and 2050	2050 and 2100	S550e
den Elzen *et al.* (2008)	10	Direct costs 2020 and 2050	2050	S450e and S550e
Leimbach (2003)	11	Consumption loss 1990–2045 and 2050–2100	2025 and 2100	2 °C
Manne *et al.* (1995), Manne and Richels (1995)	5	NPV GDP	2030 and 2200	Several explored
Nakicenovic and Riahi (2003)	5	GDP 2020, 2050 and 2100	2050 and 2100	S400c, S450c[e]
Persson *et al.* (2006)	5	NPV direct	2050	S450c
Peterson and Klepper (2007)	12	Welfare 2030	2050	S550e

[a] Consistent with a 450 to 550 ppm carbon dioxide stabilization goal (Blanchard 2002).
[b] Net Present Value: lifetime change in costs discounted to present values.
[c] Corresponds to a concentration stabilization level of 550 ppm carbon dioxide-equivalent.
[d] Stabilization greenhouse gas concentration level at 550 ppm carbon dioxide-equivalent.
[e] Stabilization concentration level at 400 and 450 ppm carbon dioxide, which corresponds to a greenhouse gas concentration level of 500 and 550 ppm carbon dioxide-equivalent (den Elzen *et al.* 2003).

Criqui *et al.* (2003), den Elzen *et al.* (2005) and Leimbach (2003) look more closely at the effects of a convergence year of 2100 instead of 2050 or 2025. All three studies conclude that delaying the convergence year reduces costs substantially for industrialized countries. On the other hand, convergence in 2100 leads to much higher costs for developing countries, especially for India and countries in Africa.

Table 4.4 *Regional costs compared to global average costs (1)*

Study	USA/ North America	European Union/ enlarged European Union	FSU/ Russia	Middle East/ Middle East and N Africa	Latin/ South America	Africa/ sub-Saharan Africa	China/ East Asia	India/ South Asia
den Elzen and Lucas (2005), den Elzen *et al.* (2005)[a]	2	2	3	3	2	0	1	0
den Elzen *et al.* (2008)[b]	2	2	3	2	1	0	1	0
Böhringer and Welsch (2004, 2006), Böhringer and Helm (2008)[c]	3	2	3	0	0	0	1	0
Criqui *et al.* (2003)[d]	1	1	0	3	1	0	0	0
Peterson and Klepper (2007)[e]	1	1	3	3	2	0	0	0
Blanchard (2002)[f]	3	2	3	3	NA	0	2	0
Leimbach (2003)[g]	2	1	3	0	NA	0	1	0
Persson *et al.* (2006)[h]	NA	NA	NA	0	0	0	1	0
Bollen *et al.* (2005)[i]	2	2	3	3	3	NA	NA	NA
Bollen *et al.* (2004)[j]	2	2	3	3	NA	NA	NA	NA

The table shows regional costs compared to global average costs for the 'contraction and convergence' regime with emission trading and convergence in 2050 (exceptions: Bollen *et al.* 2005 convergence in 2024; Leimbach 2003 convergence in 2025) for a greenhouse gas concentration stabilization target of 550 ppm carbon dioxide-equivalent or 450 carbon dioxide ppm.

Legend: 0 = no costs or gains, 1 = costs less than global average, 2 = costs between global average and twice the global average, 3 = costs more than twice the global average; NA = not available.

[a] Direct costs in 2025.

[b] Direct costs in 2050.

[c] Net Present Value of change in income.

[d] Change in welfare in 2025.

[e] Change in welfare in 2030.

[f] Direct costs in 2030.

[g] Consumption loss in the period 1990–2045.

[h] Net Present Value of direct costs.

[i] Income in 2020. In this study there is one Rest of World region with large benefits, explaining the fact that all regions reported here incur higher costs than the global average.

[j] Income in 2040. In this study there is one Rest of World region with large benefits, explaining the fact that all regions reported here incur higher costs than the global average.

Other emission allocation regimes.

Of the studies that analyse other universal emission allocation regimes, den Elzen and Lucas (2005) is the most comprehensive. In total, they analyse nine universal regimes with full and gradual participation with respect to regional abatement costs. This can be used to explore whether there are also large variations in the distribution of costs for other universal regimes. Table 4.5 summarizes their results.

All regimes analysed by den Elzen and Lucas are subject to substantial cost differences between regions. Interestingly, the Middle East, the former Soviet Union, Canada and Oceania incur high costs no matter what the regime. The regimes with the smallest cost differences between regions are Triptych, 'multi-stage' with differentiated reductions and the Brazilian proposal on historic responsibility (but this strongly depends on the parameter settings).

Rose *et al.* (1998) also compare several universal allocation regimes. They find the largest cost differences in the equal per capita allocation regime. In this regime, the costs for industrialized countries are especially high; this can be expected, since these countries currently have the highest per capita emissions. They also analyse an outcome-based allocation regime called horizontal equity, in which abatements costs are required to be an equal proportion of GDP for all. By definition, there are no cost differences between countries in such a regime.

The large cost differences for almost all allocation regimes are confirmed by other studies that analyse a single emission allocation regime (Blanchard 2002; Bollen *et al.* 2004; Böhringer and Löschel 2005; Wicke and Böhringer 2005 Böhringer and Welsch 2006; Persson *et al.* 2006; Rive *et al.* 2006; Vaillancourt and Waaub 2006; Peterson and Klepper 2007; Böhringer and Helm 2008). The large variation in the distribution of costs in almost every regime analysed will likely pose significant problems for full participation, even if average global costs are modest. At first glance, the horizontal equity allocation regime seems promising for achieving full participation, since in this regime every country incurs proportionally the same costs, figured as a share of GDP. Nevertheless, this regime is unlikely to achieve full participation, for two reasons. First, for many countries, especially developing ones, it might be unfair that they will have to pay the same costs as industrialized countries – even when calculated as a share of GDP. Second, in every universal regime the problem of free-riding remains (Carraro and Siniscalco 1998; Barrett 1999; Carraro 2000; Tol 2001; Dellink *et al.* 2005; Finus *et al.* 2005; 2006; Eyckmans and Finus 2007).

Universal regimes without a global emission target

We analyse cost projections for four universal regimes without a predefined global emission target (Table 4.2).

Table 4.5 *Abatement costs as per cent of GDP for nine regimes*

Region	GC	CSE	AP	MS	C&C	TT	BP	GF	MCC
Canada	3	3	3	3	3	2	2	2	2
USA	3	3	3	3	2	2	1	1	1
OECD Europe	2	2	2	2	2	2	2	2	1
Eastern Europe	3	2	1	1	1	2	1	1	1
FSU	3	3	1	3	3	3	3	2	2
Oceania	3	3	3	3	2	2	2	2	2
Japan	2	2	2	2	2	2	2	2	1
Central America	1	1	2	2	1	1	2	3	2
South America	2	2	2	2	2	1	2	3	2
Northern Africa	0	0	0	0	1	0	1	0	2
Western Africa	0	0	0	0	0	0	0	0	2
Eastern Africa	0	0	0	0	0	0	0	0	0
Southern Africa	2	1	0	0	3	1	0	0	3
Middle East	3	3	3	3	3	3	3	3	3
South Asia	0	0	0	0	0	0	0	0	1
East Asia	1	1	1	1	1	2	1	2	2
South East Asia	0	0	0	0	1	1	2	1	2

The table shows abatement costs as per cent of GDP in Purchasing Power Parity terms for nine regimes (all allowing for emission trading) in 2025 for greenhouse gas concentration stabilization at 550 ppm carbon dioxide-equivalent.
Legend: 0 = no costs or gains, 1 = costs less than global average, 2 = costs between global average and twice the global average, 3 = costs more than twice the global average; GC: Global preference score compromise, CSE: Equal per capita allocation, AP: Ability to pay, MS: 'multi-stage' with differentiated reductions, C&C: Contraction and convergence, TT: Triptych, BP: Historic responsibility: Brazilian proposal, GF: Grandfathering, MCC: Multi-criteria.
Source: based on den Elzen and Lucas (2005).

Dynamic targets: emissions intensity

Blanchard (2002) analyses the economic consequences of a universal regime based on dynamic targets. This regime defines reduction targets as the ratio of carbon dioxide emissions to GDP. Although there is no global emission reduction target, Blanchard set the dynamic emission intensity targets at levels that stabilize carbon dioxide concentrations at 450 to 550 ppm. Emission intensity targets for industrialized countries are set at a reduction rate of approximately 2 per cent annually from business as usual, while developing countries have to improve their emissions intensity by 0.5 per cent annually. Due to these stricter targets for industrialized countries, the abatement costs in industrialized countries in 2030 are much higher than in developing countries. Abatement costs as share of GDP in industrialized countries range from twice the global average in the European Union to more than six times the global average in countries of the former Soviet Union.

Regimes based on cooperation

Two other comparable universal regimes without a fixed global emission target are analysed by Buchner and Carraro (2003, 2004): Enhanced Global Cooperation and Technology Research and Development. Both regimes focus on cooperation between regions. In the Enhanced Global Cooperation regime, all countries cooperate in such a way that their joint welfare is maximized, and reduce emissions by an additional 10 per cent compared to this 'optimal path'. The Technology Research and Development regime focuses on global cooperation on technical innovation and diffusion, rather than maximization of joint welfare. Both of these regimes are sensitive to parameter settings like the discount rate and to uncertainties like the estimated damages of climate change, and might therefore be difficult to implement. With the parameters chosen in their study (which were deduced from the FEEM–RICE model), costs are low for all regions, but the environmental effectiveness is also very low. The disadvantage of the Technology Research and Development regime is that – as a consequence of the intensified research and development efforts – production and therefore emissions increase. In other words, for a technology regime to be successful, a carbon-free direction of technology development must to be clearly specified (Alfsen *et al.*, this volume, Chapter 13; Knopf and Edenhofer 2010).

Global carbon emission tax

Finally, the implementation of a global carbon emission tax is perhaps the most straightforward universal regime without a predefined emission target. Many studies have used a global carbon tax as a means to achieve emission reductions, but the regional results of such a regime have not regularly been reported. Table 4.6 summarizes the results of a global carbon emission tax from three studies, extended with our own calculations from the IMAGE framework and the ENV-Linkages model (see Section 4.2.1).[3] The carbon tax in these studies was raised over time to reach a certain concentration stabilization level, with the exception of Bollen *et al.* (2005), who set the carbon emission tax at a constant €20 per tonne of carbon dioxide.

The various studies report similar results for most regions. All studies project that costs as share of GDP would be somewhat less than the global average for the European Union and the United States, and higher, or much higher, for the Middle East, the former Soviet Union, East Asia and Africa. Costs tend to be higher in developing countries because the burdens of a tax regime are carried mostly by those regions with high carbon intensity or with high opportunities to reduce emissions. In

[3] Not included in the table are the results by Manne *et al.* (1995), who analysed a low carbon tax starting at USD 1 per tonne, increasing at 5 per cent per year. Results are reported for five regions only and are modest, as can be expected from such a low carbon tax.

Table 4.6 *Regional costs compared to global average costs (2)*

	USA	European Union/ enlarged European Union	FSU/ Russia	Middle East/ Middle East and N Africa	Latin/ South America	Africa/ sub-Saharan Africa	China/ East Asia	India/ South Asia
Peterson and Klepper (2007): welfare effects in 2030	1	1	3	3	2	3	3	1
Vaillancourt *et al.* (2008): discounted direct costs	1	1	2	2	2	2	2	1
Based on IMAGE framework: direct costs in 2050	1	1	2	2	1	2	2	2
Based on ENV-Linkages: GDP loss in 2050	1	1	3	2	1	1	2	3
Bollen *et al.* (2005): income loss in 2020	1	1	3	3	2	NA	NA	NA

The table shows regional costs compared to global average costs for a carbon emission tax of €20 per tonne (Bollen *et al.* 2005) and an increasing carbon emission tax in order to reach a greenhouse gas concentration stabilization level of 450 ppm carbon dioxide-equivalent (IMAGE and ENV-Linkages) or 550 ppm carbon dioxide-equivalent (other studies).
Legend: 1 = costs less than global average, 2 = costs between global average and twice the global average, 3 = costs more than twice the global average; NA = not available.

theory, a differentiated tax could equalize the cost burden among countries, although this would complicate the carbon tax implementation.

4.3.2 Fragmented regimes

Table 13.2 of the Working Group III contribution to the IPCC's Fourth Assessment Report (IPCC 2007: 770–773) mentions five fragmented regimes,

Table 4.7 *Economic consequences of fragmented climate regimes*

Name	Short description	Evaluated by
European Union only	Only European Union sets emission targets	Bollen *et al.* (2005), den Elzen *et al.* (2007b), European Commission (2007), Russ *et al.* (2005)
Industrialized countries technological cooperation	Replacement of international cooperation on emission reductions with international cooperation between industrialized countries on technological innovation and diffusion	Buchner and Carraro (2004)
Developed minus US	Developed countries except the United States set emission targets, rest of world does not	Böhringer and Löschel (2005)
Industrialized countries only	Only industrialized countries set emission targets	Bollen *et al.* (2005; 2005), Böhringer and Löschel (2005), Russ *et al.* (2005)
Industrialized countries carbon tax	Only industrialized countries levy a carbon tax	Bollen *et al.* (2005)
Without Asia and Africa	Only developing Asian and African countries do not set emission targets	Bollen *et al.* (2005)
Political willingness	Regional emission constraints on levels considered to be politically acceptable according to a number of research institutes	den Elzen *et al.* (2007a)
Fully fragmented	A palette of internationally fragmented climate policies	Boeters *et al.* (2007)

The table shows fragmented climate regimes that have been analysed regarding their economic consequences, from low to high participation.

which are rather generally defined. Most quantitative studies on fragmented regimes make assumptions about the level of participation: only one climate regime is adopted, but not all countries participate in this regime. The participating countries adopt reduction targets based on expert judgements about what they might be willing to do, or reduction targets would be set at such levels in order to reach a global emission target. Only two studies analyse the costs of fragmented regimes with several different climate agreements (Boeters *et al.* 2007; den Elzen *et al.* 2007a). We will discuss a range of fragmented regimes from low to high participation (see Table 4.7).

Emission targets for the European Union only

The regime with the lowest participation analysed is one in which only the European Union sets emission targets. According to this analysis, if the European Union were to set a target of 20 per cent emission reduction in 2020 or 2025 (compared to 1990), and joint implementation and the clean development mechanism are available, abatement costs and welfare losses for the European Union would be very small – less than 0.3 per cent in 2020 (Russ *et al.* 2005; den Elzen *et al.* 2007b; European Commission 2007, 2008). Two studies look at the implications of a 30 per cent reduction target for the European Union only. Bollen *et al.* (2005) estimates income losses for the European Union of such a target of 2 per cent in 2020; a study by the European Commission (2007) finds GDP losses of 0.9 per cent in 2025. Even with the more stringent target of a 30 per cent emission reduction, the effectiveness of such a regime on a global scale is very low: by 2020, the global emission reduction would be less than 5 per cent compared to the no-climate policy case.

Technological cooperation between industrialized countries

This fragmented regime is slightly different than the rest as the focus is not on emission reduction targets, but on technological cooperation between industrialized countries. The advantage of such a coalition is that the consequences can be relatively easily assessed by decision-makers. Buchner and Carraro (2004) assess this regime with the same assumptions as in their Enhanced Global Cooperation regime (see Section 4.1.2). The results are similar: environmental effectiveness is very low, because global emissions and even the emission/output ratio increase in this regime. This is the result of production increases due to intensified research and development efforts (as in the universal Technology Research and Development regime). Emissions per unit of output also increase, because the overall impact of accumulated research and development expenditure on economic growth is larger than the impact of accumulated research and development on emission abatement.

Emission targets for industrialized countries only

Böhringer and Löschel (2005) analyse the possibility of all industrialized countries except the United States setting emission targets. Although the United States and developing countries do not set emission targets, the study assumes that these regions can sell project-based emission reductions to the reducing countries (leading to mutual benefits). The global emission reduction target is set at 10 per cent below baseline in 2020. Different allocation rules are used to allocate this global reduction to Australia and New Zealand, Canada, the European Union, the former Soviet Union and Japan. The sovereignty rule (also called grandfathering, that is, reduction obligations based on current emissions) and the polluter-pays principle (that is, reduction obligations based on past emissions) lead to emission reductions of 30–38 per cent compared to baseline in 2020 in each of the regions. The allocation rule ability-to-pay (that is, reduction obligations based on welfare), on the other hand, leads to an emission reduction target for the former Soviet Union by only 5 per cent, whereas Japan has to reduce emissions by 66 per cent. Figure 4.2a summarizes the abatement costs (measured in consumption loss).

A fragmented climate regime in which all industrialized countries participate (thus including the United States) leads to similar results. The results of Böhringer and Löschel (2005) are shown in Figure 4.2b. The participation of the United States leads to small reductions of costs for other industrialized countries. Bollen *et al.* (2005) analyse a similar regime, but with a more stringent target and allocation based on per capita emission convergence in 2024. Consistent with the results of Böhringer and Löschel, this leads to high costs for the former Soviet Union. In contrast with that study, however, they conclude that there are benefits for the Middle East and that the European Union and United States will incur higher costs. The latter can be explained by the more stringent target and the early convergence year (which is less beneficial for industrialized countries, see Section 4.3.1). Their main conclusion is that switching from a global coalition (universal regime) to a smaller coalition of industrialized countries (fragmented regime) more than doubles the cost of the European Union objective of 2 °C, even with the possibility of the clean development mechanism in its current form. With such a regime, developing countries (except energy exporting countries) would benefit and thus effectively become free-riders. This might create an obstacle for establishing a coalition in which only industrialized countries participate.

Carbon emission tax for industrialized countries only

Besides setting emission targets, industrialized countries also could levy a carbon tax. Like the universal Global Carbon Emission Tax, Bollen *et al.* (2005) analyse the carbon tax for industrialized countries with a tax of €10 and €20 per tonne carbon dioxide. Whereas a global carbon tax of €20 per tonne carbon dioxide would reduce emissions by 25 per cent compared to baseline, a carbon tax limited to industrialized countries would reduce emissions only by 10 per cent. The costs are distributed differently as well; as expected, industrialized countries (especially the former Soviet Union) carry the burden in this regime.

Emission targets for all countries except developing countries in Africa and Asia

Bollen *et al.* (2004) analyse a fragmented regime in which only developing African and Asian countries refuse to join a climate coalition. They assume that the rest of

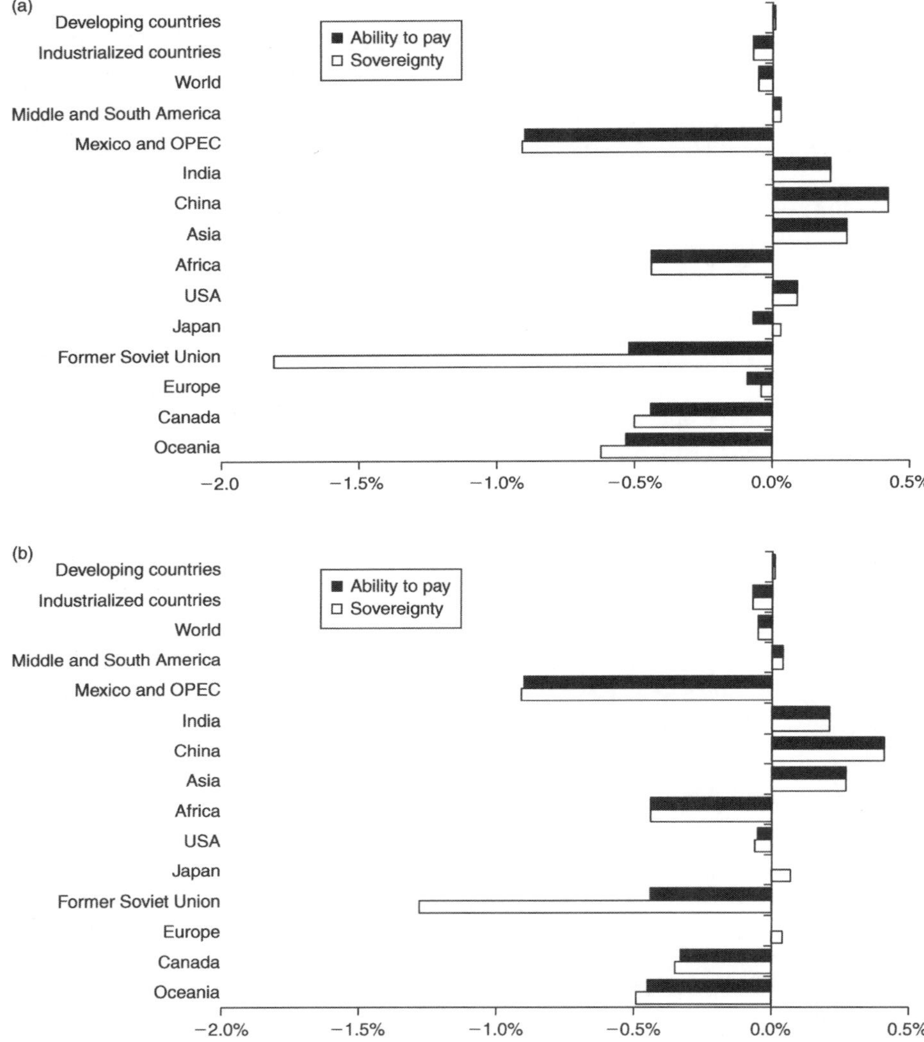

Figure 4.2 Change in consumption in 2020 relative to business-as-usual. This analysis applies in a regime where only industrialized countries set emission targets, allocated based on the sovereignty rule or ability to pay, so that global emissions are reduced by 10 per cent compared to business as usual. (a) Coalition without the United States; (b) coalition with the United States. *Source:* based on Böhringer and Löschel (2005).

the world sets an emission target of 30 per cent below 1990, allocated according to 'contraction and convergence'. The EU objective could be reached with such a regime, but the global income loss in 2020 would be almost three times higher compared to a universal 'contraction and convergence' regime.

Political willingness scenario

The Political willingness scenario, an outcome of the South–North dialogue, is an example of a fragmented regime with large participation (den Elzen *et al.* 2007a). In this proposal, emission reduction targets for different regions are set at levels based on an assessment by a number of research institutes involved in the South–North dialogue proposal. This scenario neither establishes a global emission target, nor requires regions to base their emission reduction targets on a universal regime. This proposal resembles the fragmented bottom–up or multifaceted approach, in which each country creates its own initial proposal relating to what it might be able to commit to (Reinstein 2004). According to the proposal, the European Union reduces emissions by 30 per cent in 2020 as compared to 1990; other industrialized countries by 15 per cent; newly industrialized countries by 30 per cent; and rapidly industrializing countries by 10 per cent. Developing countries continue their baseline emissions until 2020. It is implicitly assumed that all regions will comply because of political will. The costs of the political willingness scenario as a share of GDP would be similar among industrialized countries, while most developing countries would gain from financial transfers from emission trading. With the global emission reduction by 2020 resulting from this scenario, the stabilization at a concentration of 500 ppm carbon dioxide-equivalent is kept just within reach.

A fully fragmented regime compared with a universal regime

Boeters *et al.* (2007) analyses yet another fragmented regime. In contrast to the previous regimes, here even the type of goals varies by region. The United States, Australia and Canada would focus on technology improvement, while the European Union and Japan would continue with emission targets combined with emissions trading. Fast-developing countries would not set emission targets, but would invest mainly in local air quality. For a more detailed description of this fragmented regime, see Boeters *et al.* (2007). Table 4.8 compares this fragmented regime with a universal 'multi-stage' regime, also analysed in the same study.

As the table shows, the global costs of the universal 'multi-stage' regime and the fragmented regime are the same, even though the 'multi-stage' regime achieves much higher emission reductions. Another important conclusion is that no country involved in the 'multi-stage' regime is worse off as compared to the proposed fragmented regime, with the notable exception of the United States. In other words, there might be an incentive for the United States not to join a universal regime (however, allocation rules other than the ones of the 'multi-stage' proposal can limit the costs for the United States; see Table 4.5).

Table 4.8 *Comparison of a fully fragmented regime with a universal regime*

	Target (2020, relative to 1990)		Emission price (€/tonne carbon dioxide)		Change in national income (by 2020)	
	Universal	Fragmented	Universal	Fragmented	Universal	Fragmented
Industrialized countries	−20 per cent	−1 per cent	24	25	−0.3 per cent	−0.3 per cent
European Union-25	−23 per cent	−15 per cent	24	51	−0.4 per cent	−0.6 per cent
USA	−24 per cent	+25 per cent	24	14	−0.3 per cent	−0.0 per cent
FSU	−13 per cent	−25 per cent	24	7	+0.8 per cent	−0.7 per cent
Rest OECD	−16 per cent	+3 per cent	24	34	−0.4 per cent	−0.4 per cent
Non-Annex I	+100 per cent	+147 per cent	–	–	0.0 per cent	0.0 per cent
China	+101 per cent	+105 per cent	24	2	+0.4 per cent	0.0 per cent
India	+210 per cent	+203 per cent	24	4	+0.1 per cent	0.0 per cent
World	+28 per cent	+45 per cent	–	–	−0.2 per cent	−0.2 per cent

The table shows targets, emission prices and national income changes for a fully fragmented regime compared to a universal regime based on a multi-stage emission allocation rule.
Source: Boeters *et al.* (2007).

4.4 Conclusions and policy recommendations

This chapter reviewed several quantitative studies about the costs and environmental effectiveness of different universal and fragmented regimes. A large number of studies exist that qualitatively discuss different climate regimes or discuss only the emission reductions resulting from these regimes. The number of studies also discussing regional economic consequences is much more limited. In the studies that do provide costs analyses, different tools are used, such as macroeconomic models. We find that, in general, studies agree on what regions will experience high, medium or low costs under different regimes, even when using different tools.

The chief conclusions from the review of universal regimes are that (1) theoretically, binding universal regimes with high emission reduction targets or a carbon tax can achieve ambitious reduction targets at relatively low costs; and (2) almost all examined allocation regimes result in significant cost differences (in terms of costs as share of GDP) between regions. The last conclusion is obviously a major challenge for forming international coalitions. There is no single formula for emission allocation that satisfies all possible country conditions, as there is a generic conflict between a simple transparent formula and incorporating many national circumstances. As an example, both energy system and general equilibrium models indicate that Africa and South Asia would benefit from a universal climate regime based on 'contraction and convergence' with 2050 as the convergence year, while the former Soviet Union and Middle East are projected to incur high costs. In that context, for any real-world agreements, the outcomes of specific allocation rules will at best serve as a starting point for negotiations. The cost differences could be smaller after a negotiation process, which would increase the likelihood of accepting such a proposal.

From the studies that have analysed fragmented regimes, we learn that in general, it is more cost-effective to reduce emissions in a universal regime than in a fragmented regime. One reason is that with lower participation, it is more difficult for the participating regions to reach a certain global emission target (they need to compensate the higher emissions of the non-participating regions). In addition, even with high participation fragmentation implies that emission reductions are not made where it is cheapest to do so (no emission trading is usually possible between regions participating in different agreements). However, despite the higher overall costs, a fragmented regime consisting of multiple agreements could be more feasible to attain. This is mainly because individual countries have an incentive to free-ride on a universal regime.

There are many criteria to assess climate regimes. The current study mainly focused on environmental effectiveness and cost efficiency (and for instance not on institutional feasibility). Based on the long-term advantages of universal regimes for these criteria, but also the difficulties in establishing such regimes, one may argue that some kind of transition from a fragmented regime to a universal regime could provide the best possibility to achieve strong emission reductions. To simplify negotiations, a transitional, ambitious, fragmented regime consisting of all major emitting countries could be established in the short term. Such a coalition could provide the basis for a larger, universal regime in the long term. Transfer schemes or other interlinkages (Flachsland *et al.*, this volume, Chapter 5; and Zelli and van Asselt, this volume, Chapter 6) might help to achieve such a universal regime.

In terms of the mapping criteria applied in this volume, the political dimension of this recommendation is purely policy-based, since the institutional settings and type of actors involved were not subject of our analysis. This means that we do not have concrete recommendations about the international institutional environment, the type of actors involved in decision-making and the mode of governance applied to implement the policy suggestions. This notwithstanding, our recommendation of transitional fragmented regime could be interpreted as a cross-institutional solution which involves the UN climate regime along with several other institutions.

References

Bakkes, J. and P. Bosch (eds.) 2008. *Background Report to the OECD Environmental Outlook to 2030*. Bilthoven: Netherlands Environmental Assessment Agency, and Paris: Organization for Economic Co-operation and Development.

Barrett, S. 1999. 'A theory of full international cooperation', *Journal of Theoretical Politics* **11**: 519–541.

Bell, W., J. van Ham, J.-E. Parry, J. Drexhage and P. Dickey 2005. *Canada in a Post-2012 World: A Qualitative Assessment of Domestic and International Perspectives*. Winnipeg: International Institute for Sustainable Development.

Berk, M. M. and M. G. J. den Elzen 2001. 'Options for differentiation of future commitments in climate policy: how to realise timely participation to meet stringent climate goals?', *Climate Policy* **1**: 465–480.

Blanchard, O. 2002. 'Scenarios for differentiating commitments', in K. A. Baumert (ed.), *Building on the Kyoto Protocol: Options for Protecting the Climate*. Washington, DC: World Resources Institute, pp. 203–222.

Blok, K., N. Höhne, A. Torvanger and R. Janzic 2005. *Towards a Post-2012 Climate Change Regime*. Brussels: 3E.

Bodansky, D. 2004. *International Climate Efforts beyond 2012: A Survey of Approaches*. Arlington, VA: Pew Center on Global Climate Change.

Boeters, S., M. G. J. den Elzen, T. Manders, P. Veenendaal and G. Verweij 2007. *Post-2012 Climate Policy Scenarios*. Bilthoven: Netherlands Environmental Assessment Agency.

Böhringer, C. and C. Helm 2008. 'On the fair division of greenhouse gas abatement cost', *Resource and Energy Economics* **30**: 260–276.

Böhringer, C. and A. Löschel 2005. 'Climate policy beyond Kyoto: quo vadis? A computable general equilibrium analysis based on expert judgments', *Kyklos* **28**: 467–493.

Böhringer, C. and H. Welsch 2004. 'Contraction and convergence of carbon emissions: an intertemporal multi-region CGE analysis', *Journal of Policy Modeling* **26**: 21–39.

Böhringer, C. and H. Welsch 2006. 'Burden sharing in a greenhouse: egalitarianism and sovereignty reconciled', *Applied Economics* **38**: 981–996.

Bollen, J., T. Manders and P. Veenendaal 2004. *How Much Does a 30 Per Cent Emission Reduction Cost? Macroeconomic Effects of Post-Kyoto Climate Policy in 2020*. The Hague: Netherlands Bureau for Economic Policy Analysis.

Bollen, J., T. Manders and P. Veenendaal 2005. *Caps and Fences in Climate Change Policies: Trade-Offs in Shaping Post-Kyoto*. Bilthoven: Netherlands Environmental Assessment Agency.

Buchner, B. and C. Carraro 2003. 'Future commitments and participation incentives: some climate policy scenarios', paper presented at the Berlin Meeting of the Second ESRI (Economic and Social Research Institute of the Government of Japan) Collaboration Project on Environmental Issues, Berlin, 15–16 October 2003.

Buchner, B. and C. Carraro 2004. *Economic and Environmental Effectiveness of a Technology-Based Climate Protocol*, Working Paper No. 61.04. Milan: FEEM (Fondazione Eni Enrico Mattei).

Carraro, C. 2000. 'The economics of coalition formation', in J. Gupta and M. Grubb (eds.), *Climate Change and European Leadership: A Sustainable Role for Europe*? Dordrecht: Kluwer Academic Publishers, pp. 135–156.

Carraro, C. and D. Siniscalco 1998. 'International environmental agreements: incentives and political economy', *European Economic Review* **42**: 561–572.

Criqui, P., A. Kitous, M. Berk, M. G. J. den Elzen, B. Eickhout, P. Lucas, D. P. van Vuuren, N. Kouvaritakis and D. Vanregemorter 2003. *Greenhouse gas Reduction Pathways in the UNFCCC Process up to 2025*, Technical Report B4-3040/2001/325703/MARE/E.1. Grenoble: DG Environment.

Dellink, R., M. Finus and N. Olieman 2005. *Coalition Formation under Uncertainty: The Stability Likelihood of an International Climate Agreement*, Working Paper No. 98.2005. Milan: FEEM (Fondazione Eni Enrico Mattei).

den Elzen, M. G. J. and P. Lucas 2005. 'The FAIR model: a tool to analyse environmental and costs implications of climate regimes', *Environmental Modeling and Assessment* **10**: 115–134.

den Elzen, M. G. J., M. Berk, P. Lucas, B. Eickhout and D. P. van Vuuren 2003. *Exploring Climate Regimes for Differentiation of Commitments to Achieve the EU Climate Target*. Bilthoven: Netherlands Environmental Assessment Agency.

den Elzen, M. G. J., N. Höhne, B. Brouns, H. Winkler and H. E. Ott 2007a. 'Differentiation of countries' future commitments in a post-2012 climate regime: an assessment of the "South–North Dialogue" proposal', *Environmental Science and Policy* **10**: 185–203.

den Elzen, M. G. J., P. Lucas and A. Gijsen 2007b. *Exploring European Countries' Emission Reduction Targets, Abatement Costs and Measures Needed under the New EU Reduction Objectives*. Bilthoven: Netherlands Environmental Assessment Agency.

den Elzen, M. G. J., P. Lucas and D. P. van Vuuren 2005. 'Abatement costs of post-Kyoto climate regimes', *Energy Policy* **33**: 2138–2151.

den Elzen, M. G. J., P. Lucas and D. P. van Vuuren 2008. 'Regional abatement action and costs under allocation schemes for emission allowances for achieving low CO_2-equivalent concentrations', *Climate Change* **90**: 243–268.

European Commission 2007. *Limiting Global Climate Change to 2° Celsius: The Way Ahead for 2020 and Beyond*. Brussels: European Council.

Eyckmans, J. and M. Finus 2007. 'Measures to enhance the success of global climate treaties', *International Environmental Agreements: Politics, Law and Economics* **7**: 73–97.

Finus, M., J.-C. Altamirano-Cabrera and E. C. van Ierland 2005. 'The effect of membership rules and voting schemes on the success of international climate agreements', *Public Choice* **125**: 95–127.

Finus, M., E. C. van Ierland and R. Dellink 2006. 'Stability of climate coalitions in a cartel formation game', *Economics of Governance* **7**: 271–291.

Gupta, J. 1998. *Encouraging Developing Country Participation in the Climate Change Regime*, Discussion Paper E98–08. Amsterdam: Vrije Universiteit, Institute for Environmental Studies.

Höhne, N., C. Galleguillos, K. Blok, J. Harnisch and D. Phylipsen 2003. *Evolution of Commitments under the UNFCCC: Involving Newly Industrialized Economies and Developing Countries*. Cologne: Ecofys.

IPCC 2007. *Climate Change 2007: Mitigation of Climate Change. Contribution of Working Group III to the Fourth Assessment Report of the Intergovernmental Panel on Climate Change*. Geneva: IPCC.

Knopf, B. and O. Edenhofer 2010 (in press). 'The economics of low stabilisation: implications for technological change and policy', in M. Hulme and H. Neufeldt (eds.), *Making Climate Change Work for Us: European Perspectives on Adaptation and Mitigation Strategies*. Cambridge, UK: Cambridge University Press.

Leimbach, M. 2003. 'Equity and carbon emissions trading: a model analysis', *Energy Policy* **31**: 1033–1044.

Manne, A. and R. Richels 1995. 'The greenhouse debate: economic efficiency, burden sharing and hedging strategies', *Energy Journal* **16**: 1–37.

Manne, A., R. Mendelsohn and R. Richels 1995. 'MERGE: a model for evaluating regional and global effects of GHG reduction policies', *Energy Policy* **23**: 17–34.

Nakicenovic, N. and K. Riahi 2003. *Model Runs with MESSAGE in the Context of the Further Development of the Kyoto Protocol*. Berlin: German Advisory Council on Global Change.

Persson, T. A., C. Azar and K. Lindgren 2006. ' Allocation of CO2 emission permits: economic incentives for emission reductions in developing countries', *Energy Policy* **34**: 1889–1899.

Peterson, S. and G. Klepper 2007. *Distribution Matters: Taxes vs. Emissions Trading in Post Kyoto Climate Regimes*, Working Paper No. 1380. Kiel: Institute for World Economics.

Philibert, C. 2005. *Approaches for Future International Co-operation*. Paris: Organization for Economic Co-operation and Development, and International Energy Agency.

Reinstein, R. A. 2004. ' A possible way forward on climate change', *Mitigation and Adaptation Strategies for Global Change* **9**: 245–309.

Rive, N., A. Torvanger and J. S. Fuglestvedt 2006. 'Climate agreements based on responsibility for global warming: periodic updating, policy choices, and regional costs', *Global Environmental Change* **16**: 182–194.

Rose, A., B. Stevens, J. Edmonds and M. Wise 1998. 'International equity and differentiation in global warming policy: an application to tradeable emission permits', *Environmental and Resource Economics* **12**: 25–51.

Russ, P., J. C. Ciscar and L. Szabo 2005. *Analysis of Post-2012 Climate Policy Scenarios with Limited Participation*. Brussels: European Commission, Joint Research Centre and Institute for Prospective Technological Studies.

Tol, R. S. J. 2001. 'Climate coalitions in an integrated assessment model', *Computational Economics* **18**: 159–172.

Vaillancourt, K. and J.-P. Waaub 2006. 'A decision aid tool for equity issues analysis in emission permit allocations', *Climate Policy* **5**: 487–501.

Vaillancourt, K., R. Loulou and A. Kanudia 2008. 'The role of abatement costs in GHG permit allocations: a global stabilization scenario analysis', *Environmental Modeling and Assessment* **13**: 169–179.

van Vuuren, D. P., M. G. J. den Elzen, P. L. Lucas, B. Eickhout, B. J. Strengers, B. van Ruijven, S. Wonink and R. van Houdt 2007. 'Stabilizing greenhouse gas concentrations at low levels: an assessment of reduction strategies and costs', *Climatic Change* **81**: 119–159.

Wicke, L. and C. Böhringer 2005. *Cost Impacts of a 'Beyond Kyoto' Global Cap and Trade Scheme*. Berlin: Institut für Umweltmanagement.

5

Developing the international carbon market beyond 2012: options and the costs of delay

CHRISTIAN FLACHSLAND, ROBERT MARSCHINSKI, OTTMAR EDENHOFER, MARIAN LEIMBACH AND LAVINIA BAUMSTARK

5.1 Introduction

Emission trading has become one of the most important policy instruments for controlling greenhouse gas emissions. The Kyoto Protocol introduced an intergovernmental emissions trading system that runs from 2008 to 2012 and in which countries accepted economy-wide caps and the possibility to trade so-called Assigned Amount Units. Participation of developing countries is possible through the Clean Development Mechanism (CDM) (Stripple and Lövbrand, this volume, Chapter 11). With its International Transaction Log the climate secretariat provides the institutional infrastructure for these trading mechanisms. Complementing the Kyoto trading system, the European Union has established a company-level EU emissions trading scheme in 2005, with its second trading period running in parallel to the Kyoto system (van Asselt 2010). The EU emissions trading scheme regulates about 10 000 facilities that currently emit around 2 gigatonnes carbon dioxide per year (Skjærseth and Wettestad 2008).

In addition to these developments, an increasing number of industrialized countries are implementing or planning to set up national cap-and-trade systems, including Australia, New Zealand, the United States, Canada and Japan. On the sub-national level, the Regional Greenhouse Gas Initiative, the Western Climate Initiative and the Midwestern Greenhouse Gas Accord have emerged in the United States, while the Japanese provinces of Kyoto and Tokyo consider the introduction of regional emissions trading. In 2007, several of these national and sub-national initiatives and the European Union inaugurated the International Carbon Action Partnership with the explicit aim of exploring options for linking the domestic trading systems in the broader perspective of creating a global carbon market (Bergfelder 2008).[1]

[1] See www.icapcarbonaction.com/declaration.htm. Members of the International Carbon Action Partnership are the EU Commission and several EU Member States, US states from the Regional Greenhouse Gas Initiative and the Western Climate Initiative, and Australia, New Zealand and Norway. Japan, the Tokyo Metropolitan Government and the Ukraine are observers.

Global Climate Governance Beyond 2012: Architecture, Agency and Adaptation, eds. F. Biermann, P. Pattberg and F. Zelli. Published by Cambridge University Press.

These developments can be understood as manifestations of two different approaches towards emissions trading. First, there is the top–down approach, characterized by centralized multilateral decision-making and embodied in negotiations of the climate convention. Second, there is the bottom–up approach, associated with decentralized decision-making of individual nations or sub-national entities that implement emissions trading systems unilaterally, bilaterally or plurilaterally (Zapfel and Vainio 2002). These processes yield two different types of institutional architectures for international emissions trading. The backbone of top–down architectures is emissions trading between governments, while bottom–up architectures rest upon the implementation and possible linking of regional systems, based on company-level emissions trading.

This chapter analyses different architectures for international emissions trading and quantifies changes of global and regional mitigation costs when delaying the implementation of a global carbon market. In the context of this volume, our analysis serves as a case study on the 'architecture' appraisal question about the relative performance of differently fragmented policy architectures for addressing global climate change (Biermann *et al.*, this volume, Chapter 2). Our contribution is complemented by the review of Hof *et al.* (this volume, Chapter 4), which provides a meta-analysis of quantitative assessments of fragmented architectures within integrated assessment models.

Section 5.2 describes our methodology, and Section 5.3 analyses carbon market architectures and provides an economic assessment of delaying the introduction of a global carbon market. Section 5.4 concludes and maps the policy options under consideration.

5.2 Methodology

Our contribution combines traditional qualitative economic policy analysis and computer-based quantitative modelling to assess different structures and the cost of delay of international carbon market architectures. This section introduces some key terminology and describes the employed quantitative model REMIND-R.

Discussions about emissions trading systems use a distinct terminology that draws on a number of terms and concepts. Hence we first clarify some basic terms as employed in this chapter. 'Cap-and-trade systems' set a binding, absolute cap on total emissions but allow for certificates to be traded among the covered entities, which are either nations or companies. The Kyoto Protocol trading system for industrialized countries is an example for cap-and-trade at the governmental level, while the EU emissions trading scheme operates at the company level. In contrast, 'credit systems' define a certain baseline such as a business-as-usual projection or a relative target, and only allow emissions reductions that go beyond this baseline to be used as sellable credits, often referred to as 'offsets'. In this chapter, credit systems refer to non-binding systems, meaning that there is no penalty if the baseline is exceeded. The CDM and Joint Implementation mechanisms established under the Kyoto Protocol are examples of such non-binding baseline-and-credit systems. The terms 'carbon market' and 'emissions trading system' are used interchangeably to refer to both cap-and-trade and credit systems.

The policy domain under investigation is the international emissions trading system. The term 'emissions trading architecture' is used to denote the overarching structure of relations between emissions trading systems that are implemented all over the world. The implementation of and interaction with additional climate policy instruments such as standards and taxes is omitted to focus the analysis on the structure of worldwide emissions trading initiatives.

Complementing the qualitative analysis the REMIND-R model is applied to quantify changes in mitigation costs when delaying the implementation of a global carbon market. REMIND-R is a multi-regional integrated energy–economy–climate growth model, featuring trade in goods, resources and emission allowances (for a detailed description, see Leimbach *et al.* 2008). It is implemented in five-year time-steps. Land use, land-use change and forestry emissions are represented by an exogenous scenario. In climate policy scenarios, a temperature target like the 2 °C constrains the energy–economy system through an integrated climate module. The model then identifies intertemporally and interregionally optimal mitigation trajectories as well as optimal energy sector investment strategies. Damages from climate change are not taken into account.

5.3 Analysis

This section first identifies and analyses different emissions trading architectures characterized by varying degrees of fragmentation (Section 5.3.1). Then, a quantitative assessment of the costs of delaying the introduction of a comprehensive global carbon market is presented in Section 5.3.2.

5.3.1 Emissions trading architectures

From an economic theory point of view, a global company-level emissions trading system for all sectors[2] and regions may be regarded as the 'ideal' theoretical

[2] It is difficult to include some sectors to emissions trading due to uncertainties in the monitoring of emissions, for example in the agriculture sector or for N_2O emissions from stationary sources. However, the Lieberman–Warner proposal for a United States emissions trading system, for example, envisages coverage of 84 per cent of all United States emissions (World Bank 2008). In particular, emissions from small sources such as in the transport or building sector can be integrated upstream (Hargrave 2000).

benchmark for an environmentally and cost-effective global emissions trading architecture.[3] It would ensure broad coverage and incentivize emission reductions wherever they are cheapest, rule out leakage effects and minimize concerns for market power. However, it will be politically difficult to implement such a system at least in the short to mid term due to the unprecedented level of international economic policy coordination required, distributional questions, transaction costs of company-level emissions trading (for example related to monitoring emissions) and questions of legal enforcement (Victor 2007). Thus, alternative approaches for international emissions trading beyond 2012 are under discussion. Subsequently we attempt a summary of this debate by distinguishing two top–down and three bottom–up architectures for international emissions trading.

Top–down

A *global government-level cap-and-trade architecture* implies that every country adopts a well-defined and limited greenhouse gas emissions budget for its entire economy, and that emission allowances can be traded between governments. This kind of architecture would for example emerge if all countries assumed Annex B status in the Kyoto Protocol.

By contrast, a *Kyoto II architecture* would continue the current Kyoto Protocol approach beyond 2012. Only a limited group of countries, for example industrialized countries, implements a cap-and-trade system on the level of governments. All other countries, for example developing countries, could participate by means of trade without binding caps as represented by the current CDM approach or another mechanism. Given the critique of the CDM in its current form, for example with regard to additionality and transaction costs (for example Michaelowa *et al.* 2003; Schneider 2007), various reform proposals are currently discussed (for example Schmidt *et al.* 2006; see Stripple and Lövbrand, this volume, Chapter 11 for an overview).

For any government trading scheme, additional domestic policy instruments are required to translate the international price signal to domestic actors, for example domestic cap-and-trade, taxes, standards and technology support schemes. This study abstracts from such domestic policies to focus the analysis on the structure of international emissions trading.

[3] Babiker *et al.* (2004) have questioned whether international emission trading is always beneficial for all participating countries. Evidently, in a first-best setting this would be the case due to a simple gains-from-trade argument. However, in some second-best constellations, such as in the presence of tax interaction and terms-of-trade effects, international emissions trading among companies can lead to welfare reductions for permit-selling countries.

Bottom–up

Fragmented markets characterize the situation of two or more independent cap-and-trade systems at the national, supra- or sub-national level that do not have any intentional linkages between them. Even though international trade in goods already induces a certain tendency towards permit price convergence across different emissions trading systems (Copeland and Taylor 2005), allowance prices will in general vary and thus prevent a cost-effective outcome. The degree of inefficiency increases – ceteris paribus – in proportion to the allowance price differential between carbon markets.

If at least two regional cap-and-trade systems accept credits from the same baseline-and-credit scheme, an *indirect link* between them emerges (Egenhofer 2007; Jaffe and Stavins 2008). Depending on the supply curve for credits, cap levels, marginal abatement cost curves and quantity limits on the import of credits, indirect linking will lead to a complete or incomplete convergence of the allowance price in indirectly linked cap-and-trade markets. Figure 5.1 illustrates the underlying mechanics for two cases.

For each case two periods are compared. In the first period t, only cap-and-trade system A accepts credits from the baseline-and-credit scheme C, while cap-and-trade system B operates in autarky. In period $t+1$, system B also allows the import of permits from the baseline-and-credit scheme C, thereby establishing an indirect link between systems A and B.

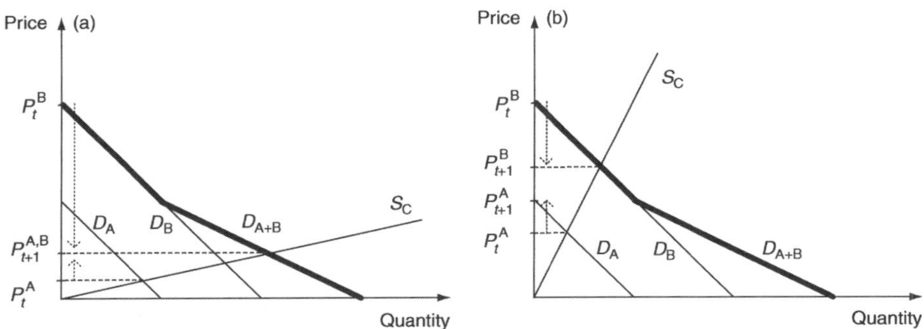

Figure 5.1 Price convergence when trading systems are linked indirectly through credits. D_A and D_B are credit demand curves for systems A and B, and D_{A+B} is the aggregate demand curve. S_C is the supply curve for credits. The price level in system A prior to system B's joining of the credit market is P_t^A. The autarky price in system B without any linkage is P_t^B. The price levels in A and B after the entry of B into the credit market are P_{t+1}^A and P_{t+1}^B, respectively. The arrows indicate the direction of price changes resulting from indirect linking. *Source:* based on Flachsland *et al.* (2009a).

Figure 5.1a illustrates the case of complete price convergence due to the indirect link. The price in system A increases from P_t^A to the new equilibrium price P_{t+1}^A (= P_{t+1}^B), while the price level in B decreases from P_t^B to P_{t+1}^B (= P_{t+1}^A).

In Figure 5.1b, price convergence is incomplete because of the steep credit supply curve S_C. A steep supply curve can be interpreted to represent low availability of low cost emission reductions relative to the demand by cap-and-trade systems, for example due to limited scope or high transaction costs of credit systems. When entering the market for credits, system B buys credits at a market clearing price P_{t+1}^B which exceeds the maximal willingness to pay of system A. The latter then resorts to domestic abatement only, leading to a new and different internal allowance price P_{t+1}^A. Here, indirect linking brings about partial price convergence as the allowance price level in A increases, while it decreases in B.

Therefore, developing country baseline-and-credit schemes that clear the way for large-scale investment opportunities into abatement, for example in the power sector of developing countries, would be conducive to international allowance price harmonization in a market architecture characterized by indirect links between regional trading schemes.

Finally, *direct linking* occurs whenever two or more regional emissions trading systems mutually recognize each others' allowances, that is, they accept emission certificates issued in other systems as valid for compliance within their own system (Haites and Mullins 2001). A direct linking architecture is thus established through a concerted linking-decision of different regional trading systems (Tangen and Hasselknippe 2005; Tuerk *et al.* 2009; Victor 2007; Jaffe and Stavins 2008).[4] Evidently, an immediate consequence of linking is the formation of a common emissions price.[5]

The benefit of enhanced cost-effectiveness comes, however, at the cost of contagiousness: once two emissions trading systems are linked, changes in the design or regulatory features in one system that influence the price formation automatically diffuse into all other systems. (Depending on the level of price convergence, this will also be the case for indirectly linked systems.) For instance, if only one country decides to adopt a price ceiling in form of a price cap, the entire linked market is in effect capped at the same price. Arbitrage trading will ensure that prices in each system do not exceed the price ceiling. Thus, there is a partial loss of control for domestic regulators over their own system. Relevant design issues with implications for the whole linked carbon market include: the setting and modification of emission

4 Only bilateral linkages are considered. A unilateral link is established if cap-and-trade system A accepts allowances from another system B for compliance but not vice versa. In such a system, the allowance price in A would remain at or below the price level of B.

5 The permit price might differ by a constant factor if systems use different units of measurement, for example metric and short tons. The latter unit is in fact the case for the Regional Greenhouse Gas Initiative.

caps; upper and lower bounds for permit prices; links to baseline-and-credit schemes, for example CDM; banking and borrowing provisions; compatible registries; rules for monitoring, reporting and verification of emissions; and penalties and enforcement of compliance (in more detail International Energy Agency 2005; Jaffe and Stavins 2007; Flachsland *et al.* 2008).

To address these issues, institutional provisions in the form of linking agreements and some form of consultation and cooperation over market regulation are required (Edenhofer *et al.* 2007; Mehling 2007; Flachsland *et al.* 2008; Mace *et al.* 2008). As a first step in that direction, several countries and regions with existing or emerging regional cap-and-trade systems established the International Carbon Action Partnership in 2007. Part of its mandate is to assess barriers to linking and work out solutions where such impediments may exist (Bergfelder 2008).

Ultimately, the decision to link or not to link two regional cap-and-trade systems will depend on each linking partner's perceived balance of benefits and drawbacks (Flachsland *et al.* 2009b). Clearly, the major economic benefit from bilateral linking is the efficiency gain from enabling trade across systems with different marginal abatement costs, that is, allowance prices (Anger 2008). Also, in particular smaller carbon markets can benefit from improved liquidity when linking to larger systems. In addition, concerns over competitiveness resulting from different carbon price levels are eliminated among bilateral linking partners. Apart from economic advantages, bilateral linking promises political benefits, including the stabilization of the expectation of an international carbon market if this is a political objective.[6] Also, bilateral linking offers an opportunity for committing a country to climate policy, as cancelling direct links will entail costs that make dismantling of carbon trading schemes less likely.

However, caveats to full bilateral linking need to be taken into account as well. Regarding distributional issues, linking partners need to recognize each other's post-linking mitigation burden, analogously to an agreement in a multilateral framework. Also, if linking leads to significant changes in the domestic incidence of mitigation costs in one or several regions, this may be a barrier to linking if this violates sensible distributional objectives, for instance if some domestic groups suffer significant welfare losses. Regarding the design of cap-and-trade systems, the setting of key parameters as outlined above reflects an implicit or explicit prioritization of policy objectives of an emissions trading system (for example emission reductions, some level of domestic abatement, incentives for technology research and development, minimization of domestic abatement costs). As design parameters of systems 'mix' when linking directly, economic outcomes of the joint trading system may change. If linking leads to unacceptable violation of one or several

[6] www.icapcarbonaction.com/declaration.htm.

partners' policy priorities (for example a reduced incentive for research and development due to a much lower carbon price), this represents a barrier to direct linking. Different views regarding cost containment measures such as price caps will be particularly difficult to reconcile. In general, one can expect that regions with close political ties will be able to resolve the challenges of coordinating the economic policy better and sooner than regions with a less profound history of joint policy-making.

Regarding the time horizon for full bilateral links, it is likely that systems will first want to observe each other's performance and governance mode (European Commission 2007). As this implies stand-alone runs of systems that have mostly not yet been implemented, we expect first bilateral links between OECD schemes to emerge within the time span 2015 to 2020. As proposed by the European Commission (2009) and German foreign and environmental ministers Steinmeier and Gabriel,[7] a transatlantic link between a United States federal cap-and-trade system and the EU emissions trading scheme not only eliminates competitiveness concerns between the two largest economies in the world, but would also create a strong signal concerning the robust prospect for the emergence of a global carbon market.

Discussion

Drawing on the characterization by Biermann *et al.* (this volume, Chapter 2), top–down approaches are cases of synergistic fragmentation, while direct and indirect linking correspond to cooperative fragmentation. 'Fragmented markets' display some features of conflictive fragmentation.

Both top–down architectures under discussion would comprise all countries as part of the same core institution, that is, the international emissions trading regime, which would provide common principles. Different trading mechanisms suiting country's circumstances may be specified in the case of Kyoto II.

A comprehensive and globally coordinated system of directly linked trading systems as well as an architecture marked by indirect links between trading systems can be regarded as examples for cooperative fragmentation. Both approaches feature more or less loosely integrated different regional institutions and decision-making procedures.

Finally, in fragmented carbon markets without any connection, where each system rests on its own institutional infrastructure, strongly differing views regarding appropriate market design (for example emission caps, or price caps) may prevail. This implies a case of conflictive fragmentation.

[7] *Frankfurter Allgemeine Zeitung*, 29 September 2008.

The remainder of this section compares the five carbon market architectures with respect to three criteria – environmental effectiveness, cost effectiveness and political feasibility – in order to identify their relative strengths and weaknesses.[8]

First, *environmental effectiveness* refers to the capability of an emissions trading architecture to bring about significant reductions in global emissions. Its potential for doing so depends, first of all, on the share of global emissions that are actually covered by the emissions trading regime. Also, leakage effects[9] and reduction targets need to be taken into account. Because this study abstracts from specific reduction targets of trading schemes to focus on their broader properties, one should more accurately speak of the *potential* environmental effectiveness of a trading architecture.

Against this backdrop, a top–down architecture with global cap-and-trade obviously offers the best possibility for significant cuts in global emissions, as it features the highest coverage and excludes carbon leakage. On the other end of the spectrum, a bottom–up architecture consisting of fragmented markets is unlikely to significantly curb global emissions, as it can be expected that only few countries will implement trading schemes with ambitious reduction targets that remain unconnected to major baseline-and-credit schemes like the CDM. The situation is less definite for the other, 'intermediate' architectures: with a sufficient number of committed participants, the indirect linking and especially the direct linking approach may come close to the environmental effectiveness of a Kyoto II architecture. In fact, while a bottom–up approach may be more likely to start out with lower initial emissions coverage, it can expand step by step.

Thereby, it gradually increases the share of global emissions that it covers. Eventually, it may resemble our theoretical benchmark configuration of global company-level trading system. Both the Kyoto II and all of the bottom–up schemes will have to face the challenge of controlling emissions leakage. In fact, this will be the case for any architecture that does not cover all major economic sectors and regions. However, Kyoto II as well as the formal and indirect linking architectures can be extended to provide economic incentives for emission control to third countries in the form of appropriately designed trading mechanisms. In general, short-term concerns over leakage can be mitigated if most or all of those countries that are close trade competitors participate in a linked carbon market with a single permit price.

[8] This operationalization is meant to support our policy analysis of different carbon market architectures and thus differs in some respects from that employed in numerical integrated assessments that focus on quantitative indicators such as GDP. See for example Hof *et al.* (this volume, Chapter 4).

[9] Leakage occurs if the regulation of emission-intensive industries in one country leads to an expansion of the output of those industries in other, less regulated countries, due to a shift in comparative advantage. The impact of this effect will depend on a number of factors, including the size of the carbon price differential, the trade exposure of affected sectors, and the relative importance of the expected persistence of the cost gap for investment decisions. In general, the available evidence suggests that this effect would not be a serious problem in most sectors, at least in the short to mid term (Stern 2007: 253–266; Neuhoff 2008).

Our second criterion, *cost effectiveness*, requires the minimization of the costs of achieving a given emissions reduction target. It is well known that cost effectiveness depends on the equalization of marginal abatement costs across regions and sectors (for example Tietenberg 2003). In theory, an emissions trading system realizes this by creating a uniform price signal reflecting marginal abatement costs across all covered entities. In practice, however, the emerging emissions price under a permit trading scheme may deviate from marginal abatement costs, in particular if (1) one or more actors possess market power[10] or (2) regulators trade on behalf of firms but lack full information on the abatement costs incurred by the latter (Kerr 2000).

Among participating countries, top–down architectures promise a complete equalization of the permit price. However, there are concerns over market power[11] and doubts about the proper revelation of domestic marginal abatement costs, reducing the cost effectiveness of these architectures. By contrast, bottom–up approaches lead to permit price equalization only in the formal linking case or, under certain conditions, in the indirect linking case. However, the price signal may be more robust since company-level trading systems are better suited to resolve the information asymmetry between governments and companies and are less prone to market power distortions.

Finally, the question of *political feasibility* is related to participation levels and the requirements for coordination and burden-sharing, as well as transaction costs of a trading architecture. Evidently, to establish an integrated trading architecture, players need to agree on a common regulatory framework that will be more difficult to achieve with a larger number of players. For example, some actors will favour higher emissions prices to foster technological development, while others may focus on cost containment by implementing price caps. With an increasing number of participants, such differences will become both more probable and more difficult to reconcile. Regarding burden-sharing, regional emission caps have distributional implications, as allocations represent each player's cost-free endowment and thus determine regional costs. In consequence, bargaining over burden sharing becomes a strategic game where self-interested players have an incentive to free-ride on the mitigation efforts of others by implementing targets with low stringency (Helm 2003; Rehdanz and Tol 2005). A rising number of players that need to agree over caps in trading architectures negatively affect political feasibility. Finally, trading architectures can be compared in terms of the transaction costs that arise from creating the necessary institutional structures. High transaction costs reduce political feasibility.

[10] A case in point would be Russia's bargaining power with its large amounts of 'hot air' within the Kyoto trading framework. See also Böhringer and Löschel (2003).

[11] In 2004, the biggest five emitters in absolute terms – the United States, China, Russia, Japan and India – accounted for 51 per cent of global carbon dioxide emissions, which would give them considerable market power in a government-level system (Climate Analysis Indicators Tool of the World Resource Institute; http://cait.wri.org).

Top–down approaches resemble 'all-or-nothing' options with respect to political feasibility: without international consensus on burden sharing, complete political standstill is imminent. This constitutes a very tangible threat, given that countries with vested interests can rather easily block any kind of agreement. Similarly, agreement on the design details of the trading and accounting system will be more difficult to achieve than for bottom–up approaches with fewer participants. In fact, the direct linking approach will always enable cooperating regions to jointly reduce emissions in a cost-effective manner, even in the absence of a global accord on burden-sharing and regulatory design. Indirect linking provides less certainty regarding permit price harmonization, but implies even less need for political coordination (Jaffe and Stavins 2008). Transaction costs of top–down architectures are relatively high, because a larger number of players need to implement the institutional infrastructure required for participating in the common carbon market.

To sum up, the choice between integrated top–down and fragmented bottom–up architectures corresponds to a trade-off between high environmental effectiveness favouring top–down approaches, and political feasibility, where bottom–up processes fare better. The picture is less clear-cut for cost effectiveness, but direct bilateral links of regional cap-and-trade systems promise superior economic performance.

In the context of ambitious climate policy goals such as the EU 2 °C objective (Jordan and Rayner 2010), global emissions need to stabilize and start declining within the next two decades. Also, in view of global business-as-usual greenhouse gas projections, large amounts of emissions need to be mitigated. Thus, if carbon markets become the major tool for coordinating emissions reductions in a future international climate regime, top–down architectures appear quasi-indispensable to meet this challenge as they enable to instantly implement globally significant reduction targets. Moreover, their major weakness, low political feasibility due to the need to resolve the burden-sharing issue, can in a way be understood as strength: the equity issue as the very crux of the climate problem is addressed at once, which keeps up the pressure on negotiators, and prevents procrastination up to a point in time where low stabilization becomes unfeasible. Thus, within this long-term perspective bottom–up architectures appear as imperfect substitutes of top–down approaches, representing a fallback option if a global agreement cannot be achieved instantly. Consequently, they would mainly serve to bring new momentum to the currently stagnant efforts to establish a global, integrated system.

On the other hand, the two approaches can be viewed as complementary in the sense that bottom–up architectures may serve as essential building blocks for more comprehensive top–down architectures. This way, efficient regional company-level carbon markets can already be put into place, while the delicate question of burden-sharing is deferred for some time. For example, it is conceivable that after the Kyoto Protocol's expiry in 2012 a group of countries willing to adopt binding economy-wide caps proceed with the Kyoto Protocol's intergovernmental cap-and-trade system, and formally link their emerging domestic trading systems within this overarching structure. By devolving intergovernmental permit trading to the company level, the economic performance of the international carbon market would be improved. In fact, this is the approach adopted by the European Union in the first commitment period of the Kyoto Protocol 2008–2012, where international allowance trades across companies within the EU emissions trading scheme are mirrored by transfers of Kyoto allowances in country's registries (Ellerman 2008; van Asselt 2010). Within participating countries, sectors remaining outside the domestic company-level systems would be covered by different instruments such as standards, and governments could trade on behalf of these remaining sectors on the international carbon market.

But unlike the Kyoto scheme, this architecture should be designed as an open system, where countries can join by linking up their domestic emissions trading system whenever they feel ready, or whenever the political momentum in the country has reached a sufficient level (as was the case with Australia and the Kyoto Protocol; see Keohane and Raustiala 2008). Such an approach could be environmentally and economically more effective than pure bottom–up approaches, while being less prone to political deadlock than the top–down approach.

5.3.2 The costs of delay

The previous section have shown that there are several options for further developing the international carbon market, possibly in a step-by-step fashion. Applying the energy–economy simulation model REMIND-R (see Section 5.2 for a brief description), this section quantifies the economic costs of delaying the introduction of a global carbon market when the 2 °C objective of the European Union is pursued.

To calculate the costs of delayed action, the following three scenarios are compared. In the 'default' policy scenario, the increase in global temperature is limited to 2 °C, allowing for temporary overshooting to a maximum of 2.1 °C. A comprehensive global carbon market comprising all world regions and sectors is introduced in 2010, and allowances are distributed according to the contraction and convergence rule (Meyer 2004). Assuming full market efficiency in the permit market, the model can be interpreted to represent our theoretical benchmark of a global company-level trading system.

In the delay scenarios, the global carbon market is implemented at a later date. Before the carbon market is introduced, behaviour of regions is characterized as follows: up to one time-step before carbon market implementation, regions invest into their energy systems as they would find optimal if they would jointly strive for a 3 °C target. That is, they modify their investment behaviour relative to business-as-usual, but they do not consider a 2 °C objective feasible at this stage. Only if they anticipate global consensus

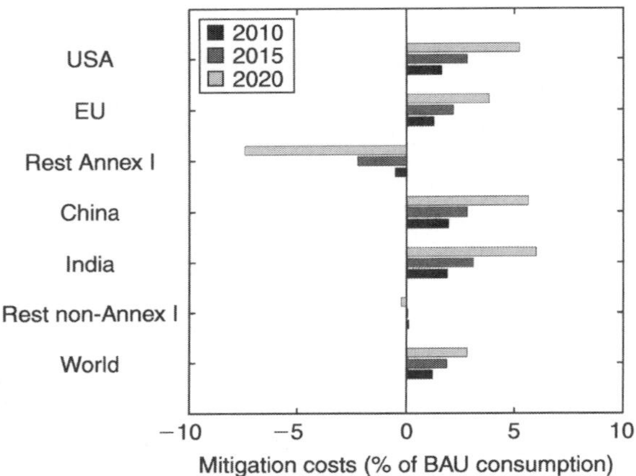

Figure 5.2 Global and regional consumption losses when implementing a comprehensive global carbon market by 2010, 2015 or 2020.

on the 2 °C objective and an accordingly designed global trading scheme, they would adjust their investment strategies. Again, allocation follows the contraction and convergence rule. The two delay scenarios 'delay 2015' and 'delay 2020' are characterized by the inception of the global carbon market in 2015 and 2020, respectively.

Figure 5.2 shows the increase in global and regional consumption losses for these three scenarios. Clearly, global costs increase with the length of the delay: by 45 per cent in the 'delay 2015' case relative to the default policy scenario (from 1.3 per cent to 1.9 per cent) and by 115 per cent in the 'delay 2020' scenario (from 1.3 per cent to 2.8 per cent). If the carbon market is initiated later (that is, 2025 and after), the model becomes infeasible, that is, 2 °C cannot be achieved. Thus, while some delay will not dramatically increase global costs, the window of opportunity for ambitious climate protection is rather small.

Regional costs would increase considerably for the European Union, in particular in case of a longer delay (by 70 per cent and 300 per cent for the 2015 and 2020 scenarios, respectively), and also for the United States. A similar picture emerges for China and India. By contrast, impacts on the rest of the developing countries are very small (Figure 5.2).

The substantial net gains from delayed implementation for the rest of the industrialized countries result from the increased requirement and value of emissions reductions from biomass energy production coupled with carbon capture and storage. This option is particularly relevant for Russia.

The main reason for these increases in global and most regional costs can be seen from Figure 5.3a–c. The later the world community starts pursuing the more

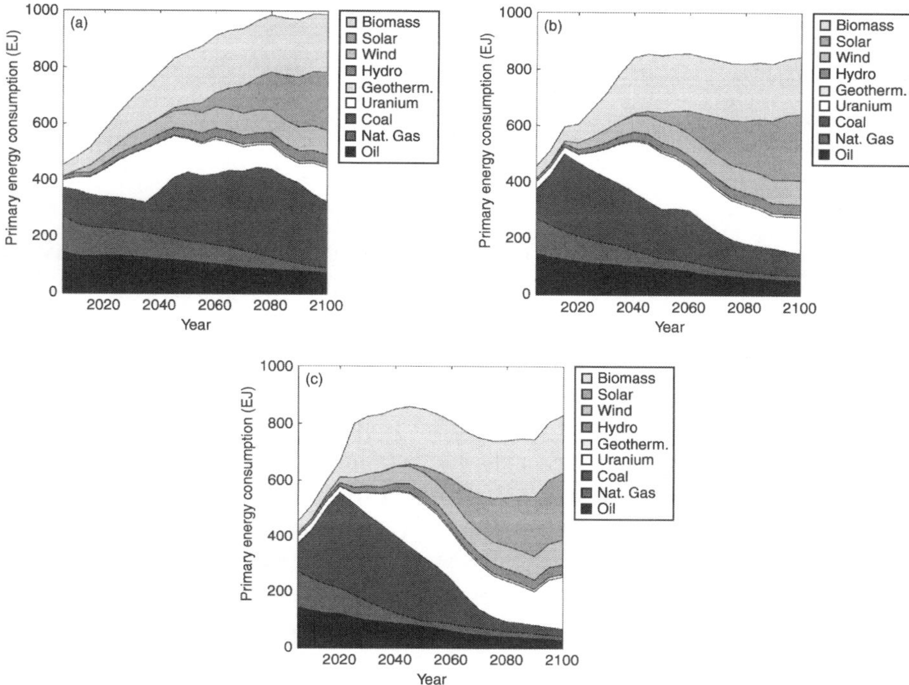

Figure 5.3 Primary energy consumption in the global energy system 2000–2100. Scenario (a), 'default' policy; scenario (b), 'delay 2015'; scenario (c), 'delay 2020'.

ambitious 2 °C target, the more coal-based generation capacity is implemented in the energy system prior to inception of the trading system. The resulting path dependency commits the energy system to high emissions for several decades, making the achievement of ambitious reduction targets difficult and, when the delay lasts too long, impossible.

The main finding of our analysis is that delayed action will increase costs of achieving ambitious reduction. This is in line with results of other integrated assessment models analysing second-best climate policy regimes (for an overview see Hof *et al.*, this volume, Chapter 4; see also Bosetti *et al.* 2008; Edmonds *et al.* 2008). In addition, these studies show that regional fragmentation, that is, only some region participate in an international carbon market equalizing marginal abatement costs, further increases the costs of climate protection. Also, they reveal that longer delays in implementing some policy instrument that sets corresponding targets and prices will make ambitious goals such as the 2 °C objective unfeasible.

These findings underline the need for swift internationally coordinated action if ambitious reduction targets shall be achieved at low cost. While bottom–up strategies for implementing a global price on carbon may represent viable fallback options, delay of action has its price. Conversely, this means that achieving agreement on global pricing of carbon across regions in a timely manner has considerable and quantifiable benefits for the world community.

5.4 Conclusions and policy recommendations

This study identified two top–down and three bottom–up architectures for the future international carbon market beyond 2012. Within the typology developed by Biermann *et al.* (this volume, Chapter 2), the top–down regimes under investigation correspond to the case of synergistic fragmentation, while the direct and indirect linking cases are examples for cooperative fragmentation. The 'fragmented markets' architecture displays some features of conflictive fragmentation.

Summarizing key features of the architectures under consideration, a top–down *government-level cap-and-trade architecture* with caps for every country in the world is rather unlikely to emerge in the short term due to the substantial distributional and institutional challenges, and there are also concerns over its economic efficiency for example due to market power. A *Kyoto II* system with government-level cap-and-trade for industrialized regions and non-binding trading mechanisms such as a reformed CDM for other countries is more feasible as it does not implement binding caps for developing countries, but still faces the challenge of resolving the burden-sharing question among major economies (Stripple and Lövbrand, this volume, Chapter 11). Also, the inherent efficiency concerns related to any government-based carbon market prevail. Both in presence and absence of Kyoto II, the bottom–up *direct linking of regional cap-and-trade systems*, for example of the EU emissions trading scheme and other emerging schemes, represents an option for developing international company-level carbon markets and promises high economic efficiency. In absence of direct links, *indirect links among regional cap-and-trade systems through CDM* work towards allowance price convergence in regional cap-and-trade systems, thus enhancing their efficiency. Finally, cap-and-trade markets may remain fully *fragmented* if none of these links are implemented.

We find that the choice between integrated top–down and fragmented bottom–up architectures corresponds to a trade-off between high environmental effectiveness – favouring top–down approaches – and political feasibility, on which bottom–up processes fare better. The picture is less clear-cut for cost effectiveness; however, one can expect superior economic performance from bilateral links of regional cap-and-trade systems.

Mapping the findings of this study along the four criteria used throughout this volume, the following conclusions emerge:

Political dimension As they rely on modalities and transactions on the one hand and a respective architecture on the other, carbon markets are located at the interface between the political software (that is, policies) and hardware (that is, institutions). We found that a combination of the Kyoto II approach and an architecture directly linking the various markets will fare best as regards environmental effectiveness and economic efficiency. Kyoto II enables timely multilateral coordination of regional emission budgets in the context of a global policy target, while linking cap-and-trade systems in this framework creates an efficient international market.

International institutional aspects The climate convention will very probably remain the main forum for addressing issues of global burden-sharing, which are relevant in all of the trading architectures under consideration. In particular, resolving burden-sharing issues is a necessary condition for implementing top–down approaches. The climate secretariat's International Transaction Log constitutes the infrastructure for the Kyoto Protocol trading mechanisms, and may also serve this purpose in a future top–down architecture. It is conceivable, however, that a regulatory body for an international carbon market may evolve independently of existing institutions, or be associated with a different institution. In a multilateral bottom–up process comprising the EU Commission, national governments from the European Union, Australia and New Zealand as well as federal states in the United States, the International Carbon Action Partnership may further evolve as a platform for coordinating and eventually governing bottom–up linking activities. As long as only a few direct bilateral links materialize for example between the EU emissions trading scheme and a future United States emissions trading system, bilateral coordination schemes for market governance may suffice. As soon as additional actors join the multilateral scheme for example in an OECD-wide cap-and-trade system, more comprehensive arrangements for policy coordination will be desirable.

Type of actors As the set-up of mandatory emissions trading schemes is essentially a regulatory task, governments will remain key actors in setting up the international emissions trading architecture. Trading schemes should ideally operate at the company level to ensure their economic efficiency.

Mode of governance The options we discussed imply both top–down and bottom–up modes. International negotiations on the set-up of an international carbon market architecture between governments will entail a process of bargaining, while domestic implementation of trading schemes vis-à-vis companies entails a strong hierarchical dimension.

To conclude, in case of a prolonged stalemate over burden-sharing in global negotiations, bottom–up processes such as indirect and direct bilateral links as examples of cooperative fragmentation can serve as fallback options enabling the further development of international emissions trading until equity issues can be resolved. For the 2 °C stabilization objective, integrated assessment model calculations show such delays in the implementation of a comprehensive and efficient system to cause extra costs that appear tolerable if the delay does not last beyond

2020. In institutional terms, timely implementation of a comprehensive yet efficient carbon market can be achieved by directly linking regional trading systems in the context of a comprehensive Kyoto II agreement.

References

Anger, N. 2008. 'Emissions trading beyond Europe: linking schemes in a post-Kyoto world', *Energy Economics* **30**: 2028–2049.

Asselt, H. van 2010 (in press). 'Emissions trading: the enthusiastic adoption of an alien instrument?', in A. Jordan, D. Huitema, H. van Asselt, F. Berkhout and T. Rayner (eds.), *Climate Change Policy in the European Union: Confronting the Dilemmas of Mitigation and Adaptation*. Cambridge, UK: Cambridge University Press.

Babiker, M., J. Reilly and L. Viguier 2004. 'Is international emissions trading always beneficial?', *Energy Journal* **25**: 33–56.

Bergfelder, M. 2008. 'In the market. ICAP – the International Carbon Action Partnership: building a global carbon market from the bottom-up', *Carbon and Climate Law Review* **2**: 202–203.

Böhringer, C. and A. Löschel 2003. 'Market power and hot air in international emissions trading: the impacts of US withdrawal from the Kyoto Protocol', *Applied Economics* **35**: 651–663.

Bosetti, V., C. Carraro, A. Sgobbi and M. Tavoni 2008. *Delayed Action and Uncertain Targets: How Much Will Climate Policy Cost?*, Working Paper No. 69.2008. Milan: FEEM (Fondazione Eni Enrico Mattei).

Copeland, B. R. and M. S. Taylor 2005. 'Free trade and global warming: a trade theory view of the Kyoto Protocol', *Journal of Environmental Economics and Management* **49**: 205–234.

Edmonds, J., L. Clarke, J. Lurz and M. Wise 2008. 'Stabilizing CO_2 concentrations with incomplete international cooperation', *Climate Policy* **8**: 355–376.

Egenhofer, C. 2007. 'The making of the EU emissions trading scheme: status, prospect and implications for business', *European Management Journal* **25**: 453–463.

Ellerman, A. D. 2008. *The EU Emission Trading Scheme: A Prototype Global System*?, Discussion Paper No. 08–02. Cambridge, MA: Harvard Project on International Climate Agreements.

European Commission 2007. *Final Report of the 4th meeting of the European Climate Change Programme Working Group on Emissions Trading on the Review of the EU–ETS on Linking with Emissions Trading Schemes of Third Countries*, 14–15 June 2007. Brussels: European Commission.

European Commission 2009. *Towards a Comprehensive Climate Change Agreement in Copenhagen*, Communication from the Commission to the European Parliament, the Council, the European Economic and Social Committee and the Committee of the Regions, Doc. No. COM(2009) 39 final. Brussels: European Commission.

Flachsland, C., O. Edenhofer, M. Jakob and J. Steckel 2008. *Developing the International Carbon Market: Linking Options for the EU–ETS*, Report to the Policy Planning Staff in the Federal Foreign Office. Potsdam: Potsdam Institute for Climate Impact Research.

Flachsland, C., R. Marschinski and O. Edenhofer 2009a. 'Global trading versus linking. Architectures for international emissions trading', *Energy Policy* **5**: 1637–1647.

Flachsland, C., R. Marschinski and O. Edenhofer 2009b. 'To link or not to link: benefits and disadvantages of linking cap-and-trade systems', *Climate Policy* **9**: 358–372.

Haites, E. and F. Mullins 2001. *Linking Domestic and Industry Greenhouse Gas Emission Trading Systems*. Palo Alto, CA: Electric Power Research Institute. www.iea.org/textbase/papers/2001/epri.pdf.

Hargrave, T. 2000. *An upstream/Downstream Hybrid Approach to Greenhouse Gas Emissions Trading*. Washington, DC: Center for Clean Air Policy.

Helm, C. 2003. 'International emissions trading with endogenous allowance choices', *Journal of Public Economics* **87**: 2737–2747.

International Energy Agency 2005. *Act Locally, Trade Globally: Emissions Trading for Climate Policy*. Paris: International Energy Agency/OECD.

Jaffe, J. and R. N. Stavins 2007. *Linking Tradable Permit Systems for Greenhouse Gas Emissions: Opportunities, Implications and Challenges*, IETA (International Emissions Trading Association) Report on linking GHG emissions trading systems. Geneva: IETA.

Jaffe, J. and R. N. Stavins 2008. *Linkage of Tradable Permit Systems in International Climate Policy Architecture*, Discussion Paper No. 2008–07. Cambridge, MA: Harvard Project on International Climate Agreements.

Jordan, A. and T. Rayner 2010 (in press). 'The evolution of climate change policy in the European Union: a historical overview', in A. Jordan, D. Huitema, H. van Asselt, F. Berkhout and T. Rayner (eds.), *Climate Change Policy in the European Union: Confronting the Dilemmas of Mitigation and Adaptation*. Cambridge, UK: Cambridge University Press.

Keohane, R. and K. Raustiala 2008. *Toward a Post-Kyoto Climate Change Architecture: A Political Analysis*, Discussion Paper No. 08–01. Cambridge, MA: Harvard Project on International Climate Agreements.

Kerr, S. 2000. 'Domestic greenhouse gas regulation and international emissions grading', in S. Kerr (ed.), *Global Emissions Trading: Key Issues for Industrialized Countries*. Cheltenham, UK: Edward Elgar, pp. 131–156.

Leimbach, M., N. Bauer, L. Baumstark and O. Edenhofer 2008. *Mitigation Costs in a Globalized World: Climate Policy Analysis with REMIND-R*, PIK Working Paper. Potsdam: Potsdam Institute for Climate Impact Research.

Mace, M. J., I. Millar, C. Schwarte, J. Anderson, D. Broekhoff, R. Bradley, C. Bowyer and R. Heilmayr 2008. *Analysis of the Legal and Organisational Issues Arising in Linking the EU Emissions Trading Scheme to Other Existing and Emerging Emissions Trading Schemes*, Study commissioned by the European Commission – DG-Environment, Climate Change and Air, Final Report. London: Foundation for International Environmental Law and Development.

Mehling, M. 2007. 'Bridging the transatlantic divide: legal aspects of a link between regional carbon markets in Europe and the United States', *Sustainable Development Law and Policy* **7**: 47–52.

Meyer, A. 2004. 'Briefing: contraction and convergence', *Engineering Sustainability* **157**: 189–192.

Michaelowa, A., M. Stronzik, F. Eckermann and A. Hunt 2003. 'Transaction costs of the Kyoto mechanisms', *Climate Policy* **3**: 261–278.

Neuhoff, K. 2008. *Tackling Carbon: How to Price Carbon for Climate Policy*. Cambridge, UK: University of Cambridge Electricity Policy Research Group. www.econ.cam.ac.uk/eprg/TSEC/2/neuhoff230508.pdf.

Rehdanz, K. and R. Tol 2005. 'Unilateral regulation of bilateral trade in greenhouse gas emission permits', *Ecological Economics* **54**: 397–416.

Schmidt, J., N. Helme, J. Lee and M. Houdashelt 2006. *Sector-Based Approach to the Post-2012 Climate Change Policy Architecture*. Washington, DC: Center for Clean Air Policy.

Schneider, L. 2007. *Is the CDM Fulfilling its Environmental and Sustainable Development Objectives? An Evaluation of the CDM and Options for Improvement*, Report prepared for WWF. Freiburg, Germany: Öko-Institut.

Skjærseth, J. B. and J. Wettestad 2008. *EU Emissions trading: Initiation, Decision-Making and Implementation*. Aldershot, UK: Ashgate.

Stern, N. 2007. *The Economics of Climate Change: The Stern Review*. Cambridge, UK: Cambridge University Press.

Tangen, K. and H. Hasselknippe 2005. 'Converging markets', *International Environmental Agreements: Politics, Law and Economics* **5**: 47–64.

Tietenberg, T. 2003. *Environmental and Natural Resource Economics*, 6th edn. Reading, MA: Addison Wesley.

Tuerk, A., M. Mehling, C. Flachsland and W. Sterk 2009. 'Linking carbon markets: concepts, case studies and pathways', *Climate Policy* **9**: 341–357.

Victor, D. 2007. 'Fragmented carbon markets and reluctant nations: implications for the design of effective architectures', in J. Aldy and R. N. Stavins (eds.), *Architectures for Agreement: Addressing Global Climate Change in the Post-Kyoto World*. Cambridge, UK: Cambridge University Press, pp. 133–160.

World Bank 2008. *State and Trends of the Carbon Market 2008*. Washington, DC: World Bank.

Zapfel, P. and M. Vainio 2002. *Pathways to European Greenhouse Gas Emissions Trading: History and Misconceptions*, Working Paper No. 85.2002. Milan: FEEM (Fondazione Eni Enrico Mattei).

6

The overlap between the UN climate regime and the World Trade Organization: lessons for climate governance beyond 2012

FARIBORZ ZELLI AND HARRO VAN ASSELT

6.1 Introduction

In Chapter 2, Biermann *et al.* discuss the increasing fragmentation of international climate governance. In this chapter, we focus on one element of this fragmentation, namely the overlap between the UN climate regime and the World Trade Organization (WTO). With a view to the appraisal question for the 'architecture' theme of this volume, we hold that this overlap not only implies benefits, but may also entail significant drawbacks for the development and implementation of the UN climate regime. This raises the question how this overlap can be addressed beyond 2012. Our main argument is that, when developing future strategies for managing this interlinkage, policy-makers should draw lessons from the past, that is, from the potential negative effects of this overlap, and from shortcomings of previous management approaches.

In Section 6.2, we introduce our methodology. Section 6.3 introduces major issues on which the two regimes overlap and respective management approaches, which have hardly yielded significant results. In Section 6.4, we discuss policy options that may be suitable to address these unresolved issues and debates in the future. We argue that appropriate strategies need to take into account core reasons for the observed interlinkages and for previous management failures: the constellation of strategic interests and the partial lack of consensual knowledge on climate–trade overlaps. We therefore suggest bringing in further expertise on climate–trade interlinkages – for example through a separate chapter in the next assessment report of the Intergovernmental Panel on Climate Change (IPCC) – as well as strategic issue-linking, for example regarding negotiations on biofuels and the transfer of climate-friendly technologies.

6.2 Methodology

While other chapters on 'architecture' in this volume – in particular Hof *et al.* (Chapter 4) and Flachsland *et al.* (Chapter 5) – have built on quantitative analyses,

Global Climate Governance Beyond 2012: Architecture, Agency and Adaptation, eds. F. Biermann, P. Pattberg and F. Zelli. Published by Cambridge University Press.

we applied methods that pertain to the two other sets of methodologies introduced by Biermann *et al.* in Chapter 1: qualitative policy assessment and participatory assessment.

First, we assessed the international law literature on the overlap between the UN climate regime and the WTO. We then analysed past interplay management by studying conference bulletins as well as the related international relations literature. The findings of this analysis are presented in Section 6.3.

Based on our insights in the current overlaps and shortcomings of management approaches, we explored how climate and trade policies could overlap in future climate governance and how these overlaps could be addressed in both regimes more appropriately. To this end, we co-organized an international workshop in September 2008 (ADAM and UNEP 2008; UNEP and ADAM 2009). More than 70 participants – from the WTO, the UN and other international organizations, from embassies and country missions, and from NGOs and academia – took part in the workshop. Keynote speakers and discussants provided briefing notes and lectures on major issues of overlap, such as flexibility mechanisms, incentive mechanisms and climate-friendly technologies, unilateral and multilateral trade-related climate measures, and respective management approaches in both regimes. In three extensive breakout group sessions, participants further discussed policy recommendations for future governance. Core results of this workshop are presented in Section 6.4.

6.3 Analysis

6.3.1 Overlaps between the UN climate regime and the world trade regime

Scholars from various disciplines have scrutinized the interplay between the UN climate regime and the world trade regime (see, for example, Brack *et al.* 2000; Chambers 2001a; Brewer 2003, 2004; Charnovitz 2003; Stokke 2004; Frankel 2005; Cosbey and Tarasofsky 2007; Howse and Eliason 2009). These authors have identified a range of overlapping issues that fall into the jurisdictional scope of both regimes, while disagreeing about the synergetic or conflictive nature of each issue. By overlaps, we do not merely refer to legal intersections between both regimes, but also to the corresponding negotiations, including the positional differences between countries and country groups on contested issues. This chapter cannot present an exhaustive overview of these overlaps or of the arguments made about their potential implications. Instead, we present merely a synopsis of the most important aspects.

Flexibility mechanisms

One overlap that has not yet been clarified concerns one of the climate regime's core elements, namely emissions trading. Tradable allowances and credits have been established under the Kyoto Protocol, in particular under the three flexible mechanisms: the Clean Development Mechanism, Joint Implementation, and international emissions trading (also Flachsland *et al.*, this volume, Chapter 5; and Stripple and Lövbrand, this volume, Chapter 11). Article 17 on international emissions trading 'implicitly prevents Parties not included in Annex B [that is, developing countries] from acquiring, issuing, or transferring emissions allowances under the Protocol' (Werksman 2001: 170). This restriction could be considered a form of trade discrimination since it effectively excludes the large majority of developing countries as well as third parties to the Kyoto Protocol from emissions trading.

However, this imbalance would only amount to a violation of WTO non-discrimination principles (that is, most-favoured-nation treatment or national treatment) if emission credits could be defined as either 'goods or products' under the General Agreement on Tariffs and Trade (GATT) or 'services' under the General Agreement on Trade in Services (GATS). Yet a classification of emission allowances as good or services is controversial and difficult to accomplish, as neither the GATT nor the GATS provides definitions for goods or services (Kim 2001: 252). Some scholars have advised against such an interpretation (for example Werksman 2001: 155f and 164). They argue that economic or financial value alone does not automatically constitute a definition as goods or services – similar to other entities such as electricity, oil or money which also do not fall under GATT or GATS requirements (Kim 2001: 252ff). On the other hand, Voigt (2008: 58) points to evidence that the GATT covers intangible goods such as electricity. Given this ambiguity – which can only be clarified by treaty amendments or case law – one cannot definitely decide whether the case of trade restrictions in emission allowances constitutes a direct regime conflict (Brewer 2003: 337).

Even when denying such a direct nature, there are further implications of emissions trading that at least point to indirect incompatibility. Given the abstract phrasing of article 17 – notwithstanding its concretization in the Marrakech Accords – various design options might be chosen for an emissions trading scheme. Depending on the design of such a scheme, measures taken in the artificial market of emission allowances might also affect the trade in goods and services in existing markets. For instance, 'brokerage, consulting and insurance services associated with emissions trading could be considered commercial services with the normal meaning of the term', and thus potentially fall under GATS rules (Brewer 2003: 337; Martin 2007). In fact, given the variety of services that can be involved in emissions trading, many scholars expect the GATS rather than the GATT to be applicable (Petsonk 1999: 203ff; Cosbey and Tarasofsky 2007: 24).

Another important design option that might collide with WTO law is the allocation of emission allowances. Certain allocation methods could be interpreted as the favourable treatment of a domestic industry over foreign competitors, in particular if a domestic emissions trading scheme stipulates the free-of-charge distribution of allowances. In fact, such a free allocation of financial assets might be classified as a subsidy (Howse and Eliason 2009). However, the WTO Agreement on Subsidies and Countervailing Measures prohibits subsidies that are specific to an enterprise or industry, or subsidies that 'can bring adverse effects to the interests of other Members' (article 1). The free allocation of emission allowances might hence be challenged under this agreement, which is thus more stringent on acceptable design options than the UN climate regime itself.

Trade-related policies and measures

Articles 2.1.a and 2.2 of the Kyoto Protocol list various policies and measures by which industrialized countries can achieve emission limitations. These policies and measures include, for instance, research, development and use of renewable energy and climate-friendly technologies; reduction or phasing out of fiscal incentives, tax and duty exemptions, and subsidies in all greenhouse gas emitting sectors; and limiting and reducing greenhouse gas emissions in the transport sector. This notwithstanding, the protocol lacks specificity. For instance, it does not spell out concrete steps or targets to achieve the listed items. Since policies and measures are not concretely defined, it is more problematic for the climate regime to claim jurisdiction over such behaviour, let alone to entitle a regime body or mechanism to address it. Hence, policies and measures are not covered by the UN climate regime's compliance system and dispute settlement procedure (Kommerskollegium 2004: 83).

On the other hand, it is not ruled out that parties apply certain trade-distorting and not fully WTO-compliant measures. This can be termed an 'indirect conflict' with WTO rules: there is no immediate rule collision with the WTO, but the Kyoto Protocol's permissive rules on policies and measures might induce such behaviour (Vranes 2006; Zelli 2008). There is a whole range of overlaps and possible clashes due to the potential implementation of certain fiscal measures (subsidies, tariffs or border taxes), regulatory measures (standards, technical regulations and labelling), and government procurement practices. Industrialized countries might consider such measures to flank their greenhouse gas emission reduction activities or to protect domestic industries that are adversely affected by the implementation of climate policies – in other words: to level the playing field between regulated domestic industries and unregulated foreign competitors (Frankel 2005: 15). In the following, we briefly reflect on some of these measures.

Subsidies to firms for climate-friendly products, research, development or export might be forbidden under the WTO Agreement on Subsidies and Countervailing Measures (Santarius *et al.* 2004: 25). The key questions are how specific a subsidy is (does it only concern selected industries?) and what injury it might cause to others (van Asselt and Biermann 2007: 501).

Governments might also choose to put burdens on energy-inefficient foreign companies by imposing tariffs or taxes on their greenhouse gas-intensive imports. One uncertainty about WTO-compatibility of such measures relates to the question of product-related processes and production methods. Energy input tariffs do not apply to the end-use of a product, but to its 'embedded carbon' (that is, greenhouse gases emitted during the product's life cycle). Yet if a WTO panel – in a possible future dispute – only considered end-uses, such process-related taxes would be found to violate the national treatment principle under GATT article III(2) which demands similar taxing for 'like products' (Biermann and Brohm 2005: 291).

In the same vein, marginal taxes on energy-intensive goods from countries that are not party to the Kyoto Protocol or do not take 'comparable' climate change action might violate both the national treatment and most-favoured-nation principles of the GATT. Such border adjustment measures might become a reality soon. A number of industrialized countries have considered offsetting measures at the border to complement an emissions trading scheme – most recently the United States Congress, the French government and the European Commission (van Asselt *et al.* 2009). Experts have led long debates on border adjustment measures, either holding that, under certain circumstances, they could be defended and sustained under WTO law (for example Biermann and Brohm 2005; Ismer and Neuhoff 2007; Pauwelyn 2007) or rather warning against their protectionist implications and possible violation of the GATT (for example Bhagwati and Mavroidis 2007; Quick 2008).

Furthermore, any product standards, labels or technical regulations that establish minimum requirements for goods on the basis of their energy or greenhouse gas intensity during production or use, might conflict with the national treatment principle under the GATT or with the Agreement on Technical Barriers to Trade (Santarius *et al.* 2004: 25).

The climate regime also permits certain government procurement policies – that is, government purchases of goods and services – that might create tensions with the WTO Agreement on Government Procurement. Yet altogether, while subsidies, tariffs and border adjustment measures might be more prone to a legal challenge, government procurement, labelling and standards (at least voluntary ones) are rather unlikely to collide with WTO rules (van Asselt *et al.* 2006: 224; van Asselt and Biermann 2007: 502).

In sum, there is a whole array of 'unresolved issues' (Brewer 2003: 228). These indirect overlaps have been acknowledged from both sides. The WTO secretariat has referred to potential tensions arising from policies and measures in various notes on the relationship with the UN climate regime (Brewer 2003: 334ff). Furthermore, article 3(5) of the climate convention holds that '[m]easures taken to combat climate change, including unilateral ones, should not constitute a means of arbitrary or unjustifiable discrimination or a disguised restriction on international trade'. Likewise, article 2(3) of the Kyoto Protocol signals negotiators' concerns (Linnér 2006: 285), asking parties to 'minimize adverse effects' when implementing policies and measures. These clauses could be interpreted as subtle priority clauses indicating a prevalence of the WTO. In turn, there is no comparable WTO clause in favour of climate-related measures – notwithstanding references to the principle of sustainable development in the preamble of the Agreement Establishing the WTO and the general exceptions to WTO non-discrimination principles, for example under GATT article XX.

Transfer of climate-friendly goods, services and technologies

While the two previous examples rather point to potential conflicts between the UN climate regime and the WTO, there are also win–win constellations, in particular the removal of trade barriers in favour of climate-friendly goods or services, and the development and transfer of low-emission technologies. Article 4(5) of the climate convention states that 'developed country Parties … shall take all practicable steps to promote, facilitate and finance, as appropriate, the transfer of, or access to, environmentally sound technologies and know-how to other Parties, particularly developing country Parties, to enable them to implement the provisions of the Convention'. This provision is based on a broad understanding of technology transfer, which includes capacity-building – in terms of human resources and knowledge bases – in the receiving countries (Knopf and Edenhofer 2010; also Alfsen *et al.*, this volume, Chapter 13; Shrivastava and Goel, this volume, Chapter 8; Winkler, this volume, Chapter 7). However, instead of facilitating knowledge transfer and capacity-building, companies in industrialized countries have much higher incentives to build new technologies completely in-house to secure maximum profits and reduce investors' risks. Only once the technology is fully developed, they might insist on the rules of trade liberalization, asking receiving countries to reduce respective import barriers.

What role does WTO law play in this constellation? On the one hand, the Agreement on Trade-Related Aspects of Intellectual Property Rights (TRIPS) strengthens the position of technology developers, since it opposes national sovereignty – and subsequent protectionism – over intellectual property rights. Moreover, the most-favoured-nation provisions for goods (under GATT) and services (under GATS) guarantee that certain measures which facilitate technology transfer towards selected countries (for example as granted by the United States in several bilateral and multilateral treaties) are expanded to all WTO members (Brewer 2008b). On the other hand, the TRIPS agreement might render the acquisition of technologies more costly, to the disadvantage of developing countries (Littleton 2008: 7–14; Alfsen *et al.*, this volume, Chapter 13; Shrivastava and Goel, this volume, Chapter 8). Moreover, the WTO Agreement on Trade-Related Investment Measures can constrain the ability of acquiring countries' governments to act by excluding the use of certain interventions, for example by not allowing enforcement of performance requirements on multinational corporations (Subbarao 2008: 14).

Similarly to the two earlier discussed topics, there is hence much uncertainty about the exact implications of WTO rules for climate-related policies and vice versa.

6.3.2 Management approaches and their shortcomings

While acknowledging these various overlaps, governments have failed to develop appropriate management approaches to enhance synergies or to tackle potentially negative implications. Negotiations on these issues have largely taken place under the umbrella of the WTO and have either ended in narrow mandates or in stalled debates.

The European Union and Switzerland submitted proposals in the WTO Committee on Trade and Environment as early as 1996. They called for an 'environmental window' in favour of multilateral environmental agreements. Such a window might take the form of a savings clause, such as an extension of the environmental exceptions under article XX GATT, or even the adoption of a whole new WTO agreement on the relationship with multilateral environmental agreements. Both of these tools intended to grant certain environmental treaty rules a permanent waiver with regard to the WTO principles on non-discrimination (Sampson 2001: 74; Santarius *et al.* 2004: 15–16; Zelli 2007). For the UN climate regime, this could have implied waivers for any restrictions on the trade in emission allowances or for the implementation of certain trade-restrictive policies and measures. However, such proposals met considerable resistance by other parties, especially developing countries who feared that such 'carte blanche' was a disguise for green protectionism (Neumann 2002: 330).

After these failed attempts, the European Union and Switzerland tried to reinvigorate discussions in the WTO Committee on Trade and Environment at the Doha ministerial conference in 2001. In fact, an explicit mandate for clarification of the relationship between the WTO and multilateral environmental agreements was one of the European Union's '"must haves" for launching a new WTO round' (Haverkamp 2001: 5). This mandate was included in paragraph 31(i) of the Doha Development Agenda, to be debated in the special session of the WTO Committee on Trade and Environment. However, the mandate was restricted in three ways. First, due to resistance by the United States, Australia and most developing countries, the formula of paragraph 31(i) was narrowed to the applicability of *existing* WTO rules with regard to *existing* treaty rules – that is, leaving out any decisions of the conference of the parties. This interpretation significantly limits prospects for any legal concessions to a future climate agreement, since the conferences of the parties are supposed to flesh out the rather abstract regulations of such an agreement on tradable allowances, policies and measures and other issues. Second, the same countries also achieved that another controversial question was 'carved out from the mandate's scope' (Palmer and Tarasofsky 2007: 14), namely the application of trade measures in multilateral environmental agreements to third parties. Third, debates in the special session of the WTO Committee on Trade and Environment soon got stuck in technical controversies about the scope and procedure of the negotiations. In the end, only specific and mandatory provisions were discussed. This implies an exclusion of trade-related policies and measures, since the list of policies and measures under article 2(1a) of the Kyoto Protocol is only indicative. Even if discussions in the special session of the WTO Committee on Trade and Environment yielded any legal concessions, they would not apply to policies and measures. This makes it more likely that trade-restrictive policies and measures will be challenged in the WTO dispute settlement system (Eckersley 2004: 36–38).

A look at other Doha Round discussions – on environmental goods and services (under paragraph 31[iii]), market access (32[i]) and eco-labelling (32[iii] of the Doha Ministerial Declaration) – further confirms that developing countries have often been the most determined opponents of any concessions for multilateral environmental agreements in general, and the UN climate regime in particular. These debates concern the third overlapping issue identified above (climate-friendly goods, services and technologies) but also are relevant for policies and measures. '[W]hereas many industrial countries regarded [measures such as eco-taxes or eco-labelling] as important environmental instruments, developing countries feared restrictions on their market access' (Santarius *et al.* 2004: 14). For instance, in 2002 Saudi Arabia tabled a proposal in the WTO Committee on Trade and Environment and the Doha Round's Non-Agricultural Market Access Negotiating Group. In line with OPEC strategies in the UN climate regime on adverse impacts of policies and measures (articles 3[14] of the Kyoto Protocol and 4[18] of the convention), this proposal called for the removal of energy-related subsidies in industrialized countries. Most remarkably, the proposal did not only target subsidies for the coal or nuclear sectors, but also for climate-friendly sectors like renewable energies (Yamin and Depledge 2004: 256).[1] But also non-OPEC members have carried the debate on adverse impacts into the WTO. A proposal by India has criticized the negative effects of environmental measures on market access for products from least-developed countries and other developing countries (Harashima 2008: 27).[2] So far, these developing country proposals have been rejected by both the European Union and the United States.

Developing countries were far more sceptical about trade liberalization once discussions addressed access to their own markets. In the environmental goods

[1] Doc. No. TN/TE/W/9. www.jmcti.org/2000round/com/doha/tn/tn_te_w_009.pdf.
[2] Doc. No. WT/CTE/W/207. http://commerce.nic.in/wt_cte_W207.pdf.

and services debate under paragraph 31(iii) of the Doha Ministerial Declaration, they have strongly criticized a 'list approach' suggested by industrialized countries. In their approach, the European Union and the United States listed a large number of environmental goods and services, including climate-friendly ones, for which trade barriers should be removed or reduced. Developing countries, in turn, held that the 'list approach' was just a disguise for a different purpose: since many of the listed goods had multiple uses, the approach rather sought to secure wide-ranging access to developing country markets (Jha 2008: 2ff). Therefore, India proposed a much narrower 'project approach', which only allows temporary trade liberalization for goods and services associated with an approved environmental project (ICTSD 2007: 12ff). Moreover, India and OPEC members demanded a relaxation of intellectual property standards under the TRIPS agreement in order to support transfer of specific climate-friendly technologies (ICTSD 2008a: 6). Due to this stand-off between North and South, discussions on environmental goods and services shared the fate of other debates in the WTO Committee on Trade and Environment and made no major progress (World Bank 2008: 75ff). In addition to the dual-use problem, parties to the special session of the WTO Committee on Trade and Environment also disagreed on the definition of environmental goods – whether they should be based on environmental end-use or also on the environmental production process (World Bank 2008: 75ff).

The WTO-internal debates on overlaps with environmental regimes have thus largely ended in negotiation stalemates. These stalemates are tied to the overall slow progress of the Doha Round. Since the Doha Development Agenda constitutes a 'single undertaking', progress on the 'trade and environment' mandate under paragraphs 31 and 32 depends on the success of talks on other Doha items, especially on tariff and subsidy cuts in the industrial goods and agriculture negotiations. An agreement on these issues 'would give delegates a sense of what products to include in the liberalisation agreement and would provide a more solid idea of the potential extent of any tariff cuts' (ICTSD 2008b). To take another illustration: debates on paragraph 31(ii) on permanent observer status of the climate convention and other multilateral environmental agreements have been hijacked by an overarching conflict among WTO members. To date, members of the Arab League and countries sympathetic to the League have blocked any applications for WTO observer status by international organizations or regimes. They thereby retaliate against the denial of observer status to the Arab League (Palmer and Tarasofsky 2007: 44).

Many of these overlaps and shortcomings of management approaches have negatively affected the jurisdictional scope and rule-development of the UN climate regime. This is due in part to the higher degree of delegation in the WTO, where the dispute settlement bodies issued important interpretations of WTO rules on the overlap of trade and environment. Some scholars have argued that the shadow of

WTO law and its stronger dispute settlement system may provoke anticipative or 'chill effects' (Stilwell and Tuerk 1999; Eckersley 2004). In order not to risk any legal challenge before the WTO dispute settlement mechanism, parties to the UN climate regime might refrain from the elaboration or implementation of more ambitious trade-relevant climate protection measures (Oberthür 2006: 57). Moreover, they might even refrain from developing more concrete provisions for the UN climate regime itself (Gehring and Oberthür 2006: 314–316).

Indeed, since the adoption of the Kyoto Protocol, negotiators have avoided any trade-restrictive modalities. For instance, the list of policies and measures has remained purely indicative and non-exhaustive. A mandatory coordinated set of policies and measures – for example with trade restrictions similar to those of the Montreal Protocol on Substances that Deplete the Ozone Layer or of the Convention on International Trade in Endangered Species of Wild Fauna and Flora – has not evolved. Furthermore, while a relatively strong compliance mechanism was established, trade restrictions were largely omitted. Non-compliance may lead to exclusion from emissions trading and reduction of the cap in the next commitment period. Yet, although proposed by the European Union, non-compliance does not entail financial penalties or a loss of carbon credits, nor does it include any other trade sanctions (Stokke 2004: 352). Finally, there has been no significant elaboration of the dispute settlement procedure of the UN climate regime, leaving the possibility that countries bring a case before the WTO dispute settlement mechanism (Chambers 2001b: 103). In other words, the UN climate regime has no ultimate clout over cases that immediately affect its jurisdiction. By the same token, negotiators have not reached any legal concessions of WTO rules in favour of the UN climate regime. The various deadlocks and restrictions we discussed above rather point to a legal prevalence of the WTO on the overlapping issues. For instance, leaving disputes between parties and non-parties de facto under the jurisdiction of the WTO dispute settlement mechanism makes it more likely that third parties to the UN climate regime will challenge trade-related climate measures.

In sum, it is evident that the legal status quo implies a lack of clarity, which has negative effects for the development and implementation of the UN climate regime rather than for the WTO. This raises the question how these overlaps can be addressed while avoiding the deadlocks in which many management approaches have ended. We address this question now.

6.4 Conclusions and policy recommendations

This chapter has shown, first, that there are still many unresolved issues and debates in the climate–trade overlap that need to be addressed. Management approaches in both the climate regime and the WTO have not resulted in cooperative and pragmatic solutions to the relationship between the two regimes. What is more, the world trade regime may result in a potential chill effect, which might harm future climate policies in the long run.

It is important to understand the reasons for the current overlaps and the shortcomings of previous management approaches. A first potential reason concerns the strategic constellation of parties. In both the climate and trade negotiations, powerful coalitions of countries influence the outcome of climate–trade interactions. In the UN climate regime, these blocks include the European Union, the Umbrella Group (a loose coalition of the United States, Canada and other non-European industrialized countries like Australia, Canada, Japan and Russia), and the Group of 77 and China, which in turn embraces a diverse range of nations ranging from OPEC members to least-developed countries and small-island developing states. In the WTO, groups are not identical and much more differentiated (especially developing country groups), but roughly follow this threefold pattern with different levels of in-group cohesion. The various interests of these coalitions can in part help explain the observed state of overlap between both regimes. This holds in particular for the 'WTO-compliant' development of the UN climate regime (that is, its largely market-based mechanisms), and the poor outcome of negotiations in the WTO Committee on Trade and Environment on the overlap between the WTO and multilateral environmental agreements under paragraph 31(i).

Another potential reason is likely to be the high uncertainties about climate–trade overlaps. Although a large body of research exists on the overlaps between the climate regime and the WTO, the IPCC reports do not even include a comprehensive analysis. Instead, with the exception of a 2000 special report on technology transfer, passages on the overlaps with the WTO are rather dispersed throughout the IPCC's assessment reports. This reflects the controversy and uncertainty about many of the climate–trade overlaps, for example the aforementioned debates about the WTO-compliance of policies and measures like subsidies or border adjustment measures, or the lack of clarity about the benefits of TRIPS relaxations for North–South technology transfer.

We conclude this chapter therefore with some preliminary policy recommendations for addressing this particular aspect of the institutional fragmentation of global climate governance.

With regard to the lack of consensual knowledge, it is important to bring in further expertise to inform discussions on climate–trade issues and to move discussions away from mostly considering unilateral trade measures such as border adjustment measures. It is crucial to first gather more evidence on the implications of such measures, including their environmental and economic effects, chances to discipline such measures in multilateral agreements, and indirect impact on climate negotiations (for example in light of perceptions by developing countries). Such evidence could be provided by a separate chapter in the fifth assessment report of the IPCC, either on climate–trade overlaps in general, or on unilateral and multilateral trade-related approaches in particular. Moreover, the WTO Committee on Trade and Environment could open up to regular scientific advice on environmental matters, for instance, by establishing a standing advisory body.

Given the change in expertise on these overlaps, one pragmatic option for several overlap issues is to build the uncertainty into respective strategies. For instance, one option to handle border adjustment measures could be a flexible system based on multilateral discussions. Such a system could address critical issues in the design of border adjustment measures, such as sectoral and country coverage (taking into account country's common but differentiated responsibilities), and setting appropriate levels for border adjustments (Climate Strategies 2008). A flexible expertise-based approach might also be an option suitable for another major issue of climate–trade overlap, the removal of trade barriers for climate-friendly goods and services. Instead of a fixed list of climate-friendly goods and services, the United States and the European Union could propose a 'living list', which can be amended based on further scientific input. For instance, building on carbon life-cycle analyses of goods and services, sustainability criteria for the removal of trade barriers could be developed.

As regards coordination among different country coalitions, informal forums or dialogues might be a more suitable starting point to discuss management attempts first, as they are less prone to political sensitivities such as fear of protectionism. Several of these dialogues have already been initiated, such as the Major Economies Process on Energy Security and Climate Change, and the Informal Trade Ministers Dialogue on Climate Change Issues during the thirteenth conference of the parties in Bali. However, it is important to arrange the dialogue across ministries, that is, between government representatives for environment, trade and development. Such a dialogue could provide a platform to discuss overlap questions outside of the WTO.

Finally, policy recommendations should accommodate the strategic interests of the involved country coalitions and the constellation of these interests. To this end, delegates in both institutions should further explore opportunities for issue linking – more than has been the case. Issue linking implies that countries or country coalitions consider aspects from related debates in their strategies. This can result in proposals for coordination, side-payments or even induce package deals, all of which has been practised in international politics in general and international trade in particular (ultimately in the form of the WTO which links a wide range of issues) – and has even found its way into recommendations of the IPCC (2001: 624–627). The most noteworthy example of a constructed link among climate and trade interests is Russia's ratification of the Kyoto Protocol in November 2004, which secured the protocol's entry into force (Henry and Sundstrom 2007).

The underlying intuition of 'tactical issue linkage' (Haas 1980; Folmer *et al.* 1993; Cesar and De Zeeuw 1996) – or even package deals – is that they can solve asymmetries among countries, each country gaining on a different issue, thereby making the agreement profitable to all participants (IPCC 2001: 626ff). In terms of game theory, such tactical issue-linkage can connect two separate bargaining situations, creating a new pay-off matrix with altered preferences, that is, an overall constellation which is more conducive to cooperation. Combining climate and trade issues in an overall deal might hence produce new bargaining chips and provide new leverage to deadlocked negotiations (Zürn 1990: 166–173).

This notwithstanding, package deals are far from being a panacea. While the potential number of tactical issue-linkages between climate and trade issues is infinite, most of these linkages are neither feasible nor sensible. Caveats one needs to consider include the nature of the linked issues. As climate negotiations provide a public good, that is, a good with non-excludable benefits, incentives are high to free ride. To reduce this, issue linking is sensible especially in negotiations on issues with excludable benefits, for example deals on technology transfer. Moreover, the agendas of both future climate governance and the Doha negotiations are overburdened, which slows down progress (ICTSD 2008c). Additional topics could hence easily make matters worse. The choice of topics therefore needs to guarantee balanced benefits for all parties. Moreover, in the Doha Round, trade topics tend to be more important to parties than climate concerns. This imbalance of preferences thus needs to be taken into account.

We briefly explore two examples for which it might be feasible to further integrate country strategies in the Doha Round and in the climate negotiations. First, the European Union could consider linking its position on the relationship between the WTO and multilateral environmental agreements under paragraph 31(i) of the Doha Ministerial Declaration to its position on the TRIPS Agreement. In the former debate, the European Union has asked for legal concessions under WTO law in favour of trade-related measures under multilateral environmental agreements, however meeting opposition from developing countries fearing green protectionism. But in the second debate, some developing countries have demanded concessions in favour of specific environmental concerns: Brazil, India and other countries have called for an amendment of the TRIPS agreement to reflect requirements of the Convention on Biological Diversity. Likewise, at the conference of the parties to the climate convention in Poznań in December 2008, some developing countries such as India and Pakistan have called for a relaxation of intellectual property rights standards for all climate-related technologies (Shrivastava and Goel, this volume, Chapter 8). As discussed above, they argued that such standards might increase costs for acquiring technologies (Meyer-Ohlendorf and Gerstetter 2009: 23–26). This debate was also carried into the special session of the WTO Committee on Trade and Environment. In discussions on environmental goods and services under paragraph 31(iii) of the Doha Ministerial Declaration, Saudi Arabia even suggested allowing compulsory licences for respective technologies.

With all these debates concerning some form of legal concessions under WTO law, there is potential for strategic issue linking: for instance, movement from one side on the debate between WTO and multilateral environmental agreements could trigger progress in TRIPS-related discussions. Switzerland and some EU countries have already demonstrated that some form of issue-linkage with regard to TRIPS discussions is possible; in order to reach a package deal, they suggested linking their demand for an extension of provisions on geographical indication to the proposal on a TRIPS amendment on disclosure (as required by developing countries) (Palmer and Tarasofsky 2007: 43–44).

A second option for issue linking is overlapping discussions on environmental goods and services, and biofuels. In the Doha Round, under paragraph 31(iii), the United States and the European Union asked developing countries to liberalize trade policies to allow more transfers of environmental goods and services. Yet Brazil and other developing countries criticized that the list of environmental goods and services presented by the European Union and the United States excludes biofuels. Moreover, Brazil included United States subsidies of biofuels in a dispute it filed in the WTO in 2007 (Brewer 2008a: 24). In light of this overlap, concessions from one or both sides on biofuels might help reinvigorate the debate on environmental goods and services. Such a concession could for instance come close to the aforementioned idea of a 'living list' that could include biofuels that fulfil certain sustainability criteria. These criteria could be developed under the UN climate regime, based on potential future IPCC work as suggested above. This consideration of sustainable biofuels would accommodate the interests of some developing countries and raise chances of a more comprehensive deal on trade barrier removals for environmental goods and services.

Our proposals of a scientific advisory body to the WTO Committee on Trade and Environment and of a cross-ministerial dialogue and of multilateral discussions on border adjustment measures affect the institutional core of climate and trade governance. Yet our proposals of a specific IPCC chapter or of a living list for climate-friendly goods and services and strategic issue linking are rather 'political software'. In terms of the institutional environment, only our proposal of an IPCC chapter on climate and trade relates solely to the UN climate regime. All other options either involve both regimes (issue-linking, living list of goods and services), only the WTO (scientific advisory body to the WTO Committee on Trade and Environment) or even other institutions (an informal arena for a cross-ministerial dialogue, unilateral or multilateral frameworks for border adjustment measures). Most options predominantly involve public actors and a top–down mode of governance with regard to decision-making. The major exceptions are the advisory body to the WTO Committee on Trade and Environment and an IPCC chapter, both of which bring in epistemic communities. Concerning policy implementation, however, private actors will be increasingly involved and concerned, in particular regarding a living list of goods and services and designing border adjustment measures.

Acknowledgements

We are highly indebted to Benjamin Simmons and Lutz Weischer from the Economics and Trade Branch of the United Nations Environment Programme – both for their invaluable comments and the excellent cooperation on the international workshop on post-2012 climate and trade policies in Geneva in September 2008. We would also like to thank all participants of this workshop as well as the participants of the workshop *Climate Governance Post-2012: Options for EU Policy-Making* in October 2008 in Brussels, organized by the ADAM Project and the Centre for European Policy Studies. Last but not least, this chapter benefited substantially from useful suggestions and critique by Sebastian Oberthür and Oran R. Young.

References

ADAM (Adaptation and Mitigation Strategies: Supporting European Climate Policy) and UNEP (United Nations Environment Programme) 2008. *International Workshop on Post-2012 Climate and Trade Policies*, Geneva, 8–9 September 2008. Workshop summary. www.unep.ch/etb/events/International%20Workshop%20CC%20and%20Trade%20Policies%20Sept%202008/ADAM-UNEP%20workshop%20summary.pdf.

Asselt, H. van and F. Biermann 2007. 'European emissions trading and the international competitiveness of energy-intensive industries: a legal and political evaluation of possible supporting measures', *Energy Policy* **35**: 497–506.

Asselt, H. van, T. Brewer and M. Mehling 2009. *Addressing Leakage and Competitiveness in US Climate Policy: Issues concerning Border Adjustment Measures*. Cambridge, UK: Climate Strategies.

Asselt, H. van, N. van der Grijp and F. Oosterhuis 2006. 'Greener public purchasing: opportunities for climate-friendly government procurement under WTO and EU rules', *Climate Policy* **6**: 217–229.

Bhagwati, J. and P. C. Mavroidis 2007. 'Is action against US exports for failure to sign Kyoto Protocol WTO-legal?', *World Trade Review* **6**: 299–310.

Biermann, F. and R. Brohm 2005. 'Implementing the Kyoto Protocol without the United States: the strategic role of energy tax adjustments at the border', *Climate Policy* **4**: 289–302.

Brack, D., M. Grubb and C. Windram 2000. *International Trade and Climate Change Policies*. London: Earthscan.

Brewer, T. L. 2003. 'The trade regime and the climate regime: institutional evolution and adaptation', *Climate Policy* **3**: 329–341.

Brewer, T. L. 2004. 'The WTO and the Kyoto Protocol: interaction issues', *Climate Policy* **4**: 3–12.

Brewer, T. L. 2008a. *Climate Change Policies and Trade Policies: The New Joint Agenda*, background paper for UNEP Expert Meeting, Geneva: United Nations Environment Programme.

Brewer, T. L. 2008b. 'Climate change technology transfer: a new paradigm and policy agenda', *Climate Policy* **8**: 516–526.

Cesar, H. and A. de Zeeuw 1996. 'Issue linkage in global environmental problems', in A. Xepapadeas (ed.), *Economic Policy for the Environment and Natural Resources*. Cheltenham, UK: Edward Elgar, pp. 158–173.

Chambers, W. B. (ed.) 2001a. *Inter-Linkages: The Kyoto Protocol and the International Trade and Investment Regimes*. Tokyo: United Nations University Press.

Chambers, W. B. 2001b. 'International trade law and the Kyoto Protocol: potential incompatibilities', in W. B. Chambers (ed.) *Inter-Linkages: The Kyoto Protocol and the International Trade and Investment Regimes*. Tokyo: United Nations University Press, pp. 87–118.

Charnovitz, S. 2003. 'Trade and climate: potential conflicts and synergies', in E. Diringer (ed.), *Beyond Kyoto: Advancing the International Effort against Climate Change*. Arlington, VA: Pew Center on Global Climate Change, pp. 141–170.

Climate Strategies 2008. *International Cooperation to Limit the Use of Border Adjustment*, workshop summary, South Center, Geneva, 10 September 2008. Cambridge, UK: Climate Strategies.

Cosbey, A. and R. Tarasofsky 2007. *Climate Change, Competitiveness and Trade*, a Chatham House report. London: Royal Institute of International Affairs.

Eckersley, R. 2004. 'The big chill: the WTO and multilateral environmental agreements', *Global Environmental Politics* **4**: 24–40.

Folmer, H., P. Von Mouche and S. E. Ragland 1993. 'Interconnected games and international environmental problems', *Environment and Resources Economics* **3**: 313–335.

Frankel, J. 2005. 'Climate and trade: links between the Kyoto Protocol and WTO', *Environment* **47**: 8–19.

Gehring, T. and S. Oberthür 2006. 'Comparative empirical analysis and ideal types of institutional interaction', in S. Oberthür and T. Gehring (eds.), *Institutional Interaction in Global Environmental Governance: Synergy and Conflict among International and EU Policies*. Cambridge, MA: MIT Press, pp. 307–371.

Haas, E. B. 1980. 'Why collaborate? Issue-linkage and international regimes', *World Politics* **32**: 357–402.

Harashima, Y. 2008. 'Trade and environment negotiations in the WTO: Asian perspectives', *International Environmental Agreements: Politics, Law and Economics* **8**: 17–34.

Haverkamp, J. 2001. 'The conflict between the WTO and MEAs: in the view of the U.S. government, only a theoretical problem', in Heinrich Böll Foundation (ed.), *Trade and Environment, the WTO, and MEAs: Facets of a Complex Relationship*. Washington, DC: Heinrich Böll Foundation Washington Office, pp. 5–11.

Henry, L. A. and L. Sundstrom 2007. 'Russia and the Kyoto Protocol: seeking an alignment of interests and image', *Global Environmental Politics* **7**: 47–69.

Howse, R. and A. Eliason (2009). 'Domestic and international strategies to address climate change: an overview of the WTO legal issues', in T. Cottier, O. Nartova and S. Bigdeli (eds.), *International Trade Regulation and the Mitigation of Climate Change*. Cambridge, UK: Cambridge University Press, pp. 48–94.

ICTSD (International Centre for Trade and Sustainable Development) 2007. *Trade in Environmental Goods and Services and Sustainable Development: Domestic Considerations and Strategies for WTO Negotiations*, ICTSD Policy Discussion Paper. Geneva: ICTSD.

ICTSD 2008a. 'Progress over next two weeks critical to Ministerial's prospects', *Bridges Weekly Trade News Digest* **12**(24).

ICTSD 2008b. 'WTO Panel rules against EU import ban in beef hormone case: both sides claim victory', *Bridges Weekly Trade News Digest* **12**(11).

ICTSD 2008c. 'TRIPS Council once again marked by divisions over disclosure amendment', *Bridges Weekly Trade News Digest* **12**(10).

IPCC (Intergovernmental Panel on Climate Change) 2001. *Climate Change 2001: Mitigation. Contribution of Working Group III to the Third Assessment Report of the Intergovernmental Panel on Climate Change*. Geneva: IPCC

Ismer, R. and K. Neuhoff 2007. 'Border tax adjustment: a feasible way to support stringent emission trading', *European Journal of Law and Economics* **24**: 137–164.

Jha, V. 2008. *Environmental Priorities and Trade Policies for Environmental Goods: A Reality Check*, ICTSD Issue Paper No. 7. Geneva: ICTSD.

Kim, J. A. 2001. 'Institutions in conflict? The climate change flexibility mechanisms and the multinational trading system', *Global Environmental Change* **11**: 251–255.

Knopf, B. and O. Edenhofer 2010 (in press). 'The economics of low stabilisation: implications for technological change and policy', in M. Hulme and H. Neufeldt (eds.), *Making Climate Change Work for Us: European Perspectives on Adaptation and Mitigation Strategies*. Cambridge, UK: Cambridge University Press.

Kommerskollegium (Swedish National Board of Trade) 2004. *Climate and Trade Rules: Harmony or Conflict*? Stockholm: Kommerskollegium.

Linnér, B.-O. 2006. 'Authority through synergism: the roles of climate change linkages', *European Environment* **16**: 278–289.

Littleton, M. 2008. *The TRIPS Agreement and Transfer of Climate-Change-Related Technologies to Developing Countries*, Working Paper No. 71. Doc. No. ST/ESA/2008/DWP/71. New York: United Nations Department of Economic and Social Affairs.

Martin, M. A. 2007. 'Trade law implications of restricting participation in the European Union emissions trading scheme', *Georgetown International Environmental Law Review* **19**: 437–474.

Meyer-Ohlendorf, N. and C. Gerstetter 2009. *Trade and Climate Change: Triggers or Barriers for Climate Friendly Technology Transfer and Development*, Dialogue on Globalization Occasional Paper No. 41. Berlin: Ecologic.

Neumann, J. 2002. *Die Koordination des WTO-Rechts mit anderen völkerrechtlichen Ordnungen: Konflikte des materiellen Rechts und Konkurrenzen der Streitbeilegung*. Berlin: Duncker and Humblot.

Oberthür, S. 2006. 'The climate change regime: interactions with ICAO, IMO, and the EU burden-sharing agreement', in S. Oberthür and T. Gehring (eds.), *Institutional Interaction in Global Environmental Governance: Synergy and Conflict among International and EU Policies*. Cambridge, MA: MIT Press, pp. 53–78.

Palmer, A. and R. Tarasofsky 2007. *The Doha Round and Beyond: Towards a Lasting Relationship between the WTO and the International Environmental Regime*, a Chatham House report. London: Royal Institute of International Affairs.

Pauwelyn, J. 2007. *US Federal Climate Policy and Competitiveness Concerns: The Limits and Options of International Trade Law*, Nicholas Institute Working Paper 07–02. Durham, NC: Nicholas Institute for Environmental Policy Solutions, Duke University.

Petsonk, A. 1999. 'The Kyoto Protocol and the WTO: integrating greenhouse gas emission allowance trading into the global marketplace', *Duke Environmental Law and Policy Forum* **10**: 185–220.

Quick, R. 2008. '"Border tax adjustment" in the context of emissions trading: climate protection or "naked" protectionism?', *Global Trade and Customs Journal* **3**: 163–175.

Sampson, G. P. 2001. 'WTO rules and climate change: the need for policy coherence', in W. B. Chambers (ed.), *Inter-Linkages: The Kyoto Protocol and the International Trade and Investment Regimes*. Tokyo: United Nations University Press, pp. 69–85.

Santarius, T., H. Dalkmann, M. Steigenberger and K. Vogelpohl 2004. *Balancing Trade and Environment: An Ecological Reform of the WTO as a Challenge in Sustainable Global Governance*. Wuppertal: Wuppertal Institute for Climate, Environment, and Energy.

Stilwell, M. T. and E. Tuerk 1999. *Trade Measures and Multilateral Agreements: Resolving Uncertainty and Removing the WTO Chill Factor*, discussion paper. Gland, Switzerland: WWF International.

Stokke, O. S. 2004. 'Trade measures and climate compliance: institutional interplay between WTO and the Marrakesh Accords', *International Environmental Agreements: Politics, Law and Economics* **4**: 339–357.

Subbarao, P. S. 2008. *International Technology Transfer to India An Impedimenta and Impetuous*, IIM Working Paper No. 2008–01-07. Ahmedabad: Indian Institute of Management.

UNEP (United Nations Environment Programme) and ADAM (Adaptation and Mitigation Strategies: Supporting European Climate Policy) 2010. *Climate and Trade Policies in a Post-2012 World*. Geneva: United Nations Environment Programme.

Voigt, C. 2008. 'WTO Law and international emissions trading: is there potential for conflict?', *Carbon and Climate Law Review* **2**: 54–66.

Vranes, E. 2006. 'The definition of "norm conflict" in international law and legal theory', *European Journal of International Law* **17**: 395–418.

Werksman, J. 2001. 'Greenhouse-gas emissions trading and the WTO', in W. B. Chambers (ed.), *Inter-Linkages: The Kyoto Protocol and the International Trade and Investment Regimes*. Tokyo: United Nations University Press, pp. 153–90.

World Bank 2008. *International Trade and Climate Change: Economic, Legal, and Institutional Perspectives*. Washington, DC: World Bank.

Yamin, F. and J. Depledge 2004. *The International Climate Change Regime: A Guide to Rules, Institutions and Procedures*. Cambridge, UK: Cambridge University Press.

Zelli, F. 2007. 'The World Trade Organization: free trade and its environmental impacts', in K. V. Thai, D. Rahm and J. D. Coggburn (eds.), *Handbook of Globalization and the Environment*. London: Taylor and Francis, pp. 177–216.

Zelli, F. 2008. *Regime Conflicts in Global Environmental Governance: A Framework for Analysis*, Global Governance Working Paper No. 36. Amsterdam: The Global Governance Project.

Zürn, M. 1990. 'Intra-German trade: an early East–West regime', in V. Rittberger (ed.), *International Regimes in East–West Politics*. London: Pinter, pp. 151–188.

7

An architecture for long-term climate change: North–South cooperation based on equity and common but differentiated responsibilities

HARALD WINKLER

7.1 Introduction

The pace of climate negotiations needs to step up significantly to deal with the urgency of the challenge. This chapter considers different proposals or 'packages' for a possible architecture for the future of the climate regime beyond 2012.[1] In terms of the appraisal question for Part I of this volume (Biermann *et al.*, this volume, Chapter 2), this chapter analyses proposals that fall between a state of 'cooperative fragmentation', namely the status quo of the UN climate regime, and the ideal type of 'universalism', that is, an all-encompassing regime. Unlike the contributions by Hof *et al.* (this volume, Chapter 4) and Flachsland *et al.* (this volume, Chapter 5), the chief assessment criterion in this chapter is the principle of equity. It is the first principle cited in the United Nations Framework Convention on Climate Change. Article 3.1 states: 'The parties should protect the climate system for the benefit of present and future generations of humankind, on the basis of equity and in accordance with their common but differentiated responsibilities and respective capabilities' (UNFCCC 1992).[2]

The objective of the climate convention as spelled out in article 2 is not only about stabilization of concentrations of greenhouse gases. This objective must be achieved in a way that does not prejudice sustainable development. From the perspective of developing countries, ensuring that economic development can proceed in a sustainable manner remains as relevant as ever, as do social considerations and quality of life issues such as food security (UNFCCC 1992).

The equity principle suggests that any proposal or package must also be fair. Achieving such fairness is not a simple matter, given that we live in a world with

[1] Parts of this chapter draw on previous work published in Winkler and Vorster (2007).

[2] The other principles in article 3 refer to the specific needs of developing countries; the precautionary approach; the right to promote sustainable development; sustainable economic growth; and an open international economic system.

Global Climate Governance Beyond 2012: Architecture, Agency and Adaptation, eds. F. Biermann, P. Pattberg and F. Zelli. Published by Cambridge University Press.

high levels of inequality (Agarwal and Narain 1991). Not only is the world an unequal place, but the problem of climate change itself has an unequal structural characteristic: those least responsible for the problem (the poor) are also the most vulnerable to the impacts of climate change (Jerneck and Olsson, this volume, Chapter 18).

The chapter is organized as follows. The next section briefly outlines the methodology and approaches the question of equity. The chapter then outlines different schools of thought in which different packages can be located. I elaborate two packages in more detail: the 'multi-stage' approach and the Transitional Ambitious package. I then draw conclusions for the future of the climate regime.

7.2 Methodology

The methodological approach taken in this chapter is policy analysis. In particular, I consider the potential future architecture of the climate regime from the point of view of equity. My starting point is that equity remains central, even as the framework of the convention encompasses further action in response to the greater urgency indicated by science. Equity requires that both the process and the outcome of negotiations be fair and be perceived as fair (Müller 1999). What is considered fair depends on the perspective. Some take the view that an outcome based on equal per capita emissions is fair; others point to historical responsibility for cumulative emissions, while a third view is that any proposal that undermines sustainable development cannot be fair.

Perceptions of equity may also differ over time. For instance, in Kyoto 1997, the principle of common but differentiated responsibilities and respective capabilities meant that industrialized countries took leadership through quantified emission limitation and reduction objectives (UNFCCC 1997), while developing countries continued with qualitative mitigation measures (UNFCCC 1992: article 4.1b).

Ten years later, at the thirteenth conference of the parties to the climate convention, negotiations culminated in the Bali Action Plan, which settled on four key elements: adaptation, mitigation, finance and technology. The emerging balance between mitigation, adaptation, technology and finance started to create some bargaining space. Balancing these core elements can establish a conceptual contract zone. But the devil is in the detail. Whether the future agreement passes the test of equity will depend on how these elements are fleshed out beyond the current climate negotiations.

Pursuing the objective of article 2 of the climate convention in both its sentences, in the context of latest evidence, hence requires that we raise the bar on both sides. Deeper cuts are required in the North, from all industrialized countries, including the United States. In addition, the South needs to take quantifiable mitigation actions, supported by the long-agreed assistance with financial and technological resources. To make progress on mitigation, we need to squarely address the issues of distributional equity (taking into account both historical responsibility and responsibility for the future), in order to build the trust required to make progress. Moreover, to meet the test of equity, an agreement needs to deal seriously with adaptation, especially for the poor, but ultimately for all (Jerneck and Olsson, this volume, Chapter 18).

Unlike other chapters that emphasize quantitative analysis and modelling, this chapter addresses these questions using a qualitative approach that is rooted in political science. Although no formal interviews were conducted, the insights of key informants with experience of the multilateral process have shaped the thinking of the author.

The future of the climate regime must be fair, effective, flexible and inclusive to be acceptable. That is what is required for 'equity in the greenhouse' (Ott *et al.* 2004). This chapter argues in favour of striking a core balance between the imperatives of development and climate. Hence, the elements of future climate governance should be packaged in terms of climate and development – or from a developing country perspective, even putting development first (Shrivastava and Goel, this volume, Chapter 8). I develop a set of criteria in the analytical body of the chapter and apply them to different schools of thought. This assists in identifying two central packages that I elaborate more fully.

7.3 Analysis

7.3.1 Review of proposals

Both governments and scholars have proposed a wide variety of approaches to regulate mitigation of greenhouse gas emissions in the global climate governance architecture beyond 2012 (Zelli *et al.*, this volume, Chapter 3). These approaches include fixed targets similar to the Kyoto Protocol; universal carbon taxes; allocations of emissions per capita (Gupta and Bhandari 1999; Meyer 2000; Aslam 2002); the Brazilian proposal that allocates emissions allowances in relation to the past contribution to change in temperature (Brazil 1997; Pinguelli Rosa and Kahn Ribeiro 2001; La Rovere *et al.* 2002); common but differentiated convergence (Höhne *et al.* 2003); emissions intensity (Kim and Baumert 2002; Herzog *et al.* 2006; Chung 2007); a sector-based Clean Development Mechanism (Samaniego and Figueres 2002; Sterk and Wittneben 2006; Stripple and Lövbrand, this volume, Chapter 11); technology agreements (Edmonds and Wise 1998; Knopf and Edenhofer 2010; Zelli and van Asselt, this volume, Chapter 6); various sectoral approaches (Ellis and Baron 2005; Schmidt *et al.* 2006; Ward 2006; Zelli *et al.*, this volume, Chapter 3); Triptych approach extended to the global context (Groenenberg *et al.* 2001; den Elzen *et al.* 2007; den Elzen, this volume, Chapter 12); converging markets (Tangen and Hasselknippe 2004; Victor *et al.* 2005); the safety valve approaches (Philibert 2002); greenhouse development rights (Baer *et al.* 2007); and sustainable development policies and measures (Winkler *et al.* 2002, 2007).

This list does not cover all proposals put forward. The literature is extensive enough to have produced a number of overviews of approaches, including the review and assessment by Hof *et al.* (this volume, Chapter 4), as well as overviews by Bodansky *et al.* (2004), Höhne and Lahme (2005) and Boeters *et al.* (2007) as well as a website (www.fiacc.net) collecting the various proposals.

The literature also includes an evaluation of several proposals focusing specifically on adequacy and equity (Baer and Athanasiou 2007; Shrivastava and Goel, this volume, Chapter 8). There have been processes bringing together perspectives from North and South, including the South–North dialogue (Ott *et al.* 2004); a dialogue on future action among selected negotiators (Centre for Clean Air Policy 2007); and the São Paulo Proposal for an Agreement on Future International Climate Policy (BASIC Project 2006). The Fourth Assessment Report of the IPCC assessed the proposals, and table 13.2 of the Working Group III report (IPCC 2007: 770–773) provides probably the most authoritative overview of recent proposals for international climate agreements, at least up to the cut-off date for literature assessed (mid-2006).

Relatively few of these proposals originate from developing countries. The Brazilian proposal of 1997 stands out as a major exception (Brazil 1997). The proposal took a scientific approach to burden-sharing among industrialized countries, calculating the contribution to temperature increase and hence responsibility for mitigation. By focusing on responsibility, the Brazilian proposal had a strong basis of equity. It also had a strong scientific basis, since the key factors determining temperature change are cumulative emissions, rather than annual ones.

Policy-makers from some developing countries have also favoured per capita approaches. There is an extensive literature that formulates climate regimes based on this principle. The essential equity-based argument is that each person should have the same right to use the absorptive capacity of the atmosphere.

For yet other developing countries, the key concern relating to equity relates to development (Ayers *et al.*, this volume, Chapter 17). This approach draws on article 2 of the Climate Convention, in particular that climate protection should occur in a way that 'enable[s] economic development to proceed in a sustainable manner'. More broadly, it argues that sustainable development in developing countries, including its ecological and social dimensions, are indispensable for an equitable solution, given that industrialized countries went through their process of industrialization without carbon constraints. In 2006 and 2007 parties to the climate convention exchanged opinions on long-term cooperative action in the so-called Convention Dialogue. In this setting, South Africa put forward the approach of sustainable development policies and measures (Republic of South Africa 2006), that is, policies and measures that help developing countries meet their own sustainable development goals more effectively, while creating significant co-benefits for the global climate.

There are various ways of framing the different types of architecture represented in this diversity of proposals. A joint paper of the South African and United Kingdom departments of environment presented at an informal ministerial discussion in Sweden identified the following four approaches: 'Atmosphere first'; 'Equity first'; 'Development first'; and 'Technology first' (DEAT and DEFRA 2007). These four approaches were assessed here against a set of criteria: the objectives of the institutions, their stringency, inclusion of quantified commitments, coverage of gases or sectors, policies and measures as well as their treatment of specific issues like technology, adaptation, natural ecosystems, etc. (see Table 7.1). It is unlikely that any 'pure' approach would be adopted in its entirety for a future climate governance architecture. Just as there is no single, definitive list of elements, there is not a single conception of a balanced package among parties. Negotiators will need to merge packages carefully balancing key elements and interests. Thus, it is more helpful to think of several packages along a theoretical continuum, namely those packages that might be capable of consensus – or to use another phrase, that are in the 'political contract zone'.

7.3.2 Reviewing different packages of options

Figure 7.1 places a range of scenarios on a theoretical continuum reflecting environmental effectiveness and broader participation. However, one cannot assess effectiveness based on the *form* of a package alone, without also specifying its *stringency*, in terms of the numerical targets it encompasses. For example, it is conceptually possible to construct an 'ambitious transitional' package with very stringent actions. This would be more effective than a 'multi-stage' package with very low levels of ambition for each stage. The graph suggests that, levels of ambition being equal, effectiveness increases along the arrows in Figure 7.1. The continuum stretches from the status quo to a universal, all-encompassing climate regime.

Status quo

The status quo – which corresponds with a state of 'cooperative fragmentation' in the terminology of Chapter 2 in this volume by Biermann *et al.* – is characterized by no new commitments for developing countries, no United States participation in quantified mitigation commitments and limited ambition for industrialized countries.

Table 7.1 *Typology of approaches to climate governance*

	Atmosphere first	Equity first	Development first	Technology first
Objective	Stabilizing greenhouse gas concentrations	Ensuring fairness of allocation of mitigation burdens (historic contributions)	Making development more sustainable	Development and transfer of low carbon technologies
Stringency	Agreement on 'safe' greenhouse gas concentration level or global greenhouse gas reduction targets and timeframes	Agreement on 'safe' greenhouse gas concentration level	Not a distinctive feature	Set in terms of technology goal or budgetary contribution to research, demonstration and development
Quantified greenhouse gas related commitments	Carbon budget is calculated and allocated among countries based on current and future emissions reduction potential Carbon markets are vital incentives to join the regime Trigger for participation at various stages	Carbon budget is allocated among countries according to historical responsibility Trigger for participation, but usually later than 'atmosphere first' Carbon markets vital with large flows to developing countries	Not the focus, contribution depends on number and ambition of sustainable development policies implemented Not only carbon markets	No quantified commitments, hence limited or no carbon markets
Coverage	All greenhouse gases including land-use change and forestry and international transport 80 per cent of global emissions. Minimum inclusion of 20–30 main emitters	All greenhouse gases including land-use change and forestry and international transport. Inclusion of all countries	Unlikely to cover all gases and sectors	Several technology agreements to cover all sectors. Unlikely to cover all gases and sectors

Policies and measures	Sustainable development policy and measures for countries before the trigger for e.g. deforestation and low carbon energy and transportation	Sustainable development policy and measures for countries before the trigger for e.g. deforestation and low carbon energy and transportation	Richer countries would pay the cost of implementing sustainable development policies and measures in developing countries: for example, enforcing the efficiency standards List of good/best practice policies could serve as information	(Coordinated) energy efficiency standards and renewable energy targets
Technology research and development Demonstration/ deployment Transfer	Not a distinctive feature	No obligation for additional technology transfer	Provision of finances and technology for developing countries	Cooperation to increase development, transfer and deployment among technologically advanced countries
Adaptation Human health Natural ecosystems Agriculture/forestry Water supply Coastal zones Infrastructure Extreme events	Funded from levy on market mechanisms Not distinctive as focus on prevention	Compensation of damage costs paid according to historical responsibility	Funded also through sustainable development policies and measures	Not a distinctive feature
Response measures	Funded from levy on market mechanisms Not distinctive as focus on prevention	Historically larger emitters to assist losers adjust to the transition	Tailor-made sustainable development policies and measures allow for diversification	Efforts could be geared towards technology, that is, contributing to diversification
Participation and compliance	Main 20–30 emitters must be included early on or at the outset of the agreement	Normative definition of historical responsibility for the trigger	High participation, high degree of international coordination and information exchange	Several technology agreements with different participation

Source: DEAT (Department of Environmental Affairs and Tourism, South Africa) and DEFRA (Department for Environment, Food and Rural Affairs, United Kingdom) 2007.

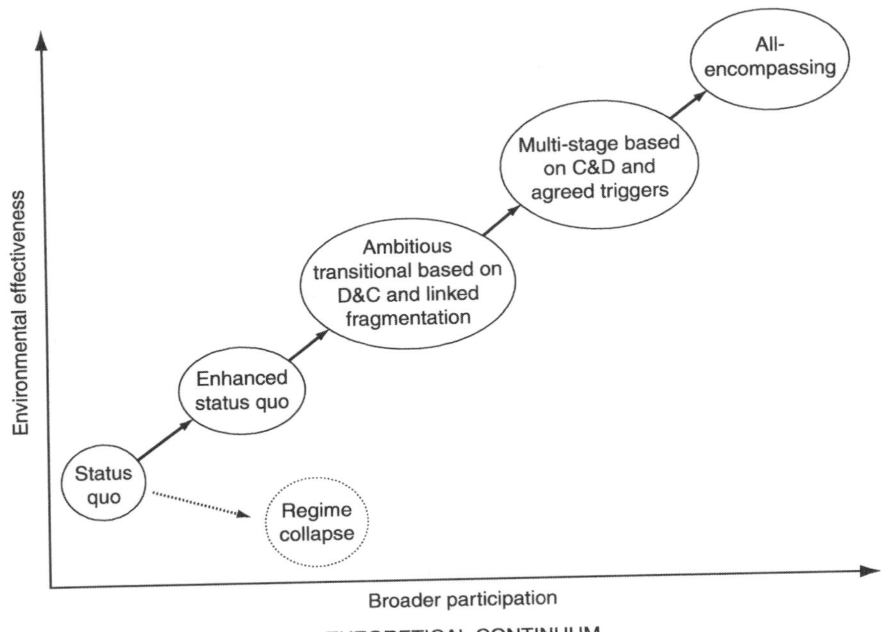

Figure 7.1 Continuum of future packages. D&C, Development and Climate; C&D, Climate and Development. *Source:* Winkler and Vorster 2007.

Enhanced status quo

A small step beyond the status quo would be 'enhanced status quo'. Under this scenario, the United States would continue to opt out of the Kyoto regime. It may take more proactive domestic action, building on various 'partnerships' and bilateral initiatives (see Watson 2007 for a complete list of these initiatives). Developing countries may scale up their action under article 4.1 of the climate convention, but it would not necessarily be quantifiable beyond participation in traditional Clean Development Mechanism projects and other voluntary domestic actions. Policies and measures to enhance technology transfer as well as financing and investment under the multilateral regime do not incentivize more stringent action by developing countries. The focus is squarely on development without a significant carbon constraint. In the absence of a comparable effort by the United States and an adequate contribution by major developing countries, the EU targets are unlikely to be more ambitious than a linear extension of their first commitment period targets under the Kyoto Protocol (EU Council 2007; Jordan *et al.* 2010). Furthermore, there is a risk in that the comparison has been shifted to efforts, which moves the indicator away from outcomes (that is, quantified emission limitation and reduction objectives), with the possibility of efforts not leading to outcomes in a verifiable manner.

Some industrialized countries might also choose to opt out of the future regime if the United States does not take comparable action to level the playing field. Without clear and visible commitments by all industrialized countries to deep and long cuts in emissions, the climate regime may not survive. There is a thin line between this scenario and regime collapse.

'Atmosphere first' approach

On one end of the spectrum shown in Figure 7.1 is a truly global climate regime, corresponding with the 'atmosphere first' approach, with a universal, all-inclusive architecture. The starting point is quantifying article 2, seeking broadened participation and then sharing the burden. Agreeing on a limit for temperature increase (for example 2 °C) or a stabilization goal (for example 450 parts per million of carbon dioxide-equivalent) is seen as the Archimedean point from which the whole world can be leveraged. A global carbon budget is calculated from this end-point and one or other formula is applied to allocate emissions to all countries. In its pure form, this approach does not consider the right of the South to develop; it thus falls outside the contract zone.

In the bargaining space between 'enhanced status quo' and an 'all-encompassing' regime – and hence between 'cooperative fragmentation' and 'universalism' – there are two possible packages. Both seek to balance development and climate. In other words, both acknowledge the right of the South to continue to grow emissions in pursuit of development. The 'multi-stage' package leans slightly more towards the climate imperative, but the various stages recognize differences in development. One might say it is based on 'climate and development'. The package called 'ambitious transitional' turns this around, basing itself on 'development and climate' (Winkler and Vorster 2007 for a detailed description of each of these two packages).

'Multi-stage' package

'Multi-stage' means that countries progress from one level of participation to another through different stages. While the definition of stages differs in various proposals (Höhne *et al.* 2003; Ott *et al.* 2004; den Elzen *et al.* 2006), the rules for successive stages are always more stringent.

Equity implies that the progression through stages be based not simply on time and current or projected emissions, but also on income levels, population size, historical responsibility and the potential to mitigate. While the urgency of the climate challenge requires more urgent mitigation action from all countries, developed and developing, simply locking developing countries into fixed stages with automatic graduation will not be a realistic political proposition. The logical consequence of such a simplistic formula is that developing countries would soon need to reduce emissions in absolute terms, *while they are still relatively poor* (Baer *et al.* 2007).

Ambitious packages will have to address sustainable production and consumption. Issues of lifestyles, among both the wealthier nations and the richer populations in developing countries, will need to remain within limits that the climate system can bear.

A fair approach to 'multi-stage' must explicitly take into account the income levels of developing countries and acknowledge that industrialized countries have built their wealth without carbon constraints. The rationale for a 'multi-stage' approach cannot simply be the perception of industrialized countries that the Kyoto Protocol is 'unfair' because developing countries have no caps. There is good reason why the first principle listed in the climate convention is equity.

The practical upshot of this distinction in rationales is how they conceive triggers for graduation. 'Multi-stage' approaches often assume automatic graduation – once a country reaches a threshold, it should graduate to the next level of commitment. Such formulations assume that 'multi-stage' is a rule-based evolution (den Elzen *et al.* 2006), that is, graduation happens automatically, according to a formula, as pre-determined rules are applied. With such an approach, 'multi-stage' becomes a 'pathway towards a global regime in which developing countries participate in a commitments regime in several stages' (Bodansky *et al.* 2004).

There is another way to frame triggers leading to increased stringency for developing country commitments. They can be agreed politically, and 'may include conditions for both developing and industrialized countries' (Winkler *et al.* 2006). Agreed triggers would imply that, after reaching a threshold, a country would need to consider whether it joins the next stage. Graduation is not automatic, simply because no sovereign country would accept the result of a formula. This is true both for developing countries, but also for industrialized countries. The São Paulo Proposal suggests that, upon reaching a trigger, a country can either graduate or opt out of the regime (BASIC Project 2006). Opting out would however have consequences, which could include trade sanctions. If countries opt out of the sequence of stages, then the 'multi-stage' package might not be much more effective than the 'ambitious transitional' approach.

'Ambitious transitional' package

An 'ambitious transitional' package assumes more urgent action by all parties. More specifically, it is premised on a larger group of countries taking on quantifiable commitments, albeit not (immediately) of the same kind. In 1997 in Kyoto, 'common but differentiated responsibilities' meant that industrialized countries took on quantified emission limitation and reduction commitments, and developing countries did not. The bar must be raised for everyone, if we are to deal with the challenge of climate change. All industrialized countries must take more stringent, binding emission reductions, and developing countries must act more urgently, in a quantifiable way and within a more empowering technology and investment framework (Alfsen *et al.*, this volume, Chapter 13).

This package assumes, however, that there is no participation by the United States in an international, legally binding cap-and-trade regime in the second commitment period, and that major developing countries do not agree to graduation under a 'multi-stage' approach. Both the United States and developing countries would do more under a more ambitious regime. This is why a transitional package may be necessary. On the other hand, this package goes further than 'enhanced status quo' by weaving together three mitigation strands into one multilateral framework:

- It links more urgent domestic action by the United States and other countries that cap-and-trade outside the regime with other domestic and/or regional cap-and-trade regimes. The link could also be to Kyoto compliance markets if methods are comparable.
- Industrialized countries as a group achieve −30 per cent reductions below 1990 levels, going beyond the −20 per cent unilateral reduction indicated by the European Union (EU Council 2007).
- It recognizes and provides incentives for enhanced mitigation action by developing countries. Although developing countries do not take on quantified mitigation targets, some do commit to meaningful quantifiable mitigation actions. This can take various forms, depending on the differing characteristics within the group of developing countries. Positive incentives enable more active leadership of the South, while taking due cognizance of the widely varying national circumstances of developing countries.

Developing countries could undertake a range of quantifiable actions, some market-based and others policies and measures. While not accepting graduation as in the multi-stage approach, they could still do so on a differentiated basis, according to their respective national circumstances. This could include sustainable development policies and measures, for example domestic targets for energy efficiency or renewable energy, cleaner use of coal, or reduced emissions from deforestation; 'no lose' sectoral targets; or traditional, programmatic, policy or sectoral Clean Development Mechanism projects.

On the demand side of the carbon market equation, long-term commitments and deeper emissions cuts by all industrialized countries will be critical to maintain price levels. These steps would also provide adequate time horizons to increase investment in low carbon economic growth in the South. Linking fragmented carbon markets or cap-and-trade regimes could further fuel demand (Flachsland *et al.*, this volume, Chapter 5).

The 'ambitious transitional' package has a stronger bottom–up character than 'multi-stage', recognizing differentiated efforts, rights and obligations. For a transitional period, it accepts some level of fragmentation, while seeking to create mechanisms to link different components of the regime. This package is 'transitional' as it assumes that the second commitment period would be a bridge to a more inclusive regime beyond 2020. The introduction of a cap-and-trade regime in the United States might build sufficient confidence within the United States, enabling the country to join a Kyoto-type regime at some point in the future, but not later than 2020. Likewise, a range of quantifiable actions taken by developing countries, including 'no lose' sectoral targets, sectoral, policy or programmatic Clean Development Mechanism, or even some categories of sustainable development policies and measures, could be linked to global, regional or national carbon markets during this transitional period (Stripple and Lövbrand, this volume, Chapter 11).

7.3.3 The future of the climate regime

Where do actual climate negotiations stand in the continuum outlined above? How might the future of the climate regime evolve over time? Having considered two packages within the conceptual contract zone, a critical factor that will determine whether there is a political contract zone, besides the balance between development and climate imperatives, is a trigger from the North. This is a fundamental political condition based on equity considerations. In addition, the South needs to take more urgent and incentivized action, structured in a way that addresses distributional issues, thereby allowing adequate space for the narrowing of the North–South development gap.

Of the two more promising packages outlined above, the 'multi-stage' package may lead to more stringent measures, since it defines a mechanism of agreed triggers by which countries move to increased levels of action. Environmental effectiveness depends critically on the level of ambition in each of the packages and broader participation, that is, the level of fragmentation.

However, taking a closer look at the course of international climate negotiations in early 2009 – with the four building blocks of adaptation, mitigation, finance and technology – political will enables at best the realm of the 'ambitious transitional'. Raising the bar on both sides, within the architecture framed by the climate convention, promises the greatest degree of equity and environmental effectiveness under current political conditions. The policy implications are that any agreement that replaces the Kyoto Protocol might still be a transition to a longer-term solution unless political conditions change dramatically.

As for mitigation, negotiations have undergone two significant shifts. Developing countries have agreed to negotiate measurable, reportable and verifiable mitigation action. 'Measurable, reportable and verifiable' can be interpreted as 'quantifiable'. Not only can the emissions implications of actions be measured, they could also be reported to the international community and be capable of verification. At the same time, technology transfer and financial resources by industrialized countries need to pass the test of being verifiable too. This similarly is a significant departure from the past, when much financing was through voluntary contributions to funds and the quantum of technology transferred was not measurable. The balance between 'measurable, reportable and verifiable' commitments by industrialized countries and 'measurable, reportable and verifiable' actions by developing countries are likely to remain central in refining the architecture of the climate regime after 2012.

Adaptation is of particular importance for developing countries (Ayers *et al.*, this volume, Chapter 17; Biermann and Boas, this volume, Chapter 14). Key issues that will need to be addressed include clarifying the costs of adaptation (Möhner and Klein 2007; Klein and Persson 2008; Hof *et al.*, this volume, Chapter 15). In this regard, the concept of incremental costs may be applicable to stand-alone adaptation activities. However, since enhancing adaptive capacity depends on development in a broader sense, the approach of funding only incremental costs may need to be re-examined. Distinguishing costs of wider efforts to address sustainable development and poverty (which include activities that respond to climate variability) from those that respond to climate change raises difficulties in attribution and disaggregation. Funding for developing countries through the financial mechanism may need to re-interpret or change the notion of 'incremental costs of global environmental benefits' encoded in the Global Environment Facility. In short, adaptation requires a knowledge base, clearer assessment of the costs of adaptation and, critically, a set of considerations to implement adaptation activities (Linneroth-Bayer *et al.* 2010).

The discussion on finance and technology evolved during 2008. Negotiations now consider both issues in an integrated way, on the means of implementation, which also includes capacity-building. The means of implementation are also considered in relation to both adaptation and mitigation. Increasing the supply of money will require innovative approaches, using public money to leverage investment from the private sector, and involving both international finance and domestic investment (Alfsen *et al.*, this volume, Chapter 13). Another issue is clarifying on what the money would be spent. Developing countries will need to conduct in-depth analysis, advancing the identification of their needs for adaptation and mitigation. They have to quantify those needs in financial terms and voice their priorities and the financial requirements in the multilateral process. In this regard, the concept of incremental costs may be applicable to stand-alone adaptation activities.

Technology for both mitigation and adaptation is critical to the future of the climate regime. This notwithstanding, investment in long-term research and development is needed at the same time. Funds have been proposed, including acquisition funds for existing technology and venture capital to get emerging technology into the market (Alfsen *et al.*, this volume, Chapter 13). Technology is now dealt with under the climate convention's subsidiary body on implementation. Moreover, in 2008 parties to the climate convention endorsed the Global Environment Facility's Poznań Strategic Programme on Technology Transfer, which aims at levering private investments that developing countries require for both mitigation and adaptation technologies. One option is that technology cooperation concentrates more on specific sectors. Likewise, debates may further attend to barriers to financing and performance indicators. In the overall discussion, the fundamental issues of intellectual property rights and trade barriers need to be addressed. This links the climate negotiations to the trade negotiations under the World Trade Organization as well as other processes; for example the World Intellectual Property Organization's development agenda (Zelli and van Asselt, this volume, Chapter 6).

Why might one categorize negotiations under these four building blocks beyond the 2009 climate talks in Copenhagen as Ambitious Transitional, rather than Multi-Stage? The answer is: because the 'trigger from the North', in the form of US re-engagement, remains too weak. The largest historical emitter must act earlier and more decisively. The convergence of two sets of expectations appears critical in establishing a political contract zone.

Re-engagement by the United States in multilateral, legally binding emission reductions remains key. The level of ambition one can expect from more proactive, incentivized leadership of the South will still depend on industrialized countries taking the lead: 'the trigger to strengthen the regime must come from the North' (Van Schalkwyk 2007). On the other hand, developing countries have increasingly faced the reality of their own growth in emissions, spelled out clearly in the Fourth Assessment Report of the IPCC (2007). From an equity point of view, there is a balance between historical responsibility by those who have emitted in the past, and responsibility for the future by those whose emissions are growing rapidly (Zammit Cutajar 2007). The scale of the mitigation challenge is so daunting, and the vulnerability of developing countries themselves so pressing that enhanced action by all countries, including developing countries is needed urgently. But developing countries continue to stress that the largest historical emitter (the United States) must act most decisively and first. The extent of developing countries mitigation action will continue to depend on the degree of support provided.

As long as these fundamental political conditions have not been met, the 'ambitious transitional' package outlined above is likely to be the only politically realistic package for the climate regime in the commitment period starting in 2013.

Rather than thinking about 'multi-stage' and 'ambitious transitional' as alternatives, however, one could understand them as part of an evolution of the regime over time. Figure 7.2 shows the two packages in the contract zone within the overall continuum. Figure 7.2 shows how it may be possible to design an 'ambitious transitional' architecture for the climate regime after 2012 – a period during which the United States will take its first steps towards integration with a truly global regime by adopting domestic quantified emission reduction commitments, industrialized countries will commit to deeper emission cuts, and developing countries will act on their differentiated responsibility by taking further quantifiable and incentivized mitigation action of a different kind. The dashed lines between the

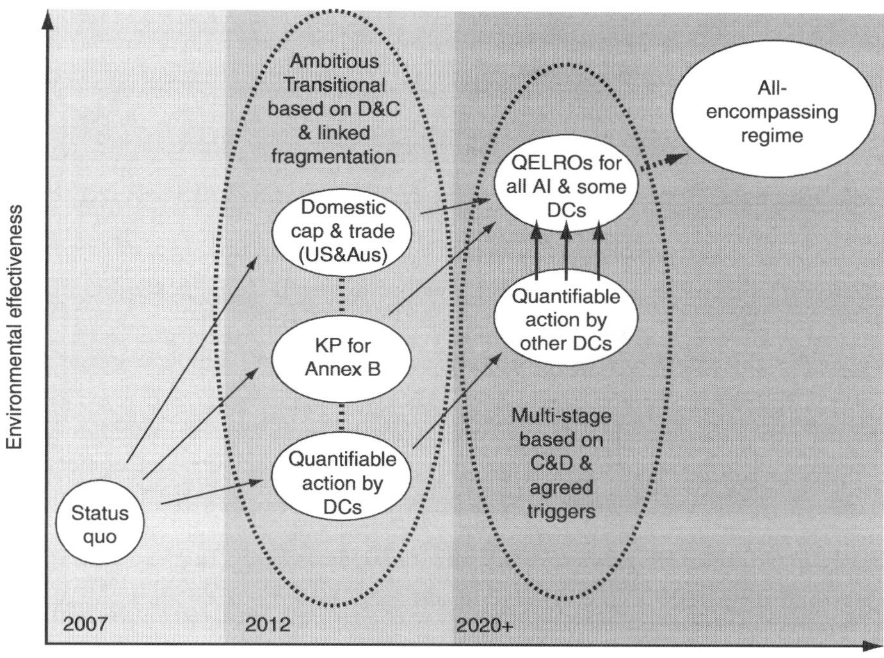

Figure 7.2 'Ambitious transitional' and 'multi-stage' packages. AI, Annex I parties; DCs, developing countries; KP, Kyoto Protocol; QELROs, quantified emission limitation and reduction objectives. *Source:* Winkler and Vorster 2007.

three elements comprising the 'ambitious transitional' package indicate that linkages are possible between the fragmented pieces. This transitional period from 2012 to 2020 will be important to build the trust and confidence required to strengthen the regime incrementally from 2020 onwards.

7.4 Conclusions and policy recommendations

In light of the IPCC Fourth Assessment Report, urgent action is needed in both developed and developing countries. Politically, differentiating between countries with different responsibilities and capabilities will remain necessary. Otherwise, it is hard to imagine any possible architecture being seen as fair and therefore acceptable.

In line with other chapters in this book, what would be the policy recommendations? The mode of governance considered in this chapter is squarely focused on the multilateral agreements under the climate convention and the Kyoto Protocol. Here, countries are the central actors. Under this umbrella, however, bottom–up approaches by developing countries could be included, such as sustainable development policies and measures, measures on reducing emissions from deforestation and forest degradation, or nationally appropriate mitigation actions. In order to implement effective action a wider range of actors will need to become involved, from both the private sector and civil society. Institutionally, the chapter assumes that the building blocks defined in Bali – mitigation, adaptation, technology and finance – could build up to a shared vision. These elements are central to institutional design of any future arrangements under the climate convention. Overall, however, this chapter focuses on policy analysis, and so the central findings are in the realm of climate policy.

The basis for climate protection must remain equity and common but differentiated responsibilities and respective capabilities. What is needed is to raise the bar on both sides of the historical divides. The climate regime after 2012 will have to see deeper cuts in emissions in the North and reduce emissions in absolute terms. The challenge that many countries in the South are facing is to make their contribution to emission reductions quantifiable and to leapfrog to a low carbon-intensity development path. This will have to be accompanied by two sets of balances – that between adaptation and mitigation, the two 'action area' building blocks; and between those two as a group and the 'means of implementation' building blocks, finance and technology transfer.

These balances will be essential to provide incentives for developing countries to raise the bar on climate action, which is needed together with greater levels of ambition from industrialized countries. Trust, as a basis for further agreements, will firstly depend on the meeting of existing commitments, and secondly on future commitments. These commitments by the North should acknowledge the North–South development deficit and address the distributional issues in the climate regime by means of a substantive and credible offer on technology and financial transfers. In short, to get to a climate deal, we will also need to strike a development deal. A clear signal is needed from the North, complemented by more proactive leadership of the South.

Acknowledgements

The author thanks the participants at the international workshop on 'Adaptation and Mitigation Options for Global Cooperation: Developing Countries' Perspectives', jointly organized by the ADAM Project and The Energy and Resources Institute in New Delhi, 25–26 February 2008, for valuable insights and comments made in response to a presentation of an earlier draft. The chapter has benefited from those comments.

References

Agarwal, A. and S. Narain 1991. *Global Warming in an Unequal World: A Case of Environmental Colonialism*. New Delhi: Center for Science and Environment.

Aslam, M. A. 2002. 'Equal per capita entitlements: a key to global participation on climate change?' in K. Baumert, O. Blanchard, S. Llosa and J. F. Perkaus (eds.), *Building on the Kyoto Protocol: Options for Protecting the Climate*. Washington, DC: World Resources Institute, pp. 175–202.

Baer, P. and T. Athanasiou 2007. *Frameworks and Proposals: A Brief, Adequacy and Equity-Based Evaluation of Some Prominent Climate Policy Frameworks and Proposals*, Global Issues Papers No. 30. Berlin: Heinrich Böll Stiftung.

Baer, P., T. Athanasiou and S. Kartha 2007. *The Greenhouse Development Rights Framework: Rationales, Mechanisms, and Initial Calculations*. Berkeley, CA: EcoEquity and Christian Aid.

BASIC (Building and Strengthening Institutional Capacity on Climate Change) Project 2006. *The São Paulo Proposal for an Agreement on Future International Climate Policy*. São Paulo: Instituto de Estudos Avancados da Universidade de São Paulo.

Bodansky, D., S. Chou and C. Jorge-Tresolini 2004. *International Climate Efforts beyond 2012*. Arlington, VA: Pew Center on Global Climate Change.

Boeters, S., M. den Elzen, A. J. G. Manders, P. J. J. Veenendaal and G. Verweij 2007. *Post-2012 Climate Policy Scenarios*, MNP (Netherlands Environmental Assessment Agency) Report 500114006/2007. Bilthoven: Netherlands Environmental Assessment Agency.

Brazil 1997. *Proposed Elements of a Protocol to the UNFCCC*, Presented by Brazil in response to the Berlin Mandate, FCCC/AGBM/1997/Misc.1/Add.3. Bonn: UNFCCC.

Center for Clean Air Policy 2007. *A Post-2012 Package: Developing Country Climate Change Strategy*, Draft July 16. Washington, DC: Center for Clean Air Policy.

Chung, R. K. 2007. 'A CER discounting scheme could save climate change regime after 2012', *Climate Policy* **7**: 171–176.

DEAT (Department of Environmental Affairs and Tourism, South Africa) and DEFRA (Department for Environment, Food and Rural Affairs, United Kingdom) 2007. *Scenarios for Future International Climate Change Policy*, Discussion paper presented at the Midnight Sun Dialogue on Climate Change, Riksgränsen, Sweden, 11–14 June 2007.

Edmonds, J. and M. Wise 1998. *Building Backstop Technologies and Policies to Implement the Framework Convention on Climate Change*. Washington, DC: Pacific Northwest National Laboratory.

Ellis, J. and R. Baron 2005. *Sectoral Crediting Mechanisms: An Initial Assessment of Electricity and Aluminium*, Doc. No. Com/Env/Epoc/Iea/Slt(2005)8. Paris: Organization for Economic Co-Operation and Development; International Energy Agency.

Elzen, M. den, N. Höhne, P. L. Lucas, S. Moltmann and T. Kuramochi 2007. *The Triptych Approach Revisited: A Staged Sectoral Approach for Climate Mitigation*, MNP Report 500114008/2007. Bilthoven: Netherlands Environmental Assessment Agency.

Elzen, M. den, P. L. Lucas, M. M. Berk, P. Criqui and A. Kitous 2006. 'Multi-stage: a rule-based evolution of future commitments under the climate change convention', *International Environmental Agreements: Politics, Law and Economics* **6**: 1–28.

EU Council 2007. *Presidency Conclusions*, 8–9 March 2007, Doc. No. 7224/07. Brussels: EU Council.

Groenenberg, H., D. Phylipsen and K. Blok 2001. 'Differentiating the burden world-wide: global burden differentiation of GHG emissions reductions based on the triptych approach – a preliminary assessment', *Energy Policy* **29**: 1007–1030.

Gupta, S. and P. M. Bhandari 1999. 'An effective allocation criterion for CO_2 emissions', *Energy Policy* **7**: 27–36.

Herzog, T., K. Baumert and J. Pershing 2006. *Target: Intensity – An Analysis of Greenhouse Gas Intensity Targets*. Washington, DC: World Resources Institute.

Höhne, N. and E. Lahme 2005. *Types of Future Commitments under the UNFCCC and the Kyoto Protocol Post-2012*. Gland, Switzerland: World Wide Fund for Nature.

Höhne, N., C. Galleguillos, K. Blok, J. Harnisch and D. Phylipsen 2003. *Evolution of Commitments under the UNFCCC: Involving Newly Industrialized Economies and Developing Countries*. Berlin: Federal Environmental Agency (Umweltbundesamt).

IPCC 2007. *Climate Change 2007: Mitigation. Contribution of Working Group III to the Fourth Assessment Report of the Intergovernmental Panel on Climate Change*. Geneva: IPCC.

Jordan, A., D. Huitema, T. Rayner and H. van Asselt 2010 (in press). 'Governing the European Union: the choices, dilemmas and frameworks of governance', in A. Jordan, D. Huitema, H. van Asselt, F. Berkhout and T. Rayner (eds.), *Climate Change Policy in the European Union: Confronting the Dilemmas of Mitigation and Adaptation*. Cambridge, UK: Cambridge University Press.

Kim, Y.-G. and K. Baumert 2002. 'Reducing uncertainty through dual-intensity targets', in K. Baumert, O. Blanchard, S. Llosa and J. F. Perkaus (eds.), *Building on the Kyoto Protocol: Options for Protecting the Climate*. Washington, DC: World Resources Institute, pp. 109–134.

Klein, R. J. T. and A. Persson 2008. *Financing Adaptation to Climate Change: Issues and Priorities*, ECP Report No 8. Brussels: CEPS.

Knopf, B. and O. Edenhofer 2010 (in press). 'The economics of low stabilisation: implications for technological change and policy', in M. Hulme and H. Neufeldt (eds.), *Making Climate Change Work for Us: European Perspectives on Adaptation and Mitigation Strategies*. Cambridge, UK: Cambridge University Press.

La Rovere, E. L., L. Valente de Macedo and K. Baumert 2002. 'The Brazilian proposal on relative responsibility for global warming', in K. Baumert, O. Blanchard, S. Llosa and J. F. Perkaus (eds.), *Building on the Kyoto Protocol: Options for Protecting the Climate*. Washington, DC: World Resources Institute, pp. 157–174.

Linneroth-Bayer, J., C. Bals and R. Mechler 2010. 'Climate insurance as part of a post-Kyoto adaptation strategy', in M. Hulme and H. Neufeldt (eds.), *Making Climate Change Work for Us: European Perspectives on Adaptation and Mitigation Strategies*. Cambridge, UK: Cambridge University Press, pp. 00–00.

Meyer, A. 2000. *Contraction and Convergence: The Global Solution to Climate Change*. Bristol, UK: Green Books and Schumacher Society.

Möhner, A. and R. J. T. Klein 2007. *The Global Environment Facility: Funding for Adaptation or Adapting to Funds?* Stockholm: Stockholm Environment Institute.

Müller, B. 1999. *Justice in Global Warming Negotiations: How to Obtain a Procedurally Fair Compromise*, 2nd revd edn. Oxford, UK: Oxford Institute for Energy Studies.

Ott, H. E., H. Winkler, B. Brouns, S. Kartha, M. J. Mace, S. Huq, Y. Kameyama, A. P. Sari, J. Pan, Y. Sokona, P. M. Bhandari, A. Kassenberg, E. L. La Rovere and A. Rahman 2004. *South–North Dialogue on Equity in the Greenhouse: A Proposal for an Adequate and Equitable Global Climate Agreement*. Eschborn, Germany: Gesellschaft für Technische Zusammenarbeit.

Philibert, C. 2002. *Fixed Targets versus More Flexible Architecture*, Revised draft note, OECD/International Energy Agency project for the Annex I Expert Group on the UNFCCC. Paris: Organization for Economic Co-Operation and Development; International Energy Agency.

Pinguelli Rosa, L. and S. Kahn Ribeiro 2001. 'The present, past, and future contributions to global warming of CO_2 emissions from fuels: a key for negotiation in the climate convention', *Climatic Change* **48**: 289–308.

RSA (Government of the Republic of South Africa) 2006. *Sustainable Development Policies and Measures: A Strategic Approach for Enhancing the Climate Regime Post-2012*, Presented at the 2nd Workshop of the Dialogue on Long-Term Cooperative Action to Address Climate Change by Enhancing Implementation of the Convention,

Nairobi, Kenya, 15–16 November. Pretoria: Department of Environmental Affairs and Tourism.

Samaniego, J. and C. Figueres 2002. 'Evolving to a sector-based Clean Development Mechanism', in K. Baumert, O. Blanchard, S. Llosa and J. F. Perkaus (eds.), *Building on the Kyoto Protocol: Options for Protecting the Climate*. Washington, DC: World Resources Institute, pp. 89–108.

Schalkwyk, M. van 2007. *Keynote address by the South African Minister of Environmental Affairs and Tourism at the G8+5 Environment Ministerial*. Potsdam, Germany.

Schmidt, J., N. Helme, J. Lee and M. Houdashelt 2006. *Sector-Based Approach to the Post-2012 Climate Change Policy Architecture*. Washington, DC: Center for Clean Air Policy.

Sterk, W. and B. Wittneben 2006. 'Enhancing the Clean Development Mechanism through sectoral approaches: definitions, applications and ways forward', *International Environmental Agreements: Politics, Law and Economics* **6**: 271–287.

Tangen, K. and H. Hasselknippe 2004. *Converging Markets*, Paper under the Post-2012 Policy Scenarios Project of FNI (Fridtjof Nansen Institute), CRIEPI (Central Research Institute of Electric Power Industry), HWWA (Hamburg Institute of International Economics) and CASS (Chinese Academy of Social Sciences). Polhøgda: Fridtjof Nansen Institute.

UNFCCC 1992. *United Nations Framework Convention on Climate Change*. New York: United Nations.

UNFCCC 1997. *Kyoto Protocol to the United Nations Framework Convention on Climate Change*. Bonn: UNFCCC Secretariat.

Victor, D. G, J. House and S. Joy 2005. 'A Madisonian approach to climate policy', *Science* **309**: 1820–1821.

Ward, M. 2006. *Climate Policy Solutions: A Sectoral Approach*. Wellington, NZ: Global Climate Change Consultancy.

Watson, H. L. 2007. Testimony of Dr. Harlan L. Watson, Special Climate Negotiator and Special Representative, U.S. Department of State, before the Committee on Foreign Affairs Subcommittee on Asia, the Pacific, and the Global Environment, United States House of Representatives, Hearing on 'The Kyoto Protocol: an Update', 11 July 2007. Washington, DC: US Government Printing Office.

Winkler, H. and S. Vorster 2007. 'Building bridges to 2020 and beyond: the road from Bali', *Climate Policy* **7**: 240–254.

Winkler, H., B. Brouns and S. Kartha 2006. 'Future mitigation commitments: differentiating among non-Annex I countries', *Climate Policy* **5**: 469–486.

Winkler, H., M. Howells and K. Baumert 2007. 'Sustainable development policies and measures: institutional issues and electrical efficiency in South Africa', *Climate Policy* **7**: 212–229.

Winkler, H., R. Spalding-Fecher, S. Mwakasonda and O. Davidson 2002. 'Sustainable development policies and measures: starting from development to tackle climate change', in K. Baumert, O. Blanchard, S. Llosa and J. F. Perkaus (eds.), *Building on the Kyoto Protocol: Options for Protecting the Climate*. Washington, DC: World Resources Institute, pp. 61–87.

Zammit Cutajar, M. 2007. *Protecting the Global Climate: 'Shared Vision – Differentiated Future'*, Presentation by Michael Zammit Cutajar, Ambassador for Climate Change, Malta, to the Midnight Sun Dialogue on Climate Change, 11–14 June 2007. Riksgränsen, Sweden.

8

Shaping the architecture of future climate governance: perspectives from the South

MANISH KUMAR SHRIVASTAVA AND NITU GOEL

8.1 Introduction

The urgent need for a globally concerted effort to address climate change has been acknowledged by all countries; yet the core challenge in designing a global climate governance architecture is to arrive at a universally acceptable multilateral agreement that will ensure that climate change is addressed in a coordinated and holistic manner (Lewis and Diringer 2007). Throughout the climate negotiations there have been disagreements among countries on specific provisions and principles (Kilaparti 1990; Redclift and Sage 1998; Rajamani 2003), resulting in 'fragmented' global climate governance architecture (Biermann *et al.*, this volume, Chapter 2; Zelli *et al.*, this volume, Chapter 3; Asselt *et al.* 2008). Such disagreements and the consequent fragmentation in global climate governance architecture, however, are not peculiar to climate negotiations. The literature on the evolution of institutions and international law, for instance, suggests that global governance architectures are in general determined and influenced by conflicting economic and political interests, differential technological and economic capabilities, jurisdictional constraints and at times cultural differences (North 1990; Cooter 1998; Boyle 1999). According to Boyle (1999), these unavoidable factors not only constrain the effectiveness of the involved institutions, but may also induce varying degrees of institutional fragmentation across the institutional architecture.[1]

Biermann *et al.* (this volume, Chapter 2) and Zelli *et al.* (this volume, Chapter 3) discuss the fragmentation in climate governance architecture in detail and argue that even though a universal architecture is more effective than a fragmented one, the former is not a feasible option. Therefore, the best option for effective global climate governance is a low degree of synergistic fragmentation, that is, detailed general

[1] Boyle uses the term 'soft law' instead of 'fragmented architecture'. Despite differences in the operational connotations of these two terms, the underlying idea – that is, flexibility as well as differential norms for different actors – remains the same.

Global Climate Governance Beyond 2012: Architecture, Agency and Adaptation, eds. F. Biermann, P. Pattberg and F. Zelli. Published by Cambridge University Press.

principles for regulating the policies in distinct yet substantially integrated institutional arrangements. This chapter looks at this proposition from the perspective of developing countries and identifies key principles for future global climate governance architecture that may encourage and enable developing countries to not only take up a greater responsibility in climate change mitigation – the core objective of any climate regime – but also achieve a sustainable development path. The analysis presented in this chapter, therefore, is mainly driven by two questions: first, what are the reasons that make developing countries hesitant, if not reluctant, to take up a greater responsibility in global efforts to mitigate climate change? Second, what kind of climate governance architecture can best address the apprehensions of developing countries?

The chapter is organized into four sections. The methodology adopted for this study is explained in Section 8.2. Section 8.3 presents our analysis that focuses on four key aspects of climate governance, namely, adaptation, mitigation, technology transfer and finance. Based on this analysis, in our conclusion, we outline key principles for a future global climate governance architecture from the point of view of developing countries.

8.2 Methodology

In order to arrive at a comprehensive understanding of the Southern perspective on climate governance, this chapter relies on two sources: the existing literature on developing countries in the context of climate change; and views by various stakeholders on the current regime and options for a future climate regime elicited at a major conference on this topic in New Delhi in 2008.

The literature dealing with the perspective of developing countries covers a wide range of issues. The survey presented in this chapter, however, concentrates on the concerns raised by developing countries with reference to their status in the present climate regime and their future role in climate change mitigation and adaptation. In addition to the academic literature, we also consulted the country submissions by developing countries on various agenda items for the fourteenth conference of the parties to the climate convention in 2008. The implicit assumption here is: by understanding the issues developing countries have with the current climate regime and their apprehensions in accepting various proposals that assign them a greater responsibility in future, we can identify the limitations of the current climate regime from a developing countries' point of view, and we subsequently derive the principles for a future climate governance architecture.

In order to collect views and suggestions from stakeholders and policy-makers, The Energy and Resources Institute (TERI) organized a two-day international workshop for the EU-funded ADAM Project on *Developing Countries' Perspectives for*

Adaptation and Mitigation Options for Global Cooperation in February 2008 in New Delhi, India. More than 80 participants, including climate policy negotiators, non-governmental organizations, representatives of industry and financial agencies, and researchers represented views from China, India, Bangladesh, Indonesia, South Africa and the United Arab Emirates. Researchers and diplomats from the United Kingdom, Germany, Norway, Switzerland and the Netherlands also participated. The workshop comprised theme-based presentations, followed by open discussion. The main themes included 'Priorities and expectations of the South', 'Upscaling and mainstreaming adaptation – technology transfer and finance', 'Reducing emissions from deforestation in developing countries', 'Sustainable forest management and reforestation', 'Technologies and finances for low carbon development', 'Role of markets and private players', and the 'Bali Action Plan and expectation of developing countries from Annex I countries'.

In our attempt to identify the apprehensions and requirements of developing countries, this chapter focuses on the synergies between issues identified in the existing literature and points highlighted by stakeholders and policy-makers in the workshop. In order to broaden the base for inputs from stakeholders, we also used reports on similar workshops conducted by other research institutes (for instance Institute for Global Environmental Strategies 2006, 2008).

8.3 Analysis

The Southern approach towards climate change can largely be summed up in three claims: First, for developing countries, development is the topmost priority (Section 8.3.1). Second, the agenda of climate change adaptation comes before mitigation (Section 8.3.2). Third, the ability of developing countries to contribute to global efforts to address the problem of climate change is constrained by the lack of access to technological and financial resources (Sections 8.3.3 and 8.3.4). These aspects are often articulated in terms of a divide between industrialized and developing countries. The analysis presented here presumes that the best strategy for arriving at a low and synergistic degree of governance fragmentation (Biermann *et al.*, this volume, Chapter 2) is to seek to bridge this divide between North and South.

8.3.1 Priority: development over climate change

The need for political action against climate change is more acute for societies that are less equipped to face climate change impacts (Jerneck and Olsson, this volume, Chapter 18; Winkler, this volume, Chapter 7). By and large, these are the societies in the developing world (Biermann and Boas 2008). More alarmingly, a large part of the population in developing countries is still deprived of basic amenities, which further reduces adaptive capacities. As a result, developing countries face the dual challenge of addressing the problem of climate change and alleviating poverty simultaneously (Oxfam 2008). While poverty makes these countries more vulnerable to climate change, poverty alleviation requires a higher energy consumption, which aggravates the problem of climate change (Asian Development Bank 2007). Thus, developing countries face a dilemma of development versus climate change, wherein the developmental objectives are perceived more urgent and tangible (Winkler *et al.* 2002; Ghosh 2008; Institute for Global Environmental Strategies 2008).

The emphasis on development over climate change is reflected, most forcefully, in the assertion of developing countries that in assigning any binding responsibility of mitigating climate change to countries, the criteria should be their per capita emissions as opposed to their total emissions (Dasgupta 2007). Clearly, this assertion puts development before growth. The North–South divide is most visible in this context, as the industrialized countries have been arguing that the advanced developing economies – India and China in particular – should also accept binding emission reduction targets because their growth rates are higher, and subsequently, they are going to be the largest emitters (International Energy Agency 2008). Evidently, the industrialized countries assign more significance to growth than development.

The South's preference for development over climate change is rooted in two aspects: the convictions that developing countries have in the context of climate change; and the constraints that they face in taking up mitigation actions on their own. The most strongly held convictions of developing countries include:

(1) The industrialized countries are responsible for climate change and must hence accept and meet even higher emission reduction targets.
(2) It is unjust to ask developing countries to take up greater mitigation responsibilities, as historically they have not contributed to climate change.
(3) Developing countries have the right to development and to use their natural resources to meet their developmental requirements. A binding emission reduction target is not acceptable: it would put a cap on the development of the South, entailing that the majority of the Southern population continue to live in poor conditions.
(4) The majority of the people in developing countries are already vulnerable to climate change. Therefore, the primary aim of developing countries is to build their adaptive capacities, which cannot be achieved without attaining a higher level of development.
(5) Since industrialized countries have benefited from carbon-intensive development, they must provide technology and financial assistance to developing countries and enable them to build adaptive capacities against climate change.

Clearly, the difference in the level of development among countries, which is largely captured by the idea of the North–South divide, is the main cause of disagreement between industrialized and developing countries. This divide has led these two groups to perceive the threat of climate change differently. While for the industrialized North, the victim is nature and humans play the role of the culprit, for the developing South, the poor are the victims and only the humans in the industrialized North are culprits (Müller 2002; Jerneck and Olsson, this volume, Chapter 18). The developing countries, thus, articulate their concerns by emphasizing adaptation and technological and financial support from industrialized countries. In the following subsections, we discuss these two aspects in detail.

8.3.2 Agenda: adaptation before mitigation

The early impacts of climate change are already materializing, as evident from the Fourth Assessment Report of the Intergovernmental Panel on Climate Change. Scientists believe that the distribution of impacts is likely to be uneven and that developing countries will face greater impacts (IPCC 2007). As a result, the immediate concern for developing countries is to build adaptive capacities. Mitigation, on the other hand, is perceived as a co-benefit of activities driven by a developmental agenda such as poverty reduction, protection of the local environment, energy security and enhancing export competitiveness (Pew Center 2002; Asian Development Bank 2007; Dasgupta 2007; Winkler, this volume, Chapter 7).

It is difficult for developing countries to undertake mitigation and adaptation activities simultaneously as they lack the necessary resources. Given their poor levels of development, accepting mitigation responsibilities would require them to divert resources from development to mitigation, which would weaken their adaptive capacities (Winkler *et al.* 2002; Dasgupta 2007). Moreover, the South perceives the need for adaptation as a burden imposed by the North, which is articulated in the demand for funds, as compensation, from industrialized countries for adaptation programmes (Müller 2002, 2008; Dasgupta 2007; Sharma and Bhadwal 2008; Ayers *et al.*, this volume, Chapter 17).

The dissatisfaction of developing countries with the current regime is also rooted in the fact that adaptation-related issues have been marginalized, as the Kyoto mechanisms are primarily concerned with mitigation (Lewis and Diringer 2007). Also, there is no specific mechanism for adaptation under the convention. During the seventh session of the conference of the parties to the climate convention in 2002, funds were established to support adaptation activities in developing countries. These funds include the Special Climate Change Fund, the Least Developing Countries Fund and the Adaptation Fund under the Kyoto Protocol. The Adaptation Fund mobilizes resources by levying a 2 per cent share of proceeds from the Clean Development Mechanism under the Kyoto Protocol. However, the volume of these funds is very small compared to the anticipated cost of adaptation in developing countries (Möhner and Klein 2007; Klein and Persson 2008; Linneroth-Bayer *et al.* 2010; Hof *et al.*, this volume, Chapter 15). Dissatisfaction among developing countries on support for adaptation and on the distribution of existing resources is evident. This leads them to demand enhanced implementation of the convention, which requires industrialized countries to help developing countries in addressing climate change.

This demand is reflected in their proposals for a separate framework on adaptation or at least a multi-window approach, i.e. a system of assessed contributions under the convention, guided by a committee or subsidiary body, assisted by complementary subsystems like insurance schemes.[2] For example, on behalf of the Africa Group, South Africa submitted a proposal for a framework for a consolidated work programme on adaptation (Third World Network 2008a). The proposal identifies the need for an institutional framework for the implementation of adaptation, which must facilitate access to the means of implementation, that is, finance, technology and capacity-building. Similarly, Bangladesh proposed the establishment of an International Adaptation Research Centre in Bangladesh and a Brazilian proposal emphasized the need for capacity building and vulnerability mapping. All these proposals assert that adaptation cannot be treated as a domestic matter and that there is a need for global support for building adaptive capacity (Third World Network 2008a). Since the priorities of the countries differ according to their vulnerabilities and socio-economic development, the principles for a new adaptation framework should be designed keeping in view the specific nature and priorities of recipient countries.

At the thirteenth conference of the parties to the climate convention in 2007, the Adaptation Fund was put into operation, and it was granted legal status at the fourteenth conference of the parties in 2008. Developing countries have welcomed these developments, yet their dissatisfaction with the inadequacy of the current and pledged levels of resources dedicated to adaptation is still visible (Third World Network 2008b, c). The South has been consistently demanding the extension of the share of proceeds from emissions trading and joint implementation for scaling up funding for adaptation. Industrialized countries are against this proposal (Ghosh 2008) despite the fact that time and again they have emphasized mobilizing resources through market-based mechanisms.

8.3.3 Constraints: technology and finance

The need for the flow of technology and financial resources from the North to the South has been stressed consistently in all important international decisions, declarations and action plans. For instance, the Rio Action Plan, the climate convention and its Kyoto Protocol as well as the Bali Action Plan from 2007 underline

[2] Such a multi-window approach has been proposed by the Alliance of Small Island States during the climate negotiations in Poznań in December 2008. http://unfccc.int/files/kyoto_protocol/application/pdf/aosisinsurance061208.pdf.

that industrialized countries should support developing countries with technology and finance. It is evident from the literature on the Southern perspectives on climate change (Müller 2002, 2008; Elliott 2004), and more so from views expressed at our developing country workshop in 2008, that the limited access to technology and financial resources is what makes developing countries hesitant towards accepting binding emission reduction targets (Winkler, this volume, Chapter 7). The report from the climate convention secretariat *Investment and Financial Flows to Address Climate Change* (2008) also concludes that the additional estimated amount of investment and financial flows needed in 2030 to address climate change is immense and much larger than the funding currently available under the convention and its protocol, but small when compared with GDP and investments.

Developing countries demand financial support mainly for two purposes: to acquire advanced and clean technologies and to undertake adaptation programmes. If the access to technologies is ensured, the demand for financial support is likely to diminish. The need and availability of financial support for adaptation have been discussed in Section 8.3.2. In this section, we elaborate on the issues concerned with technology.

The proposals by developing countries have so far emphasized technology transfer from industrialized countries to developing countries. For instance, the fourteenth conference of the parties to the climate convention in 2008 endorsed the Global Environment Facility's Poznań Strategic Programme on Technology Transfer. The aim of this programme is to scale up the level of investment by leveraging private investments that developing countries require both for mitigation and adaptation technologies. However, the experience with the existing mechanisms for enhancing technology transfer from industrialized countries to developing countries, for example through the Clean Development Mechanism, has not been encouraging (Institute for Global Environmental Strategies 2006, 2008; Olsen and Fenhann 2008; Stripple and Lövbrand, this volume, Chapter 11).

The Intergovernmental Panel on Climate Change has listed various hurdles to technology transfer, including high capital costs, limited access to capital, poor access to information, institutional and administrative difficulties in developing technology transfer contracts, lack of infrastructure to absorb riskier technologies, absence of economic incentives and the issue of intellectual property rights (Metz *et al.* 2000). In addition to endorsing the issues identified by the Intergovernmental Panel on Climate Change, other studies have concluded that the nature of and the need for technology transfer in different sectors differ (Klein *et al.* 2006). In the absence of a favourable institutional framework in the host country, technology transfer becomes particularly difficult (Chesnais 1995; Hagedoorn 1995; Gonsen 1998; Cohen 2004; Shrivastava 2007; Schneider *et al.* 2008; Knopf and Edenhofer 2010).

These studies put forth a number of theoretical propositions on technology transfer from industrialized to developing countries. They highlight that developing countries lack technological capability and supporting institutions to develop new technologies; that technological capability-building is a path-dependent learning process; that many developing countries lack resources to finance their institutional and technological capability building projects and thus need external assistance; and that a rigid system of intellectual property rights hinders technology development and transfer in developing countries. These statements are confirmed by various submissions made on technology-related issues to the climate convention by developing countries.

The question of additional investment in technology research and development and the removal of barriers to technology transfer, including intellectual property rights and finance have been contested in climate negotiations (Zelli and van Asselt, this volume, Chapter 6). For instance, developing countries demand a shift from a rigid system of intellectual property rights to a more relaxed regime, along the lines of compulsory licensing for life-saving drugs. Developed countries, on the other hand, stress the importance of intellectual property rights to ensure innovation in clean technologies (Elliott 2004).

Similarly, developing countries argued during the fourteenth conference of the parties in 2008 that raising public finance to an adequate scale was crucial for meeting the objectives of the convention. Industrialized countries, on the other hand, objected that the role of public finances is merely to enable and leverage financing for the implementation of the convention, the bulk of which should rather come from private sources (Third World Network 2008c; Alfsen *et al.*, this volume, Chapter 13).

Developing countries also recognize the importance of private sector participation, as they increasingly emphasize promoting public–private partnerships (Stripple and Pattberg, this volume, Chapter 9). The need for a favourable policy environment at the national level to encourage domestic private investment has been highlighted by many industry representatives. It is argued that spreading the risks across private and public investors through public–private partnerships would improve the involvement of private players. For instance, renewable energy has flourished in countries with supportive policies such as feed-in tariffs, developed financial markets and subsidies (Goel 2008). At present, the largest share of investment in renewable energies and energy efficiency comes from the private sector (UNFCCC 2007). Nonetheless, at the global level, developing countries assert the need for public finance for enabling technology development and transfer by removing barriers at different stages of the technology diffusion cycle.

The South argues for financial support in the form of grants rather than commercial flows, holding that commercial flows may not be sufficient to meet the requirements of developing countries, where significant incremental costs are involved. The industrialized countries, on the other hand, favour soft loans and market-based financial mechanisms that involve private actors (Third World Network 2008d). This difference in approaches is reflected in the proposals that countries have submitted to the Ad-hoc Working Group for Long-term Cooperative Action for scaling up resources to address climate change.

For instance, the proposal of the Group of 77 and China suggests mobilizing USD 201–402 billion through budgetary support (0.5–1 per cent of GDP) from industrialized countries. These funds could be raised from environmental and energy taxes in these countries, revenues from permit auctions or public budgets. Furthermore, these funds would be 'new and additional', i.e. additional to official development assistance. An important aspect of the proposal is that any funding committed outside of the convention shall not be regarded as fulfilment of commitments by industrialized countries under article 4.3 of the convention (Third World Network 2008d). In addition to the financial mechanism, a technology mechanism has been suggested. The funds thus collected would form a multilateral climate technology fund supporting research and development, capacity building as well as deployment and transfer of new technologies.

Similarly, the Mexican proposal of a world climate change fund targets the mobilization of USD 10 billion through voluntary dedicated budgetary contribution from all countries in accordance with the principle of common but differentiated responsibilities and respective capabilities. On the other hand, the Norwegian proposal expects to generate USD 15–25 billion by auctioning small portions of assigned amount units. The proposal submitted by Switzerland suggests a global carbon tax of USD 2 per tonne of carbon dioxide emissions, exempting countries with emission levels below 1.5 tonne carbon dioxide-equivalent per inhabitant. The funds collected would be divided into two parts – a multilateral adaptation fund and a national climate change fund. The former fund would be used to finance adaptation policies, while the latter would concentrate on adaptation, technology transfer and /or mitigation.

Clearly, there are differences in opinions among developing as well as industrialized countries on how money should be generated to meet the financial requirements. Developing countries attach a greater importance to public funding and a new financial architecture as opposed to the emphasis by industrialized countries on market-based mechanisms and existing institutions.

Overall, developing countries are not satisfied with the existing financial arrangements. For instance, funds allocated by the Cool Earth Partnership and the Climate Investment Funds are envisaged to be concessional loans. This is a departure from the earlier paradigm of growth-oriented grants of the World Bank. In *Illustrative Investment Programs for the Clean Technology Fund*, the World Bank (2008) outlined that the share of grant from the Climate Investment Funds in total financing would only be approximately 11–15 per cent. The rest of the amount (almost 90 per cent) will be in the form of loans that would entail an additional strain for developing countries in terms of debt burdens and increased economic dependence on international donors. According to Redman (2008), these funds have been opposed by developing countries that see them as violating the principle of common but differentiated responsibilities. The control over such funds should thus be established under the climate convention to ensure that they are used equitably and effectively and that recipient nations are involved in their design and implementation.

Altogether, there is a need to reform the current financial architecture to increase the level of funding and to ensure better distribution of funds. It is evident that technological and financial support to developing countries for mitigation and adaptation constitutes the core of international cooperation on climate change. It is also clear that mechanisms ensuring flow of technological and financial resources from industrialized countries to developing countries are critical for ensuring sustainable development. There have been efforts towards this end. However, the existing market-based mechanisms and financial arrangements have altogether been unsatisfactory in providing adequate technological and financial resources to developing countries.

8.3.4 Missing ingredient: technological capability

While the convictions of developing countries have been incorporated in the present climate regime, captured by the principle of common but differentiated responsibilities, their constraints have not been addressed adequately. If a future climate architecture equally falls short of addressing these constraints, developing countries may not be able to break free from the vicious circle of climate against development. For instance, the present climate regime has provisions for the transfer of technology and finance from industrialized countries to developing countries, while the responsibility of developing new technologies lies with industrialized countries only. While the transfer of technology and finance has been unsatisfactory so far (Tamura 2006; Institute for Global Environmental Strategies 2008), restricting responsibility of developing new technologies to industrialized countries perpetuates the need for the flow of technology from industrialized to developing countries (Shrivastava 2007). This would mean that the South would not be able to undertake mitigation activities on its own (Alfsen *et al.*, this volume, Chapter 13).

Given the existing inequalities among countries, the flow of technology and finance from industrialized to developing countries is inevitable. However, we assert that the purpose of this flow, and therefore that of the global governance architecture, should be towards reducing technological and financial inequalities rather than just acknowledging and reducing disparities among countries. The literature suggests that technology transfer as a policy option for rapid economic and technological development has not been universally successful. Only those countries have benefited from technology transfer that had certain institutional, technological and industrial prerequisites (Cooper 1973; Lall 1985; Salomon and Lebeau 1993; Cohen 2004). This implies that a North–South transfer alone may not be sufficient for narrowing the disparities. Considering that the technological needs of different countries vary and many countries do not even have the capability to identify their technological needs appropriately (Cohen 2004), the emphasis on technology transfer in the current climate regime is an incomplete package for narrowing the technological disparity.

The core problem of climate change is fossil-fuel-based development. What is thus needed is a complete shift in the development path of all countries. While industrialized countries need to reduce their per capita emissions without compromising the living standards of their people, developing countries have to make out a development path for themselves so that they can improve living conditions without overly increasing per capita emissions. This means that developing countries must think of a path that is different from what industrialized countries have followed so far. In this context, technological choices can play an important role. This not only implies that all countries should be able to identify technologies suited to their specific national contexts; in addition, they also should have the capability to produce those technologies on their own.

Developing countries are growing aware of this gap on technological capacity-building and country-driven solutions in the existing climate regime. Proposals recently submitted by the Group of 77 and China and the Third World Network demand a new institutional arrangement for technology development and transfer to the climate convention. These proposals call for establishing a global fund for research and development as well as for facilitating transfer of technology. However, given the varied national circumstances of developing countries (Kameyama 2004) and different technological needs (Salomon and Lebeau 1993; Institute for Global Environmental Strategies 2008), a universal technology development and transfer architecture may not be conducive to removing the constraints of all developing countries. The future climate architecture must therefore support the context-dependent technological capability building needs of developing countries.

The literature on technological change (Rosenberg 1976; Perez 1983, 2004; Coombs *et al.* 1987; Nelson 1987, 2002; Freeman and Perez 1988; North and Wallis 1994) suggests that the deployment of technology is essentially an entrepreneur's decision, which is, to a great extent, guided by the external economic environment, including the policy and regulatory environment (Cohen 2004). Besides, institutions at the national level play a crucial role in maintaining the trajectory of technological advancement. Therefore, it is vital that national governance architectures provide and support the technological change in a given economy. Industrial representatives of many stakeholder consultations have highlighted this need for enabling environments. If adequate incentives are provided – e.g. the removal of subsidies on non-efficient technologies, compulsory purchases from renewable energy, tax concessions on using environmentally sound technologies or other fiscal incentives, availability of cheap loan for sustainable enterprises – then the industrial sector would be more than willing to assume a greater role in addressing climate change. In summary, appropriate governance architectures at the national level should form an essential component of a low or synergistic degree of fragmentation of global climate governance.

8.4 Conclusions and policy recommendations

A major hurdle in arriving at a universally accepted climate governance architecture is the difference between industrialized and developing countries in terms of the levels of development, technological capabilities and access to finance. These differences have a long history and are likely to persist. While the current climate regime recognizes and incorporates these differences, it offers little to reduce the disparities among countries. Consequently, it seems to have locked the status quo and has proven to be inadequate in addressing the challenges of climate change in a sustainable manner.

The discussion in Section 8.3 emphasizes that from a Southern perspective the future climate regime at the global level should aim at reducing the disparities among countries – instead of merely incorporating them – and at supporting developing countries in identifying technological needs and designing adequate policies and programmes to build technological capabilities. Many scholars have argued that allowing countries to put forward policies of their choosing can enable them to tailor commitments to their domestic needs, priorities and policy cultures. Commitments are more likely to be accepted and fulfilled if they emanate from national contexts (Lewis and Diringer 2007). The current regime does not fulfil these normative requirements because in its genesis, the Kyoto Protocol with its flexibility mechanisms was basically designed 'to provide Annex-I Parties with cost-effective tools to meet their near-term emission reduction targets' (Kameyama 2004).

Considering the importance of the flow of technological capability and finance from industrialized countries to developing countries, it is critical that the global regime is guided by the national circumstances of developing countries. We suggest that the best policy option could be a *two-tier approach* with two distinct but deeply integrated components: (a) a set of institutions, policies and programmes at the national level to identify the direction of technological development within the country; and (b) a network of global institutions, financial mechanisms and technological programmes to support the institutions, policies and programmes in developing countries.

A *two-tier approach* entails that the countries are differentiated according to their specific constraints and technological requirements and then integrated through global support to eliminate the factors responsible for differentiation. The differentiation that we propose here is qualitatively different from the Japanese proposal (Government of Japan 2008). The Japanese proposal argues for emission reduction targets for relatively advanced developing countries like India and China focusing on capabilities as the basis for differentiation. It ignores the context-specific constraints and requirements of developing countries. On the contrary, we consider the constraints and requirements as the basis for differentiation among developing countries and argue for a differentiated support from industrialized countries. Such a differentiation would be more acceptable to the South than the one proposed by Japan. For instance, various special programmes for African countries and small island developing states are generally welcomed by other developing countries (Elliott 2004). In the light of the discussion in Section 8.3, the *two-tier approach* could be implemented through the following institutional arrangement at the global level.

First, an international group of experts would identify specific technological needs of countries in light of economic feasibility, impact on poverty alleviation, adaptive capacities and emission reductions. This body should comprise experts from both developing and industrialized countries. Along with scientists, engineers and social scientists, the experts should include industry representatives and policy-makers. The objective of this body should be to identify the types of technologies that are best suited in a particular country's specific conditions; to run comprehensive programmes to help developing countries build their capacities so that in future, they can identify the best suitable technological direction for themselves; and to develop a global database with information about available technologies, along with their deployment costs in different countries, potential barriers and available support programmes. This database should be freely accessible to developing countries.

Second, for those identified technology types that are not available in the world market, a network of research and development laboratories and training institutes should be built in developing countries. A respective proposal of centres of innovation and excellence has been tabled by the Indian delegation and climate negotiations in March 2009 (International Institute for Sustainable Development 2009). These centres should enable developing countries to build their own technological capability in the identified technologies. This network could be a platform where the flow of technological know-how and know-why can take place through the interactions between industrialized country experts and developing country experts. To meet the financial requirements, a global research and development fund for the development of clean technologies should be established. This fund should primarily draw contributions from industrialized countries in line with the principles of the climate convention. Other countries may contribute voluntarily. Besides, various market mechanisms could be used to mobilize revenues from carbon-intensive activities and private-sector funding. For example, a shared patent on new technologies could provide an incentive for private investors to be a part of this fund, as the risks of new investments would be lowered due to a large share of public funding. Alternatively, putting all new technologies developed by this network of research and development laboratories in the public domain may induce additional research and development investments by private actors to develop new and more efficient technologies in order to stay competitive in the future global market of technologies.

Third, for those identified technologies that are available in the world market, adequate financial and technological support should be provided to developing countries, which will enable them not only to deploy these technologies but also to modify them according to their requirements. A part of the global research and development fund could be dedicated to these tasks. In general, a combination of public and market-based funding should be institutionalized for the diffusion of existing technologies. For instance, if the resulting market-based resource mobilization at the prevailing carbon prices is not adequate to meet the total incremental cost, then the remainder must be guaranteed by public funding. A separate 'technology diffusion fund' can be envisaged for this purpose.

Fourth, the 'technology diffusion fund' and the 'global research and development fund' could encourage a differentiated system of 'hard' and 'soft' laws at the national level, requiring the large-scale business utilities to necessarily use the best available technologies while providing incentives to small- and medium-scale utilities to opt for identified technology type. For instance, in India the use of supercritical boilers has been made mandatory for so-called Ultra Mega Power Plants (in the range of 3500 megawatt capacity), while for other power plants, its use is merely suggested. To facilitate the transition to this new type of boilers, import of the respective technologies has been exempted from duties. Likewise, a supportive global regime would have enabled India to make supercritical technology mandatory for other types of power plants as well. For many countries, however, such a step is not feasible due to low levels of economic and technological development. Therefore, it is important that a global mechanism is put in place to support and facilitate such policy initiatives.

Fifth, development is the priority for developing countries, and developmental policies are also guided by other global economic agreements. Therefore, in the long run, a superseding global agreement could be established that promotes cooperative links between climate and non-climate institutional arrangements to support the national systems of 'hard' and 'soft' laws. In the same vein, a separate body under the secretariat of the climate convention could be established, tasked with supervising the evolution and performance of the abovementioned institutions should be established.

These institutional arrangements at the global level, we believe, would give a strong signal to developing countries, leading them to shed their hesitation in taking a more proactive part in global efforts to address climate change. However, substantial financial and technological support from industrialized countries is necessary to embark on such an institutional architecture. Considering that the North has been largely reluctant to provide adequate assistance – financial and technological – to the South, a two-tier approach in designing global governance architecture would require a strong political will from the industrialized world.

References

Asian Development Bank 2007. *Energy for All: Addressing the Energy, Environment, and Poverty Nexus in Asia*. Manila: Asian Development Bank.

Asselt, H. van, F. Sindico and M. A. Mehling 2008. 'Global climate change and fragmentation of international law', *Law and Policy* **30**: 423–449.

Biermann, F. and I. Boas 2008. 'Protecting climate refugees: the case for a global protocol', *Environment* **50**: 9–16.

Boyle, A. E. 1999. 'Some reflections on the relationship of treaties and soft law', *The International and Comparative Law Quarterly* **48**: 901–913.

Chesnais, F. 1995. 'Some relationships between foreign direct investment, technology, trade and competitiveness', in J. Hagedoorn (ed.), *Technical Change and the World Economy: Convergence and Divergence in Technology Strategies*. Cheltenham, UK: Edward Elgar, pp. 6–33.

Cohen, G. 2004. *Technology Transfer: Strategic Management in Developing Countries*. London: Sage.

Coombs, R., P. Saviotti and V. Walsh. 1987. *Economics and Technological Change*. London: Macmillan.

Cooper, C. (ed.) 1973. *Science, Technology and Development: The Political Economy of Technical Advance in Underdeveloped Countries*. London: Frank Cass.

Cooter, R. 1998. 'Expressive law and economics', *Journal of Legal Studies: Social Norms, Social Meanings and the Economic Analysis of Law* **27**: 585–608.

Dasgupta, C. 2007. 'Keynote lecture', *National Seminar on socio-economic impacts of Extreme Weather and Climate Change*, Kuala Lumpur, 22 June 2007.

Elliott, L. 2004. *The Global Politics of the Environment*, 2nd edn. New York: Palgrave.

Freeman, C. and C. Perez 1988. 'Structural crises of adjustment: business cycles and investment behavior', in G. Dosi, C. Feeman, R. Nelson, G. Silverberg and L. Soete (eds.), *Technical Change and Economic Theory*. London: Pinter, pp. 39–66.

Ghosh, P. 2008, 'Darkness at Poznan', *TERI4U Newsletter* No. **1**(2). www.teriin.org/newsletter/index.htm.

Goel, N. 2008. *Financing Mitigation: Case for Clean Energy Investments*, TERI viewpoint paper No. 7. New Delhi: TERI.

Gonsen, R. 1998. *Technological Capabilities in Developing Countries: Industrial Biotechnology in Mexico*. London: Macmillan.

Government of Japan. 2008. Proposal for AWG-LCA: For preparation of Chair's Document for COP 14. www.unfccc.int/files/kyoto_protocol/application/pdf/japanbap300908.pdf.

Hagedoorn, J. (ed.) 1995. *Technical Change and the World Economy: Convergence and Divergence in Technology Strategies*. Cheltenham, UK: Edward Elgar.

Institute for Global Environmental Strategies 2006. *Asian Aspirations for Climate Regime beyond 2012*. Hayama: Institute for Global Environmental Strategies.

Institute for Global Environmental Strategies 2008. *The Climate Regime beyond 2012: Reconciling Asian Priorities and Global Interests*. Hayama: Institute for Global Environmental Strategies.

International Energy Agency 2008. *Energy Technology Perspectives: Scenarios and Strategies to 2050*. Paris: International Energy Agency.

International Institute for Sustainable Development 2009. *Earth Negotiations Bulletin* 12, No. 407. www.iisd.ca/vol12.

IPCC 2007. *Climate Change 2007: Mitigation of Climate Change. Contribution of Working Group III to the Fourth Assessment Report of the Intergovernmental Panel on Climate Change*. Geneva: IPCC.

Kameyama, Y. 2004. 'The future climate regime: a regional comparison of proposals', *International Environmental Agreements: Politics, Law and Economics* **4**: 307–326.

Kilaparti, R. 1990. 'North–South issues, common heritage of mankind and global climate', *Millennium* **19**: 429–445.

Klein, R. J. T., M. Alam, I. Burton, W. W. Dougherty, K. L. Ebi, M. Fernandes, A. Huber-Lee, A. A. Rahman and C. Swartz 2006. *Application of Environmentally Sound Technologies for Adaptation to Climate Change*, UN Technical Paper. Doc. No. FCCC/TP2006/2. Bonn: UN Framework Convention on Climate Change Secretariat.

Klein, R. J. T. and A. Persson 2008. *Financing Adaptation to Climate Change: Issues and Priorities*, ECP Report No. 8. Brussels: CEPS.

Knopf, B. and O. Edenhofer 2010 (in press). 'The economics of low stabilisation: implications for technological change and policy', in M. Hulme and H. Neufeldt (eds.), *Making Climate Change Work for Us: European Perspectives on Adaptation and Mitigation Strategies*. Cambridge, UK: Cambridge University Press.

Lall, S. 1985. *Multinationals, Technology and Exports*. London: Macmillan.

Lewis, J. and E. Diringer 2007. *Policy Based Commitments in a Post-2012 Climate Framework*, Pew Center on Global Climate Change Working Paper. Arlington, VA: Pew Center on Global Climate Change.

Linneroth-Bayer, J., C. Bals and R. Mechler 2010 (in press). 'Climate insurance as part of a post-Kyoto adaptation strategy', in M. Hulme and H. Neufeldt (eds.), *Making Climate Change Work for Us: European Perspectives on Adaptation and Mitigation Strategies*. Cambridge, UK: Cambridge University Press.

Metz, B., O. R. Davidson, J. W. Martens, S. N. M. Van-Rooijen and L. V. W. McGrory, (eds.) 2000. *Methodological and Technological issues in Technology Transfer*, IPCC Special Report on Climate Change. Geneva: IPCC. www.grida.no/climate/ipcc/tectran/index.htm.

Möhner, A. and R. J. T. Klein 2007. *The Global Environment Facility: Funding for Adaptation or Adapting to Funds?* Stockholm: Stockholm Environment Institute.

Müller, B. 2002. *Equity in Climate Change: The Great Divide*. Oxford, UK: Oxford Institute for Energy Studies.

Müller, B. 2008. *International Adaptation Finance: The Need for an Innovative and Strategic Approach*. Oxford, UK: Oxford Institute for Energy Studies.

Nelson, R. R. 1987. *Understanding Technical Change*. New York: Elsevier.

Nelson, R. R. 2002. 'Bringing institutions into evolutionary growth theory', *Journal of Evolutionary Economics* **12**: 17–28.

North, D. C. 1990. *Institutions, Institutional Change and Economic Performance*. Cambridge, UK: Cambridge University Press.

North, D. C. and J. J. Wallis 1994. 'Integrating institutional change and technical change in economic history: a transaction cost approach', *Journal of Institutional and Theoretical Economics* **150**: 609–624.

Olsen, K. H. and J. Fenhann (eds.) 2008. *A Reformed CDM: Including New Mechanisms for Sustainable Development*. Roskilde: UNEP Risoe Centre.

Oxfam 2008. *Climate, Poverty and Justice: What the Poznan UN Climate Conference Needs to Deliver for a Fair and Effective Global Climate Regime*, Oxfam Briefing Paper No. 124. Oxford, UK: Oxfam.

Perez, C. 1983. 'Structural change and assimilation of new technologies in the economic and social systems', *Futures* **15**: 357–375.

Perez, C. 2004. 'Technological revolutions, paradigm shifts and socio-institutional change', in E. Reinert (ed.), *Globalization, Economic Development and Inequality: An Alternative Perspective*. Cheltenham, UK: Edward Elgar, pp. 217–242.

Pew Center on Global Climate Change 2002. *Climate Change Mitigation in Developing Countries: Brazil, China, India, Mexico, South Africa, and Turkey*. Arlington, VA: Pew Center on Global Climate Change.

Rajamani, L. 2003. 'From Stockholm to Johannesburg: the anatomy of dissonance in the international environmental dialogues', *Review of European Community and International Environmental Law* **12**: 23–32.

Redclift, M. and C. Sage 1998. 'Global environmental change and global inequality: North/South perspectives', *International Sociology* **13**: 499–516.

Redman, J. 2008. *Dirty is the New Clean: A Critique of the World Bank's Strategic Framework for Development and Climate Change*. Washington, DC: Institute for Policy Studies, Sustainable Energy and Economy Network.

Rosenberg, N. 1976. *Perspectives on Technology*. Cambridge, UK: Cambridge University Press.

Salomon, J. J. and A. Lebeau 1993. *Mirages of Development: Science and Technology for the Third World*. Boulder, CO: Lynne Rienner.

Schneider, M., A. Holzer and V. H. Hoffmann 2008. 'Understanding the CDM's contribution to technology transfer', *Energy Policy* **36**: 2930–2938.

Sharma, U. and S. Bhadwal 2008. *Adaptation Financing*, TERI viewpoint paper No. 6. New Delhi: TERI.

Shrivastava, M. K. 2007. Convergence in climate change institutions and consequences for developing countries: a case study of supercritical technology adoption by NTPC. M.Phil. dissertation, Jawaharlal Nehru University, New Delhi.

Tamura, K. 2006. 'Technology development and transfer', in A. Srinivasan. (ed.), *Asian Aspirations for Climate Regime beyond 2012*. Hamaya: Institute for Global Environmental Strategies, pp. 53–76.

Third World Network 2008a. Developing countries submit proposals for comprehensive adaptation framework. *TWN Accra News Update* No. 7, 27 August 2008 www.twnside.org.sg/title2/climate/news/TWNaccraupdate7.doc.

Third World Network 2008b. Divergence over IPR issue in technology transfer. *TWN Poznan News Update* No. 11, 10 December 2008. www.twnside.org.sg/title2/climate/news/TWNpoznanupdate11.doc.

Third World Network 2008c. Mitigation contact group discusses 'MRV'. *TWN Poznan News Update* No. 14, 11 December 2008. www.twnside.org.sg/title2/climate/news/TWNpoznanupdate14.doc.

Third World Network 2008d. G77-China propose 'enhanced financial mechanism' for UNFCCC. *TWN Info Service on Finance and Development*, 8 September 2008. www.twnside.org.sg/title2/finance/twninfofinance20080803.htm.

UNFCCC 2007. Bali Action Plan. Decision CP.13. www.unfccc.int/files/meetings/cop_13/application/pdf/cp_bali_action.pdf.

Winkler, H., R. S. Spalding-Fecher, S. Mwakasonada and O. Davidson 2002. 'Sustainable development policies and measures: starting from development to tackle climate change', in K. Baumert, O. Blanchard, S. Llosa and J. Perkaus (eds.), *Building a Climate of Trust: The Kyoto Protocol and Beyond*. Washington, DC: World Resources Institute, pp. 61–87.

World Bank 2008. *Illustrative Investment Programs for the Clean Technology Fund*, CIF/DM.2/Inf.2/Rev.1. Washington, DC: World Bank. www.siteresources.worldbank.org/INTCC/Resources/Illustrative_Investment_program_May_15_2008.pdf.

Part II

Agency

9

Agency in global climate governance: setting the stage

JOHANNES STRIPPLE AND PHILIPP PATTBERG

9.1 Introduction

It has almost become a truism that global climate governance is about more than bargaining among governments within the United Nations climate convention. For many observers of the climate negotiations, side-events have been for some time as interesting as the official negotiations. Already around the Kyoto negotiations of 1997 many important activities took place in parallel with the official negotiations among governments. The idea of side-events had diffused from the forums of non-governmental organizations that had accompanied the series of high-level UN conferences in the early 1990s.

A brief glance at the numerous side-events at any conference of the parties may serve as an illustration of the multiplicity and complexity of contemporary climate governance and, consequently, can help to illuminate from where ideas about 'agency beyond the state' have emerged. Already at the fourth conference of the parties in Buenos Aires in 1998, the insurance industry held seminars on the increased frequency and costs of extreme weather events. The Climate Action Network drew media attention to its 'Fossil of the Day Award' and organized seminars on justice and equity with regard to emissions of greenhouse gases. The Global Commons Institute presented early work on 'contraction and convergence' as an approach to stabilize atmospheric concentrations of greenhouse gases through converging per capita emissions. In several workshops, academics and policy advisors debated the implications of the newly launched Clean Development Mechanism. The nuclear power industry and the wind power industry both presented their low-carbon technologies, and the city of Curitiba held an exhibition about its public transport system as a best practice of transport-related energy efficiency. This 'social forum' of climate-related activities was crowded and energetic. It was, of course, much smaller and less organized compared to, for example, later meetings in Bali (2007) and Poznań (2008), where the daily Earth Negotiations Bulletin had a specific coverage of a sample of side-events. Nevertheless,

Global Climate Governance Beyond 2012: Architecture, Agency and Adaptation, eds. F. Biermann, P. Pattberg and F. Zelli. Published by Cambridge University Press.

the conference of the parties in Buenos Aires in 1998 already contained a social dimension, a full agenda of activities beyond the state, which has since then grown much larger and gained in popularity.

Most research on climate governance has focused on the effectiveness of the overall climate regime (Mitchell 2002; Stripple 2006). But how could we further develop our analysis to also include the activities of non-governmental actors? How could we understand the multifaceted arena of different agents, networks, levels, standards, norms and behaviour? This introductory chapter conceptualizes the transnational arena of global climate governance. We provide both theoretical insights into how this arena can be approached and empirical arguments for the importance of including this sphere of analysis into research about global climate governance. The chapter outlines ways in which the discipline of international relations has tried to capture this sphere and how new ways of thinking could enrich our analysis. The overarching argument is that current developments in global climate governance are signs of the gradual institutionalization of a transnational public sphere in world politics where norms and rules are devised and implemented independently from the intergovernmental negotiation process.

Agency in our understanding is neither fully located in the public sphere of governments and intergovernmental organizations nor in the private sphere of non-governmental organizations and business actors. It rather emerges in different places and at different times as a crucial mix of public and private resources, roles and responsibilities. It consequently transcends the crude distinction between public and private interests. Agency beyond the state is not agency without the state. It is agency that is constituted within an emergent transnational public sphere that is considerably broader than the intergovernmental system of politics.

This chapter sets the stage by introducing the context of the academic debate about non-state agency in global politics, engaging in particular with the specific contribution of the discipline of international relations.

9.2 Conceptualizing agency beyond the state: a review of the literature

As Thynne (2008: 329) noted, ‘An increasingly pertinent feature of the global public order in and beyond environmental protection and sustainability is the dynamic mixing of the public and the private, with state-based public power being exercised by state institutions alongside and along with the exercise of private power by market and civil society institutions and other actors committed to the public interest and public weal.’ The claim about ‘agency beyond the state’ is a claim that is difficult to comprehend without a few brief notes on where this question comes from and what is meant by it. Some ten years ago, Ian Hurd (1999: 404) asked, ‘Can our theories of the state accommodate a locus of authority outside the state?’

What he had in mind was that the common theories about international politics as they had developed in the last 50 years had an assumption of authority that makes it difficult to account for legitimate 'agency beyond the state'. For instance, Hedley Bull portrayed world politics as similar to Hobbes' *Leviathan*. Bull (1981: 721) stated that 'all of what Hobbes says about the life of individual men in the state of nature may be read as a description of the condition of states in relation to one another'. Bull draws, in this quote, attention to the similarity between the 'state-of-nature' and the 'state of sovereign authority'. In essence, the starting position for a theory of international politics is the existence of states and the absence of a higher level of authority in the system of states. This Hobbesian foundation has allowed for an image of the international arena as 'anarchical'. A classic metaphor is the image of billiard balls, where closed entities relate to each other only by way of force. In one of his most famous works, *The Anarchical Society* from 1977, Bull argues that the system of states could develop into a society of states through common institutions and rules (if these were recognized by the participating states).

That international cooperation is organized within a system of sovereign authorities can be seen as the crucial problematic for thinking about environmental issues. The report *Our Common Future* by the World Commission on Environment and Development (1987) opened with the statement that 'the Earth is one, but the world is not'. Hardin's seminal work 'The tragedy of the commons' (1968) has been used to capture the inherent difficulty of jointly managing a common atmosphere in a world of sovereign states (Vogler 2001). Neither of the two possible solutions to the tragedy of the commons offered by Hardin – privatization or the imposition of central authority – are feasible at the international level.

To sum up, viewed from the perspective of international politics, 'agency beyond the state' is viewed as a spatial claim about where to find legitimate authority (within the society of states). This spatial demarcation between different spheres of agency contains, in turn, an assumption of authority, to which we now turn.

Agency is often conceptualized along a public–private continuum. The distinction between the public and the private has been a crucial ordering device in social life and it continues to shape much of the debates surrounding various forms of authority and governance. While it is common to refer to a 'divide' or a 'gap' between the public and the private, such dichotomous thinking actually turns out to be, not necessarily wrong, but rather unhelpful if we are to understand how authority is being articulated and how governance is shaped through non-state actors in issue areas such as climate change. It was Hannah Arendt who drew attention to the separation of Greek life into two opposed realms: a public (the *Polis*) and a private (the household). Arendt (1958: 198), in a classic formulation, uses the *Polis* metaphorically and states that the *Polis* 'is not the city-state in its physical location; it is the organization of the people as it arises out of acting and speaking together, and its true space lies between people living together for this purpose, no matter where they happen to be'.

Beacroft underscores the centrality of Arendt's thinking for our conceptualization of politics. The 'Greek model of the *Polis* remains relevant to political theory as it highlights the centrality of the public realm for political life as a way of speaking, acting and living between human beings' (Beacroft 2007:42). For the discipline of international relations, the equation of the public, the state and the territory have had fundamental implications for how we have come to conceive of authority and governance. Authority – that is, legitimate power – has been conceptualized to be only possible *inside* the *Polis* and, hence, as illegitimate *outside* the territory, the state or the public. In sum, when viewed from within the traditional categories of politics, non-state actors as *legitimate* actors pose particular analytical problems.

The emergence of non-state actors on the international scene has sparked a wealth of literatures. Theories of regimes, globalization and global governance have all tried to define the role and relevance of non-state actors. It might even be possible to understand the development of the concept of 'governance' as signifying this phenomenon: namely an increase of private and civil actor involvement in the global arena, implying a transition from 'government' to 'governance' (Arts 2005: 1; Dingwerth and Pattberg 2006). The role of non-state actors was first accounted for in the early 1970s in writings such as Keohane and Nye's edited volume on *Transnational Relations and World Politics* from 1972. The predominant conceptualization in the book was to account for the influence of non-state actors (mostly multinational corporations) on state behaviour. This thinking evolved into a theoretical model, 'complex interdependence' (Keohane and Nye 1977), which portrayed a world where transnational activity affected states capacity to act; a world where the distinction between 'high' (security) and 'low' (trade) politics became obsolete and where military force were seen as ineffective. The essence of Keohane's approach to non-state actors is not about authority, but about changes in state power and influence. Later, in *After Hegemony: Cooperation and Discord in the World Political Economy*, Keohane (1984) moved away from the idea of providing a rather separate perspective on non-state actors in world politics and instead constructed a functional theory of regimes that could account for patterns of international cooperation. This crucial move made Realist and Liberal schools of thought united in a shared 'rationalist' research programme, premised on the condition of anarchy in the international system and oriented towards investigating international cooperation generally and specifically when, where and how regimes and institutions make a difference. This move still inspires considerable research in international environmental politics and is crucial for the way in which we conceive of agency as being 'within' or 'beyond' the state (for a recent example see Breitmeier *et al.* 2006).

Starting from a similar position as Keohane, James Rosenau put a rather different emphasis on non-state actors and authority in world politics. Writing in the wake of the unexpected demise of the bi-polar world system and the sudden end of the Cold War, Rosenau struggles to empirically and conceptually make sense of a 'turbulent world' (Rosenau and Czempiel 1992; Rosenau 1997). Crucially, for Keohane, engagement with non-state actors did not raise questions of authority, but quite the reverse for Rosenau who identified that 'governance without government' was present in many issue areas. Order is achieved through regime building and other rule-making activities without the prevalence of a state or a formal intergovernmental institution. The emergence of such new authority structures led Rosenau to identify two (separate) political worlds, one 'state-centric' consisting of 'sovereignty-bound states' and the other 'multi-centric' consisting of 'sovereignty-free' actors. *Turbulence in World Politics* (Rosenau 1990) spends a lot of time on elaborating on the elements, parameters and evolution of these two separate worlds, while the sequel *Along the Domestic–Foreign Frontier* (Rosenau 1997) accounts for non-state actors as more generic 'spheres of authority'. Spheres of authority are the building blocks of a new ontology where states are regarded as only one of many sources of authority.

Contemporary with Rosenau, Ferguson and Mansbach write in a similar broad liberal vein. However, in contrast to Rosenau, they do not anchor their analysis in a post-Cold War era of 'novelty', but instead turn to history. Ferguson and Mansbach (1996, 2004) provide a comprehensive 'remapping of global politics' in which authority is fragmented among polities with little hierarchical arrangement among them. The strength of their analysis is a nuanced reading of the state that does not adhere to a Westphalian stereotype of what capacity, legitimacy and authority means for the state. The Westphalian image of the modern state, which led to the strong association of state territory and authority, was, in their view, always a contingent outcome of place and time. The important question is hence about the state's ability to change and adapt to new conditions (Ferguson and Mansbach 2004: 142).

Two recent edited volumes (Cutler *et al.* 1999; Hall and Biersteker 2002a) build on Rosenau, Ferguson and Mansbach. These volumes are central for a new approach towards theorizing authority in world politics. Hall and Biersteker note (2002a) that traditional approaches to international politics regard states as not only the principle actors, but also the only *legitimate* actors. They argue that equating authority with government has for too long constrained an analysis of other forms of authority. However, from a theoretical standpoint, the public does not need to equal the governmental.

> Being public does not, however, imply that a state or public institution must be involved or wielding authority, even though they might participate in recognizing it in certain situations. It does imply, however, that the social recognition of authority should be publicly expressed. This opens the possibility for the emergence of private, non-state based, or non-state legitimated authority.
>
> *(Hall and Biersteker 2002b: 5)*

These two books provide many empirical examples of instances where non-state actors have acquired agency (that is, authority and capacity) to govern people and issue areas. The distinction between the public and the private is not a helpful guide to where to find, and not to find, authority. Nor does it allow making claims about where authority should, or should not, be located. It seems today that norms, rules, roles and responsibilities are becoming institutionalized beyond the confines of the state and the international society they construct. As Ruggie (2004: 521) has argued,

> the arena in which "the authoritative allocation of values in societies" now takes place increasingly reaches beyond the confines of national boundaries, and a small, but growing fraction of norms and rules governing relations among social actors of all types (states, international agencies, firms, and of civil society) are based in and pursued through transnational channels and processes.

We define this emerging space of interactions, the related norms and rules and the resulting roles and responsibilities of actors within the field of climate change as a *transnational arena of global climate governance.*[1]

The concept of global governance has been perhaps the main avenue towards conceptualizing contemporary international politics for some time, and its popularity has been facilitated by the ambiguity of the concept itself. We will not iterate the various meanings here (Dingwerth and Pattberg 2006), but what is important is where this literature speaks about steering, about purposeful activities of human collectives and about rules at all levels of human activity (Rosenau 1995). Hence, Jagers and Stripple (2003: 385) define climate governance as 'all purposeful mechanisms and measures aimed at steering social systems towards preventing, mitigating, or adapting to the risks posed by climate change'. These mechanisms and measures are only to a limited extent the result of public agency in the formal sense, but often the outcome of agency beyond the state. In contrast to much attention to international agreements, such as the climate convention, the Kyoto Protocol as well as questions about a post-2012 climate treaty, we argue that an institutionalized arena of transnational global climate governance is emerging. The multi-actor and multi-level tendencies that have been under way in world politics for a while are visible here. This broader perspective on global climate governance takes into account mechanisms with public, hybrid and private sources of authority. A number of actors deliberately form social institutions to address the problem of climate change without being forced, persuaded or funded by states and other public agencies. The transnational institutionalization of climate governance is similar to what Ruggie (2004) has called the reconstitution of a global public domain. As a domain, it does not replace states but 'embed systems of

[1] Transnational relations are defined as 'regular interactions across national boundaries when at least one actor is a non-state agent or does not operate on behalf of a national government or an international organization' (Risse-Kappen 1995: 9).

governance in broader global frameworks of social capacity and agency that did not previously exist' (Ruggie 2004: 519). The original claim about 'agency beyond the state' concerns the role and relevance of different actors. The power of individual and collective actors to change the course of events (or the outcome of processes), is seen to be increasingly located in sites beyond the state and its international organizations. What is different, compared to earlier accounts of these actors, is that these are now being embedded in a thickening social environment. Transnational climate governance is becoming institutionalized, and this has implications for the way in which agency can be understood. As Stripple and Lövbrand argue (this volume, Chapter 11) 'agency beyond the state' can be illustrated with reference to *actors*, but also to the *procedures* and mechanisms that make agency possible in the first place. It is important to look at the processes of governing, that is, the techniques and practices through which governance is accomplished (Sending and Neumann 2006: 657). The transnational arena of global climate governance has both legitimate agencies as well as resources (such as standards, measures or schemes) at their disposal. It is therefore likely that this arena will be increasingly important beyond the immediate negotiations. Consequently, academic research has to refine its analysis considerably, both conceptually and theoretically, to account for the innovation and contestation in the emerging domain of transnational global climate governance.

9.3 Conclusions

This chapter has introduced the multifarious and complex emerging transnational arena of global climate governance, in which agency is constructed, maintained and challenged not only by central governments but also by a host of other actors such as non-governmental organizations, business actors, scientists and sub-national governments.

With regard to these new actors and their agency, the following chapters address a wide range of theoretical questions as well as empirical observations. In particular, the following questions are addressed. What is the role and relevance of an increasing trend towards privatized and market-based governance mechanisms for climate change mitigation and the host of private actors, from non-governmental to business actors, that surrounds these new mechanisms in global climate governance? To what extent, and under what conditions, do private or public–private governance mechanisms produce policy outcomes that are comparable, or even superior, to (traditional) forms of intergovernmental cooperation? How are climate governance mechanisms beyond the state related to international negotiations and policy making?

These questions have been addressed with regard to a number of different empirical manisfestations of agency in climate governance. Pattberg and Stripple (2008) have proposed to map current climate governnace arrangements according to the actor constellation and the primary mode of governance employed. While actor constellations range from purely public to hybrid and exclusively private, modes of governance include networks and markets in addition to hierachical steering (in the form of government and intergovernmental organizations). The following chapters loosely draw on this distinction. While Pattberg (this volume, Chapter 10) explores networked forms of climate governance, Stripple and Lövbrand (this volume, Chapter 11) analyse the increasing marketization of climate governance. Chapter 12 by den Elzen *et al.* explores the contribution of the private sector through a revised sectoral approach, while finally Alfsen *et al.* provide an assessment of the role of public agency through international research and development agreements in Chapter 13. Each of these appraisals contributes to a better understanding of agency in climate governance.

References

Arendt, H. 1958. *The Human Condition*. Chicago, IL: University of Chicago Press.

Arts, B. 2005. 'Non-state actors in global environmental governance: new arrangements beyond the state', in M. Koenig-Archibugi and M. Zürn (eds.), *New Modes of Governance in the Global System: Exploring Publicness, Delegation and Inclusiveness*. New York: Palgrave, pp. 177–200.

Beacroft, M. 2007. 'Defining political theory: an Arendtian approach to difference in the public realm', *Politics* **27**: 40–47.

Breitmeier, H., O. R. Young and M. Zürn 2006. *Analyzing International Environmental Regimes: From Case Study to Database*. Cambridge, MA: MIT Press.

Bull, H. 1977. *The Anarchical Society: A Study of Order in World Politics*. New York: Columbia University Press.

Bull, H. 1981. 'Hobbes and the international anarchy', *Social Research* **48**: 717–738.

Cutler, A. C., V. Haufler and T. Porter 1999. *Private Authority and International Affairs*. Albany, NY: State University of New York Press.

Dingwerth, K. and P. Pattberg 2006. 'Global governance as a perspective on world politics', *Global Governance* **12**: 185–203.

Ferguson, Y. H. and R. W. Mansbach 1996. *Polities: Authority, Identities and Change*. Columbia, SC: University of South Carolina Press.

Ferguson, Y. H. and R. W. Mansbach 2004. *Remapping Global Politics: History's Revenge and Future Shock*. Cambridge, UK: Cambridge University Press.

Hall, R. B. and T. J. Biersteker 2002a. *The Emergence of Private Authority in Global Governance*. New York: Cambridge University Press.

Hall, R. B. and T. J. Biersteker 2002b. 'The emergence of private authority in the international system', in R. B. Hall and T. J. Biersteker (eds.), *The Emergence of Private Authority in Global Governance*. New York: Cambridge University Press, pp. 3–22.

Hardin, G. 1968. 'The tragedy of the commons', *Science* **162**: 1243–1248.

Hurd, I. 1999. 'Legitimacy and authority in international politics', *International Organization* **53**: 379–408.

Jagers, S. C. and J. Stripple 2003. 'Climate governance beyond the state', *Global Governance* **9**: 385–399.

Keohane, R. O. 1984. *After Hegemony: Cooperation and Discord in the World Political Economy*. Princeton, NJ: Princeton University Press.

Keohane, R. O. and J. S. Nye 1972. *Transnational Relations and World Politics*. Cambridge, MA: Harvard University Press.

Keohane, R. O. and J. S. Nye 1977. *Power and Interdependence: World Politics in Transition*. Princeton, NJ: Princeton University Press.

Mitchell, R. B. 2002. 'International environment', in W. Carlsnaes, T. Risse and B. A. Simmons (eds.), *Handbook of International Relations*. London: Sage Publications, pp. 500–516.

Pattberg, P. and J. Stripple 2008. 'Beyond the public and private divide: remapping transnational climate governance in the 21st century', *International Environmental Agreements: Politics, Law and Economics* **8**: 367–388.

Risse-Kappen, T. 1995. 'Bringing transnational relations back in: introduction', in T. Risse-Kappen (ed.), *Bringing Transnational Relations Back In: Non-State Actors, Domestic Structures and International Institutions*. Cambridge, UK: Cambridge University Press, pp. 3–35.

Rosenau, J. N. 1990. *Turbulence in World Politics*. Princeton, NJ: Princeton University Press.

Rosenau, J. N. 1995. 'Governance in the twenty-first century', *Global Governance* **1**, 13–43.

Rosenau, J. N. 1997. *Along the Domestic–Foreign Frontier: Exploring Governance in a Turbulent World*. Cambridge, UK: Cambridge University Press.

Rosenau, J. N. and E. O. Czempiel 1992. *Governance without Government: Order and Change in World Politics*. Cambridge, UK: Cambridge University Press.

Ruggie, J. G. 2004. 'Reconstituting the global public domain: issues, actors, practices', *European Journal of International Relations* **10**: 499–541.

Sending, J. O. and B. I. Neumann 2006. 'Governance to governmentality: analysing NGOs, states, and power', *International Studies Quarterly* **50**: 651–672.

Stripple, J. 2006. 'Rules for the environment: reconsidering authority in global environmental governance', *European Environment* **16**: 259–264.

Thynne, I. 2008. 'Climate change, governance and environmental services: institutional perspectives, issues and challenges', *Public Administration and Development* **28**: 327–339.

Vogler, J. 2001. 'Future directions: the atmosphere as a global commons', *Atmospheric Environment* **35**: 2427–2428.

World Commission on Environment and Development 1987. *Our Common Future*. New York: United Nations.

10

The role and relevance of networked climate governance

PHILIPP PATTBERG

10.1 Introduction

Next to governing through markets, networks have emerged as a central steering mechanism in modern governance. Climate governance is no exception. Uncounted numbers of initiatives, projects, programmes, institutions and organizations fill the realm of global climate governance, at times targeting public actors and intergovernmental decision-making processes, while on many other occasions targeting private actors such as corporations or individuals. While it is generally acknowledged that these forms of networked governance have become more widespread and pervasive, there is rather little scholarship on their performance and broader implications (exceptions are Bäckstrand 2008; Okereke *et al*. 2009).

This chapter appraises different types of networked climate governance, including public non-state networks such as the C40 global cities partnership; hybrid networks emanating from public–private sources of authority such as the more than 340 partnerships that have emerged out of the 2002 World Summit on Sustainable Development (WSSD) in Johannesburg; and networks whose authority derives from purely private sources, such as corporate social responsibility and standard-setting initiatives. I appraise these different types of networked governance in the climate realm according to three sets of criteria. First, their potential contribution to effective climate change mitigation and adaptation; second, their contribution to broader political goals such as increased participation and inclusiveness in global environmental governance; and third, their linkages and fit with the existing institutional architecture of international climate change governance. The chapter concludes with a number of policy options for strengthening the contribution of networked governance mechanisms to effective climate policies.

The research reported in this chapter is linked to the main appraisal questions in the analytical domain of agency within the ADAM Project: What is the role and relevance of an increasing trend towards privatized and market-based governance

Global Climate Governance Beyond 2012: Architecture, Agency and Adaptation, eds. F. Biermann, P. Pattberg and F. Zelli. Published by Cambridge University Press.

mechanisms for climate change mitigation and the host of private actors, from non-governmental organizations to business actors that surrounds these new mechanisms in global climate governance? Turned into more applied, problem-solving language: To what extent, and under what conditions, do private or public–private, transnational governance mechanisms produce policy outcomes that are comparable, or even superior, to (traditional) forms of intergovernmental cooperation? In answering these questions, the chapter suggests a number of long-term policy options for further discussion.

10.2 Methodology

Recent scholarly debate within the discipline of international relations has focused on the transformation of the global order from a territorial-based one to one of multiple spheres of authority in flexible and issue-specific arrangements (Stripple and Pattberg, this volume, Chapter 9; also Ruggie 1993; Rosenau 1997; Held *et al.* 1999). Reflecting debates about the organizational transformation of the modern nation state, theorists of international relations have begun to reflect on the changing nature of the Westphalian system itself (Zacher 1992). One central empirical observation is the emergence of networked forms of organization that operate under a different logic compared to other types of social organization, such as markets and hierarchies. Whereas network governance has been discussed as a complementation and gradual innovation of older forms of policy-making (for example corporatism) within the domestic context,[1] networks at the transnational and global level have been largely conceptualized as new forms of governance that potentially overcome the limitations of more traditional approaches (Börzel 1998; Benner *et al.* 2004). However, the analysis of networks in world politics substantially builds on the foundations of general network theory as it has been developed in organizational sociology and comparative politics.

Networks are defined, in their most general meaning and with reference to social processes, as the interconnection of three or more communicating entities. The structural components of such social networks include individuals and organizations, while network processes include communication and more general forms of exchange. Within political science, broadly speaking, networks are understood as interactions of organizational actors. Consequently, the concept of 'policy networks' refers to the production of public policies through a relatively stable and defined interaction of divergent actors within a policy field. Stated differently, policy networks are polycentric governance arrangements that integrate the competing interests of actors within a horizontal structure.

[1] Note, however, that Renate Mayntz (2004) for example interprets the general interest in networked governance as a reaction to real-world transformations of society through modernization and functional differentiation.

This conceptualization should be understood in contrast to the standard assumption of policy studies according to which the formulation and implementation of public policies are the sole responsibility of governments that try to transform the preferences of their voters into adequate political programmes. Organized interests of non-state actors are reflected only insofar as they deliberately address the public decision-making process. In contrast, the policy network approach reflects the transformation of policy-making within modern societies. It analyses the emergence of network governance as a reaction to a number of interconnected trends, including the increase in sub-systemic autonomy within the formerly monolithic nation state, the growth of state functions and its accompanying bureaucracies as well as the increase and further differentiation of organized civil society. Mirroring many aspects of the debate about new public management at the domestic level, the concept of network governance has been recently transferred to the global level. The appropriateness of the network approach in this context is frequently justified through reference to the changing capacity of state intervention and effective governance under the constraints of de-nationalization and accelerating globalization dynamics (Zürn 1998).

There are numerous typologies of networked governance (widely used is Börzel and Risse 2005; Börzel 1998 reviews the literature on social network research). I will rely on a simple differentiation based on the type of actors involved. This results in three fundamental types of networked governance, with many in-between states. First, public networks that exclude the state as a unitary central actor (in the form of government); second, hybrid networks, that is, involving both actors from the public and the private sphere; and finally fully private networks that incorporate both profit-making and not-for-profit logics.

This chapter offers an assessment of networked governance in the field of climate change. In particular, the following criteria are discussed: (1) the potential contribution of climate governance networks to effective climate change mitigation and adaptation; (2) their contribution to wider goals such as increased participation and improved inclusiveness; (3) and finally their linkages with the existing institutional architecture of climate governance. The first criterion is based on the assumption that climate governance networks are geared towards mitigation and adaptation and consequently, their strategies and outputs can be assessed against this goal (this perspective excludes unintended consequences and larger side effects). The second criterion relates to current debates about wider governance deficits in world politics and the potential value-added of new forms of governance in addressing those gaps (Haas 2004; Biermann *et al.* 2007). In particular, networked forms of governance are expected to broaden participation and increase the inclusion of otherwise marginalized actors as well as to provide more accountable forms of governance through increased transparency (Gupta 2008). Finally, climate governance networks are assessed according to their fit with the established architecture of global climate governance, that is, the international regime. This institutional compatibility can be synergetic, dysfunctional or neutral. This chapter will in particular scrutinize whether climate governance networks are integrated into the institutional structure provided by the climate convention and the Kyoto Protocol or rather result in an additional sphere of climate governance beyond the international.

10.3 Analysis

This section provides an appraisal of three distinct types of networked climate governance: public non-state networks (global city partnerships), public–private networks (WSSD partnerships for sustainable development) and private networks (corporate social responsibility arrangements).

10.3.1 Public non-state networks in transnational climate governance: the case of global city partnerships

Next to public–private and private networks, the cooperation of public non-state actors gains relevance in global climate governance. Cities are a prime example of public authority that transcends the dichotomy of national/international (Bulkeley and Betsill 2003). Increasingly, cities have formed cooperative arrangements to exchange information, learn from best practices and consequently mitigate carbon dioxide emissions independently from national government decisions. These developments are interesting from both the agency and architecture perspective. With regard to our analytical question of agency, city networks illustrate that the drivers of change in climate policies can no longer be equated with governments and their diplomatic corps, but have diversified to include the local as a central level of climate governance. The following section addresses their potential contribution to effective problem-solving in the area of climate change, their added value in terms of participation and inclusiveness as well as their potential fit, overlap and interlinkage with the international climate regime.

Potential contribution to effective climate governance

A prime example of a public non-state network in global climate governance is the Cities for Climate Protection (CCP) programme organized by Local Governments for Sustainability (ICLEI), an international association of local governments and national and regional local government organizations that have made a commitment to sustainable development.[2] ICLEI began working

[2] In addition to city networks, individual cities also have acquired agency in global climate governance. For example, the recent commitments of London (a reduction target of 60 per cent below 1990 levels by 2025) and Los Angeles (a reduction of 35 per cent below 1990 levels by 2035) go far beyond the Kyoto targets.

on global climate change in 1991, when it launched the Urban CO_2 Reduction Project, involving 14 municipalities in North America and Europe. This campaign, which ran until 1993, was designed to 'develop comprehensive local strategies to reduce greenhouse gas emissions and quantification methods to support such strategies' (ICLEI 1997: 5). Based on the success of the Urban CO_2 Reduction Project, ICLEI launched its CCP campaign in 1993 at the Municipal Leaders' Summit on Climate Change and the Urban Environment held at the United Nations (Betsill 2001: 395).

The CCP programme has three main goals: first, quantifiable reductions in greenhouse gas emissions; second, improvement of urban air quality; and third, the enhancement of urban quality of life and sustainability. In achieving these goals, the CCP programme is premised on the assumption that while the efforts of any single local government to reduce greenhouse gas emissions may be relatively modest, by working together local authorities can contribute to efforts to mitigate climate change (Betsill and Bulkeley 2004: 477). The approach through which the CCP's goals are expected to be reached is the so-called 'five milestones' approach to which members commit themselves. It consists of the following elements: municipalities (1) conduct a baseline emissions inventory and forecast; (2) adopt an emissions reduction target for the forecast year; (3) develop a Local Action Plan through a multi-stakeholder process (most plans also incorporate public awareness and education efforts); (4) implement policies and measures (for example energy efficiency improvements to municipal buildings and water treatment facilities, streetlight retrofits, public transit improvements, installation of renewable power applications, and methane recovery from waste management); and finally (5) monitor and verify the results. Tangible impacts of this approach are difficult to verify. ICLEI itself estimates that the United States-based CCP participants mitigate approximately 23 million tonnes of carbon dioxide annually (ICLEI 2006). Scholars have thus emphasized the 'soft' results of the CCP, such as increased access to relevant technical information and policy learning (Betsill and Bulkeley 2004: 487).

A second example of a public non-state network in transnational climate governance is the C40 network. In August 2006, the Large Cities Climate Leadership Group, a coalition of then 18 global cities, was joined by the Clinton Climate Initiative to form the C40, a partnership of 40 major cities that pledged to reduce carbon emissions and increase energy efficiency. According to Nicholas Stern, 'The C40 Cities Climate Leadership Group is a tremendous idea and a fine example of the different dimensions of international collaboration' (C40 2008a). Despite such praise, the C40 initiative is in such an early stage of implementation that an evaluation of its performance and impacts is not feasible.

Participation and inclusiveness

The emergence of public non-state actors such as cities has arguably broadened the scope of actors that possess agency in global climate governance. Local governments are key actors in tackling climate change at the international level since more than half of the world's population lives in cities, consumes around 75 per cent of the world's energy and emits approximately 80 per cent of the world's greenhouse gases (C40 2008b).

As a result, a number of networked governance arrangements have emerged in recent years that involve cities. In addition to approaches that involve small to mid-sized municipalities, global mega-cities have assumed a more central role in recent years. While a general argument can be made that city networks indeed broaden participation in global climate politics – the C40 partnership alone formally represents approximately 302 million people – what can we say about their inclusiveness? Any municipal government is able to join Cities for Climate Protection by becoming a formal signatory to a National Municipal Leaders' Declaration on Climate Change. However, participation in the CCP programme requires that a number of defined steps be taken. First, interested local governments begin participating in the CCP programme by passing a resolution pledging to reduce greenhouse gas emissions from their local government operations and throughout their communities. Each local government sets its own emission reduction target and develops a Local Action Plan outlining actions that the city will pursue to meet its target. After passing the resolution, the local government designates a staff member and an elected official to serve as the city's liaisons to ICLEI. In 2008, 692 communities in 31 countries were CCP members, with a clear bias towards Australia (196), the United States (159) and Canada (109). It is estimated that CCP members account for approximately 15 per cent of global anthropogenic greenhouse gas emissions.[3] Participation in the C40 Large Cities Climate Leadership Group is limited to large cities that have a potentially high impact on greenhouse gas emissions. A large number of C40 member cities are located in countries that have not, or only lately, ratified the Kyoto Protocol.

Linkages with the international climate governance architecture

Although public non-state networks involved in global climate governance are not formally linked to the international climate governance architecture, there are a number of potential overlaps. First, a number of global cities such as London and Los Angeles have linked their emission reduction targets to the international framework by agreeing on reductions measured against the 1990 baseline established within the Kyoto framework. By accepting this frame of reference, cities signal their

[3] For more information about CCP membership, see www.iclei.org/index.php?id=800.

willingness to contribute to international cooperation on climate change. Furthermore, by acting as a positive example and role model, cities are pressuring their federal governments to take action on climate change. As Schroeder and Bulkeley conclude concerning global cities, 'a non-nation state actor such as Los Angeles may be significant beyond its jurisdictional realm' (2008: 5). Second, city networks also exchange best practices on issues ranging from energy-efficient buildings to water and waste treatment. As a result, cities are key actors when it comes to disseminating applied knowledge and solutions to the challenge of climate change. The linkages with the international climate governance architecture are synergetic, as city networks provide an important venue for the transfer of knowledge and technology, stimulate national policies and potentially increase the willingness for international cooperation.

In sum, city networks for climate change mitigation add a crucial layer to the complexity of global climate architecture, as their individual contributions to problem-solving can no longer be subsumed under national commitments taken by states within the framework of the climate convention and the Kyoto Protocol.

10.3.2 Public–private networks in transnational climate governance: the case of WSSD partnerships for sustainable development

Public–private partnerships – that is, networks of different societal actors, including governments, international agencies, corporations, research institutions and civil society organizations – have become a cornerstone of the current global environmental order, both in discursive and material terms. At the United Nations level, partnerships have been endorsed by the former Secretary-General Kofi Annan through the establishment of the Global Compact, a voluntary partnership between corporations and the United Nations, as well as through the so-called type-2 outcomes concluded by governments at the World Summit on Sustainable Development in Johannesburg in 2002 that institutionalizes public–private implementation partnerships in issues areas ranging from biodiversity to energy. Both arrangements have been criticized for effectively privatizing parts of the policy responses to global change (Biermann *et al.* 2007; Rieth *et al.* 2007).

Public–private partnerships typically build on a voluntary agreement between actors from various sectors – governments, industry, activists, scientists or international organizations – to achieve a specific sustainability goal, in other words govern a distinct issue area. This section focuses on partnerships that are related to the WSSD process. WSSD partnerships are defined as 'specific commitments by various partners intended to contribute to and reinforce the implementation of the outcomes of intergovernmental negotiations of the WSSD (Programme of Action and the Political Declaration) and to help the further implementation of Agenda 21 and the

Millennium Development Goals (MDGs)' (Kara and Quarless 2002). The United Nations invited such partnerships to register with the secretariat of the Commission on Sustainable Development (CSD), a sub-commission of the UN Economic and Social Council. By March 2007, 323 multi-stakeholder initiatives have been listed in the CSD Partnerships Database.[4] Out of the 323 partnerships formally registered, 96 are within the primary categories of 'energy for sustainable development', 'air pollution/ atmosphere' and 'climate change' (these categories are based on the self-description of partnerships in the database). For the purpose of this chapter, I focus on those 19 partnerships that have identified climate change as their primary thematic area.

Potential contribution to effective climate governance

Assessing the potential contribution of public–private partnerships to effective climate governance is not simple. A number of problems have to be overcome. First, partnerships within the WSSD sample greatly differ in scale, scope and ambition, which makes it difficult to compare them. Second, as the long-term effects of public–private partnerships on greenhouse gas concentration levels are impossible to estimate, most attention necessarily will be directed towards their output and outcome, that is the products and activities partnerships produce and the behavioural changes that can be attributed to these outputs. Finally, reliable data on climate change partnerships are hard to find and are biased towards the most visible partnerships. To at least partially overcome these problems, the following analysis is based on a large-n dataset on WSSD partnerships (Biermann *et al.* 2007; but see also Bäckstrand 2008). The following indicators of a potentially positive effect of public–private partnerships on global climate governance are addressed. First, the geographical scope of climate change partnerships, including the countries of implementation; second, the average duration of partnerships in the issue area of climate change; and finally, the dominant function of the partnership.

With regard to the geographical scope of WSSD partnerships in the thematic area of climate change, the lack of local and national scope is noteworthy (see Table 10.1). As one might expect given the global nature of the climate problem, globally geared partnerships are frequent, performing slightly above average compared to the total partnership sample (52.6–50.8 per cent). However, given the great importance of adaptation within the issue area of climate change and the immediate relevance of sustainability at the local level, the absence of local and national partnerships from the climate sample is surprising. In fact, it underlines the frequently raised criticism that WSSD partnerships reflect given interest-structures and therefore seldom deliver additional benefits that have not already been realized in more traditional multilateral or bilateral implementation programmes.

[4] See http://webapps01.un.org/dsd/partnerships/public/welcome.do.

Table 10.1 *Geographical scope of all partnerships for sustainable development and climate-related partnerships*

	All partnerships for sustainable development registered with UN Commission on Sustainable Development, in per cent	All climate-related partnerships for sustainable development registered with UN Commission on Sustainable Development, in per cent
Global	50.8	52.6
Local	0.9	0
National	4.7	0
Regional	19.6	21.1
Subregional	24.0	26.3

Source: author's calculations.

In addition to the generic geographic scope of climate partnerships, it is important to focus on the country of implementation. As shown in Figure 10.1, the large majority of climate partnerships have several countries of implementation: 68.4 per cent have five or more countries of implementation; the range is from one to 48. By comparison, in the WSSD average the majority has less than five countries of implementation. However, of the climate change partnerships a greater percentage implement in at least one country from the OECD region (Organization for Economic Cooperation and Development), while roughly the same number implement in African states, Asian states (although fewer than in the WSSD average implement in non-OECD Asia) and Latin American countries. The overrepresentation of industrialized countries as countries of implementation within the climate partnerships sample could be interpreted as a necessary focus on high-emission countries, whereas project implementation in developing countries is of secondary interest.

A second interesting observation relates to the average duration of WSSD climate change partnerships. Given the long-term effects of climate change and the given inertia of the climate system, it seems at least plausible to assume that partnerships in the area of climate change will be either frequently open-ended or long term. In fact, our assessment of the available data shows that 37 per cent of all climate change partnerships are open-ended, compared to 28.3 per cent in the total sample. In addition, the average duration compares 6.1 to 4.9 years in favour of climate partnerships. We can tentatively conclude that climate change partnerships within the context of WSSD reflect the specific long-term nature of the climate problem in their duration. However, it is unclear whether the observed duration pattern is adequate for achieving the partnership goals and thereby contributes at least partially to solving the climate change problem.

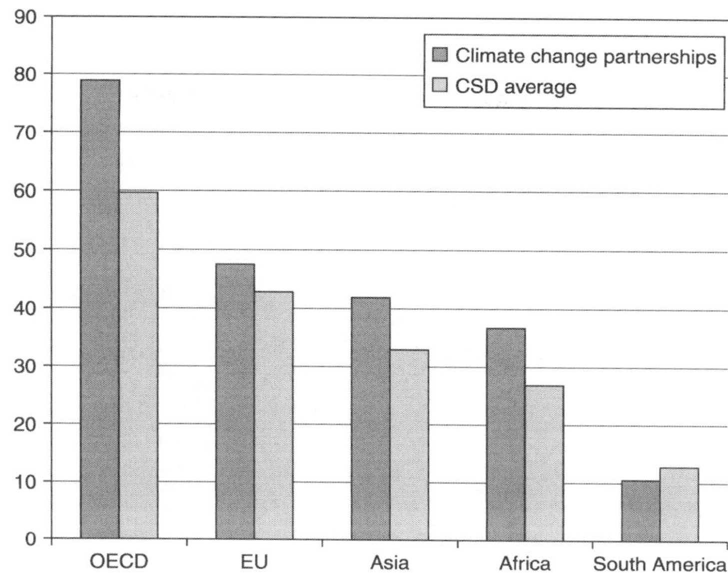

Figure 10.1 Percentage of partnerships implementing in at least one country in the region. Comparision of climate change partnerships with average of all partnerships registered with the UN Commission on Sustainable Development. *Source:* Global Sustainability Partnerships Database (on file with author).

As shown in Figure 10.2, an analysis of partnership functions as coded in the Global Sustainability Partnerships Database reveals that functions are most frequently related to facilitating knowledge-related activities, such as knowledge dissemination and production, institutional capacity building and planning, while only a limited number focuses on technical implementation or technology transfer. Both functions are important; however, the bias towards knowledge-related partnerships suggests that the effects of most public–private climate partnerships will only be realized in the long term, as capacity-building and knowledge disemination only have indirect effects on the climate system.

Participation and inclusiveness

In order to assess the added value of public–private partnerships in global climate governance with regard to participation and inclusiveness, I first discuss how far climate change partnerships are dominated by one specific type of actors; and second, whether or not partnerships are restricted in terms of participation. Turning to the question of leadership, three observations are noteworthy (Figure 10.3).[5]

[5] Leadership refers to the question of who is formally (by registration with the CSD) a lead partner within a partnership. Note that multiple lead partners per partnership are possible.

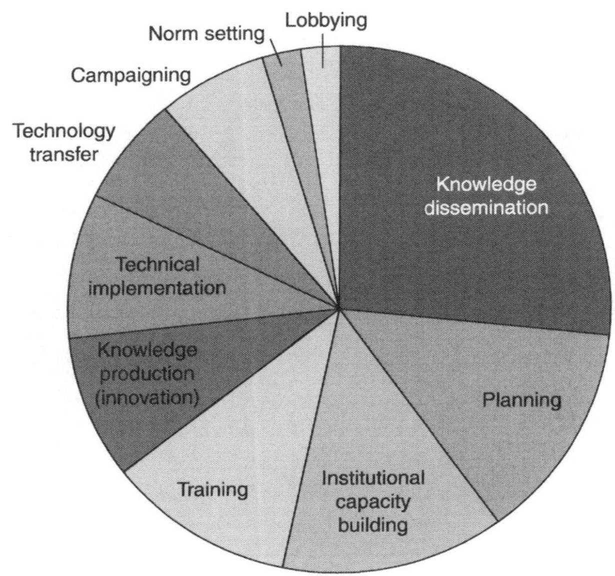

Figure 10.2 Frequency of occurrence of certain functions among climate change partnerships. *Source*: Global Sustainability Partnerships Database (on file with author).

First, leadership by United Nations agencies is less frequent in the climate change sample than in the total (8–16 per cent), while state leadership is above average by 44 to 24 per cent. This finding is consistent with the argument that the politically sensitive area of climate change is less likely to be governed by international agencies but is expected to remain under the control of governments. As a second observation, business actors are slightly overrepresented as lead-partners in the climate change sample (4–2 per cent), but are still less frequently found in leadership roles than standard arguments about business interests in climate change might suggest. One explanation could be that the advantages of participation in partnerships as a lead partner do not outweigh the costs and therefore business actors either remain absent or participate in less prominent roles. However, as the participation rate for business is higher than in the total sample, a business case for climate change might well exist. This observation is in line with the growing relevance of specific business interests in climate change, such as insurance, investors and consultancy firms. Finally, research institutions as lead partners are absent in climate change partnerships (12 per cent in the total sample), which is surprising in so far as science plays a major role in defining the problem of climate change as well as in finding solutions. The existence of the Intergovernmental Panel on Climate Change may partially explain this observation.

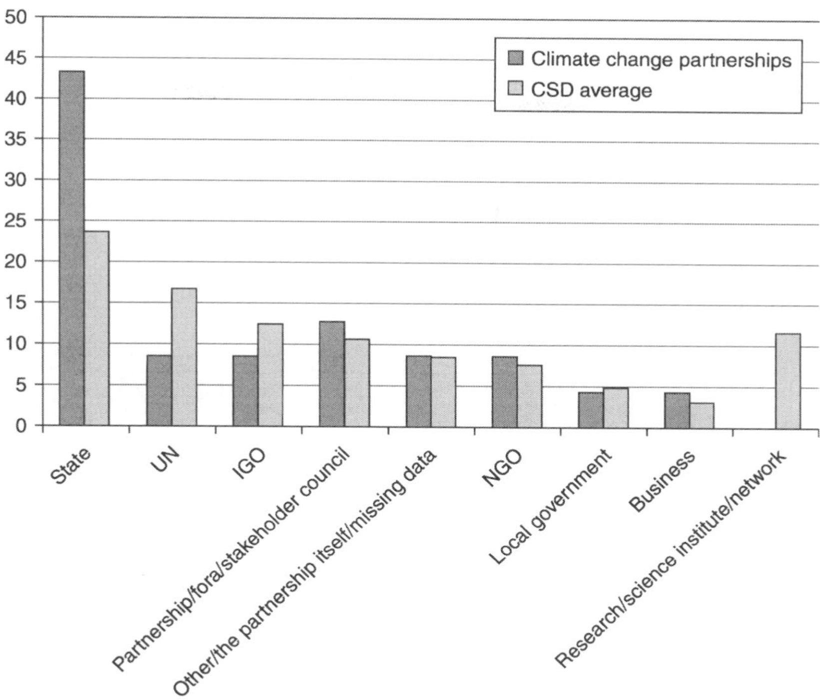

Figure 10.3 Lead partners in partnerships for sustainable development. Comparison climate change partnerships with CSD average, data in per cent. *Source:* Global Sustainability Partnerships Database (on file with author).

In addition to lead partners, the general membership of climate partnerships is also biased towards states (Figure 10.4). While general state membership is comparable between the climate and the total WSSD sample (states participate in 90 per cent of all climate partnerships compared to 85 per cent in the WSSD average), states belonging to the Group of 20 are overrepresented in the climate sample with 85–62 per cent. It is also noteworthy that NGOs participate only in close to 40 per cent of all climate partnerships while it is close to 50 per cent in the total sample.

With regard to the openness of climate partnerships, we observe that membership is generally quite restricted among climate change partnerships. Only 5.3 per cent are 'open,' while 42.1 per cent are closed and 52.6 per cent are semi-open. This means that climate partnerships either completely or partially restrict memberships to partners from a certain sector (47.4 per cent of partnerships), a specific geographical area (10.5 per cent of partnerships), a certain function (10.5 per cent) or the time-frame during which members can join (36.8 per cent).

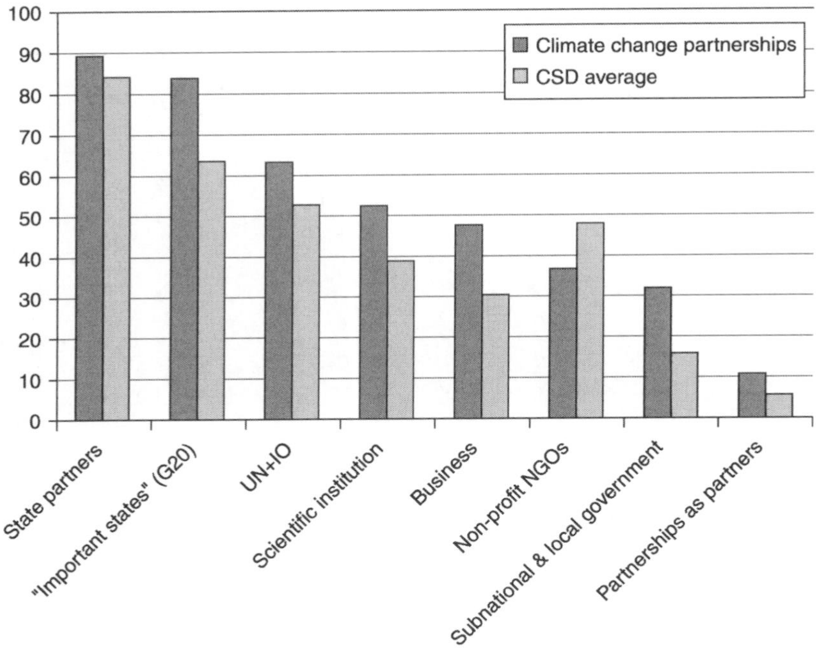

Figure 10.4 Type of members in partnerships for sustainable development. Comparison climate change partnerships with average of all partnerships registered with the Commission on Sustainable Development, shown as per cent of partnerships. *Source:* Global Sustainability Partnerships Database (on file with author).

In sum, climate partnerships are dominated by states, both in terms of leadership and general membership, while scientific institutions and business actors frequently participate but rarely take the lead. This observation is in line with the expectation that politically contested areas such as climate change politics remain, by and large, under the control of governments.

Linkages with the international climate governance architecture

As the outcome of an international agreement, WSSD partnerships are closely linked to the international arena where they are expected to contribute to the implementation of widely accepted sustainability goals such as the Millennium Development Goals and the Agenda 21. However, WSSD partnerships have a thematic focus on energy, but not on climate (the other main themes are water, health, agriculture and biodiversity). Hence, the linkages with the global climate architecture of climate convention and Kyoto Protocol are very weak as only a few individual partnerships within the climate sample attempt to link their work to the existing policy framework. For example, the Cement Sustainability Initiative has called for creation of a policy framework that would allow effective sectoral approaches within the climate convention.

Consequently, a stronger link with the existing policy framework on climate change could benefit both the WSSD partnerships, giving them guidance and a clear goal, and the international climate regime, assisting its implementation through public–private partnerships. So far, the potentially synergetic relationship between WSSD partnerships and the international climate change regime has not been further discussed or explored in negotiations.

10.3.3 Private networks in transnational climate governance: the case of corporate social responsibility

In addition to hybrid networks of public and private actors that are still embedded within the larger multilateral arena, at least partially, there are a number of policies that are beyond the state in a more concrete sense, as their authority neither derives from nor addresses public actors. Instead, they target transnational corporations and their global value- and supply-chains. Consequently, the majority of these approaches are frequently discussed under the heading of corporate social responsibility, understood as 'a concept whereby companies integrate social and environmental concerns in their business operations and in their interactions with their stakeholders on a voluntary basis' (Commission of the European Communities 2001: 6).

Potential contribution to effective climate governance

In addition to firm- and industry-level emissions reduction schemes and market-building approaches, a number of private networks have emerged that only indirectly aim at the reduction of greenhouse gas emissions, but rather focus on creating the necessary information and transparency for societal actors to assess corporate responses to climate change and thereby induce lasting behavioural change. Consequently, these benchmarking processes create a global competition among business actors to address climate change as a serious limitation to their profit-making activities. These emerging information-based governance schemes effectively institutionalize new norms at the transnational level, for example the norm to disclose corporate carbon emissions (in addition to the country-based reporting of the climate convention). I discuss the Carbon Disclosure Project as an illustrative example.[6]

[6] Another emerging non-state information-based governance scheme is the Investor Network on Climate Risk, organized by the non-profit organization Coalition for Environmentally Responsible Economies, an institutionalized cooperation of leading United States environmental organizations, social responsible investors and companies. For a more detailed analysis of Carbon Disclosure Project and the Investor Network on Climate Risk, see Enechi and Pattberg 2009.

The Carbon Disclosure Project (CDP) is an independent non-profit organization representing, as of 2008, 385 institutional investors with an investment base of approximately USD 57 trillion.[7] In 2003, it issued its first questionnaire that asks companies to disclose their past and future carbon emissions. CDP to date has issued six questionnaires to companies requiring disclosure of their carbon emissions (the questionnaire is issued annually). After 47, 59, 71 and 72 per cent response rate, 77 had responded to the sixth survey in 2008. Interestingly, sectors that have a high impact on carbon emissions, such as the electric utility sector, performed above average in the Fortune 500 index as a whole, while United States companies lagged behind European companies (60 per cent compared to 82 per cent). In addition to the regular survey, more than 1000 large corporations report on their emissions through the CDP website. The required disclosure stimulates information sharing about climate change and the related business risks between companies and stakeholders. The project goals are based on the assumption that information disclosure will motivate and facilitate meaningful dialogue among business actors, investors and the wider public to induce corporate responses to climate change. Although it is too early to assess the effectiveness of the CDP and the wider carbon disclosure discourse, arguably institutional investors have acquired agency beyond the state in global climate governance by, at least partially, institutionalizing the norm of corporate disclosure of carbon emissions and carbon reductions (see Pattberg 2006 for a general assessment of the influence of transnational corporate social responsibility schemes).

Participation and inclusiveness

Investors are at the centre of the CDP decision-making procedure. Although they are not represented individually, the institutional investors that make up the signatory investors and members of the project represent them in the decision- and rule-making process. The constituency is drawn in such a way that it includes investors across sectors and regions. The involvement of non-state actors is not explicitly stated; however, as initiators and supporters of the idea of corporate carbon disclosure they retain some influence in the decision-making process. In addition, the voluntary funding given to the CDP by some national governments has further diversified the participation to include public actors to some degree. While the institutional investors decide the content and structure of the CDP questionnaire, the responding companies are only passive participants in the project. Hence, there are two levels of participation in the governance arrangement – the responding companies and the global public are passive participants, while the institutional investors take

[7] The information used for the empirical analysis is sourced mainly from the publications of CDP. The figure is based on the CDP 2007 report available on the website of the CDP.

decisions. Although the institutional investors are assumed to represent the investors, it would have been more appropriate to allow participation of every interested member of the global public in the decision- and rule-making process of the project by making available the draft questionnaire publicly through the Internet before its final approval. In addition, the output of the arrangement (that is, the questionnaire) is intended for global use, but most institutional investors are based either in Europe or in North America. Institutional investors from other regions are not explicitly excluded from participation; their non-participation, however, suggests that they lack the capacity to do so. Hence, legitimacy could be increased by mechanisms that encourage the participation of institutional investors from other parts of the world.

Linkages with the international climate governance architecture

In addition to the complexity of agency, the architecture of global climate governance is highly fragmented (also Biermann *et al.*, this volume, Chapter 2). Within transnational climate governance, a number of approaches exist that are not linked to the international arena and therefore can hardly be integrated in or at least synchronized with the current negotiations. However, a number of interlinkages are also visible. The most obvious case are companies that have related their firm- or industry-level emissions reduction programmes to the international targets and timetables of the Kyoto Protocol. No less important is the business-NGO partnership The Climate, Community and Biodiversity Alliance that has announced the first two forestry projects to be independently certified under its Climate, Community and Biodiversity Standards.[8] The standard evaluates land-based carbon mitigation projects in forestry and thereby relates to the so-called land use, land-use change and forestry section of the Kyoto Protocol.[9] On this account, private standardization attempts to fill critical gaps in the operationalization and implementation of international agreements. In sum, the current developments in corporate social responsibility underscore the relevance of a broadened analytical perspective on global climate governance. With an increasing number of non-state actors acquiring agency beyond the state and the deepening institutionalization of non-state approaches towards climate change such as market- and information-based mechanisms, a strictly international and state-centred perspective seems no longer viable. Instead, focusing on the transnational global climate governance arena shows the importance of corporate social responsibility for effective climate politics.

[8] See www.climate-standards.org.

[9] Under article 3.3 of the Kyoto Protocol, parties decided that greenhouse gas removals and emissions through certain activities – namely afforestation and reforestation since 1990 – are accounted for in meeting the Kyoto Protocol's emission targets.

10.4 Conclusions and policy recommendations

This chapter has analysed agency beyond the state in global climate governance in a dual meaning. First, networked governance solutions addressed in this chapter are policy instruments that are driven by non-state actors, both public (such as cities) and private (such as corporations). Second, the regulative targets of these new mechanisms in global climate governance are frequently not states, but non-state actors themselves, ranging from transnational corporations to global mega-cities.

This chapter has analysed networked climate governance in the form of public non-state city networks, public–private partnerships and private corporate social responsibility schemes. This research was based on three evaluation criteria: problem-solving capacity; participation and inclusiveness; and synergies or dys-functional linkages with international climate governance. First, as for problem-solving capacity, it appeared that several obstacles prevent the realization of the full potential of public–private partnerships. In particular, the geographical bias towards global partnerships instead of local or regional ones underscores the frequent criticism that partnerships reflect pre-existing interest structures and therefore seldom deliver additional benefits that have not been realized in more traditional multilateral or bilateral implementation arrangements. With regard to city networks and corporate social responsibility schemes, the analysis highlighted that measurable results, in particular in terms of emissions reduction, are hard to find. However, the case studies also show that the emergence of norms, for example of corporate carbon disclosure, might have more significant effects in the future.

Second, regarding increased participation through public–private partnerships, this chapter highlights the overrepresentation of governments in climate partnerships as compared to the total sample of partnerships. Climate partnerships are largely dominated by states, both in terms of leadership and general membership. This finding is in line with the expectation that politically contested areas such as climate politics remain, by and large, under the control of governments. Regarding increased participation through city networks and corporate social responsibility schemes, this chapter finds that neither form of networked climate governance substantially closes the participation gap in global environmental governance. However, all three types of networked climate governance diversify the agency in climate politics and thereby lay the groundwork for increased and sustained participatory climate governance.

Finally, with regard to positive or negative linkage with international climate governance, it appears that neutral interlinkages are more frequent than negative ones; however, the potential of increased linkage of transnational and international climate governance has not yet been realized.

To address the problem of insufficient coordination among existing environmental programmes and initiatives, a number of scholars have suggested setting up a world environment organization that could integrate existing governance interventions into a coherent whole (for example Biermann 2000). A similar argument can be made in the case of transnational climate governance, where increased coordination among, for example, the climate convention and the WSSD process, could be mutually beneficial. However, similar to the discourse about a world environment organization, the possibility of administrative congestion and overburdening of already burdened bureaucracies presents a powerful caveat to such an idea. Therefore, I propose a light coordination mechanism that would deliver real benefits, thrive on limited resources and could be easily integrated into any international successor treaty of the current Kyoto Protocol. As a first step towards greater coordination between international and transnational governance mechanisms, a clearing house should be institutionalized that gathers information about existing non-state climate governance initiatives, evaluates their complementarity with international mechanisms and makes recommendations towards improved integration. The Commission on Sustainable Development could host such a clearing house that essentially provides an authoritative overview of the current landscape of transnational climate governance. As an observer to the conferences of the parties, the commission could serve as the missing link between the international negotiations and the burgeoning arena of transnational climate governance.

References

Bäckstrand, K. 2008. 'Accountability of networked climate governance: the rise of transnational climate partnerships', *Global Environmental Politics* **8**: 74–102.

Benner, T., W. H. Reinicke and J. M. Witte 2004. 'Multisectoral networks in global governance: towards a pluralistic system of accountability', *Government and Opposition* **39**: 191–210.

Betsill, M. M. 2001. 'Mitigating climate change in US cities: opportunities and obstacles', *Local Environment* **6**: 393–406.

Betsill, M. M. and H. Bulkeley 2004. 'Transnational networks and global environmental governance: the cities for climate protection program', *International Studies Quarterly* **48**: 471–493.

Biermann, F. 2000. 'The case for a world environment organization', *Environment* **42**: 22–31.

Biermann, F., P. Pattberg, S. Chan and A. Mert 2007. 'Multi-stakeholder partnerships for sustainability: des the promise hold?', in P. Glasbergen, F. Biermann and A. P. J. Mol (eds.), *Partnerships, Governance and Sustainable Development: Reflections on Theory and Practice*. Cheltenham, UK: Edward Elgar, pp. 239–260.

Börzel, T. A. 1998. 'Organizing Babylon: on the different conceptions of policy networks', *Public Administration* **76**: 253–273.

Börzel, T. A. and T. Risse 2005. 'Public–private partnerships: effective and legitimate tools of international governance', in E. Grande and L. W. Pauly (eds.), *Reconstituting*

Political Authority: Complex Sovereignty and the Foundations of Global Governance. Toronto: University of Toronto Press, pp. 195–216.

Bulkeley, H. and M. M. Betsill 2003. *Cities and Climate Change: Urban Sustainability and Global Environmental Governance*. London: Routledge.

C40 2008a. C40 large cities climate summit. C40, London. Available at http://www.c40cities.org/summit/.

C40 2008b. History of the C40. Available at http://www.c40cities.org/, 20 October 20 2008).

Commission of the European Communities 2001. *Promoting a European Framework for Corporate Social Responsibility*. Brussels: Commission of the European Communities.

Enechi, O. and P. Pattberg 2009. 'The business of transnational climate governance: legitimate, accountable and transparent?', *St Antony's International Review* **5**: 76–98.

Gupta, A. 2008. 'Transparency under scrutiny: information disclosure in global environmental governance', *Global Environmental Politics* **8**: 1–7.

Haas, P. M. 2004. 'Addressing the global governance deficit', *Global Environmental Politics* **4**: 1–15.

Held, D., A. McGrew, D. Goldblatt and A. Perraton 1999. *Global Transformations: Politics, Economics and Culture*. Stanford, CA: Stanford University Press.

ICLEI 1997. *Local Government Implementation of Climate Protection*, report to the United Nations. Toronto: ICLEI.

ICLEI 2006. *Combating Climate Change: A Comprehensive Look at Local Climate Protection Programs*. Toronto: ICLEI. Available at www.iclei-usa.org/action-center/learn-from-others/(resource)Case_Studies_Dec-2006.pdf.

Kara, J. and D. Quarless 2002. 'Guiding principles for partnerships for sustainable development ("type 2 outcomes") to be elaborated by interested parties in the context of the World Summit on Sustainable Development (WSSD)', presented at the *Fourth Summit Preparatory Committee (PREPCOM 4)*, Bali, Indonesia.

Mayntz, R. 2004. *Governance Theory als fortentwickelte Steuerungstheorie*? Köln: Max Planck Institute for the Study of Societies.

Okereke, C., H. Bulkeley and H. Schroeder 2009. 'Conceptualizing climate change governance beyond the international regime', *Global Environmental Politics* **9**: 58–78.

Pattberg, P. 2006. 'The influence of global business regulation: beyond good corporate conduct', *Business and Society Review* **111**: 241–268.

Rieth, L., M. Zimmer, R. Hamann and J. Hanks 2007. 'The UN Global Compact in Sub-Sahara Africa: decentralization and effectiveness', *Journal of Corporate Citizenship* **7**: 83–99.

Rosenau, J. N. 1997. *Along the Domestic–Foreign Frontier: Exploring Governance in a Turbulent World*. Cambridge, UK: Cambridge University Press.

Ruggie, J. G. 1993. 'Territoriality and beyond: problematizing modernity in International Relations', *International Organization* **47**: 139–174.

Schroeder, H. and H. Bulkeley 2008. *Governing Climate Change Post-2012: The Role of Global Cities – Los Angeles*, Tyndall Briefing Note No. 33. Oxford, UK: Tyndall Centre for Climate Change Research.

Zacher, M. W. 1992. 'The decaying pillars of the Westphalian temple: implications for international order and governance', in J. N. Rosenau and E. O. Czempiel (eds.), *Governance without Government: Order and Change in World Politics*. Cambridge, UK: Cambridge University Press, pp. 85–101.

Zürn, M. 1998. *Regieren jenseits des Nationalstaates*. Frankfurt am Main: Suhrkamp.

11

Carbon market governance beyond the public–private divide

JOHANNES STRIPPLE AND EVA LÖVBRAND

11.1 Introduction

Despite current disagreements over the future climate policy architecture, carbon markets represent a central feature in most proposals to move society towards a low carbon economy. The worldwide creation of carbon markets is emblematic of the marketization trend in climate governance that accelerated after Russia's ratification of the Kyoto Protocol in November 2004 – the crucial moment making the treaty legally binding. With the entry into force of the Kyoto Protocol, the inception of the EU emissions trading scheme in 2005 and the more recent emergence of regional carbon markets in North America, Australia and New Zealand, global transactions in emission reductions have become more than mere social imagining. As estimated by the World Bank (Capoor and Ambrosi 2008: 1), as much as 2983 million tonnes of carbon dioxide were traded by public and private actors in 2007 to a total value of USD 64 035 million.

In this chapter, we illustrate the agency beyond the state in contemporary carbon markets. However, in contrast to global governance studies (Rosenau 1999; Biersteker and Hall 2002), we do not conceptualize carbon market governance along the public–private continuum. Instead of asking which *entities* (for example public or private authorities) govern the carbon economy, we draw attention to the *procedures* by which carbon markets are made thinkable and operational as administrative domains in the first place. Hence, we analyse the complex body of knowledge, techniques and practices that have turned tradable carbon offsets into a governable reality. In particular, this chapter focuses on the multilateral verification, validation and certification practices of the Clean Development Mechanism (CDM) that have enabled the making of the 'certified emission reduction', and the private rules and standard-setting underpinning 'voluntary' or 'verified emission reductions'. When focusing on these 'calculative practices' (Miller and Rose 2008: 11),

Global Climate Governance Beyond 2012: Architecture, Agency and Adaptation, eds. F. Biermann, P. Pattberg and F. Zelli. Published by Cambridge University Press.

carbon market governance does not signify a retreat of the state or politics as often implied by students of private governance. Rather, in line with Lemke (2002), we argue that the carbon economy as imaginary space represents a transformation of politics and statehood that involves a replacement of formal and hierarchical techniques of government with more indirect regimes of calculation.

While approaching carbon market governance as a messy set of rules, practices, norms and authority relations that lack a single origin, driving force or systemic coherence, this chapter identifies the tension between *practices of commensuration* and *practices of differentiation* as a central market feature. On the one hand, the success of carbon markets hinges on their ability to make emission reductions generated in different locations and under very different circumstances 'the same' (MacKenzie 2009). Through a range of carbon government technologies such as standards, baseline methodologies, verification and auditing schemes, combined with certain ways of thinking (derived from climate science and resource economics), 1 tonne of avoided or sequestered carbon dioxide emissions ($1tCO_2$) is today approached as a standardized economic good. This commensuration of emission reductions allows us to compare and trade certified and verified emission reductions produced through highly diverse project activities in developing countries with the European Union allowances generated within the EU emissions trading scheme. On the other hand, as noted by MacKenzie (2009: 13), standards that seek to enhance the quality of the certified emission reductions or verified emission reductions by specifying and certifying sustainable project practice (for example the Gold Standard) represent techniques that make things *not* the same.

In our view, much of contemporary carbon market governance can be understood as a contestation between the various practices that make emission reductions the same and those that make them different. In order to understand this tension, we begin our chapter by outlining the ways of thinking necessary to approach $1tCO_2$ as a tangible and tradable commodity. Consequently, we offer a brief overview of contemporary carbon markets where this new commodity is produced and transformed into various carbon currencies. Section 11.3 examines two concrete offset contracts purchased by the authors to compensate for their air travel to Bali and San Francisco. We highlight the different calculative practices that have brought these offsets into being and made them available for sale in the web stores of the two carbon brokers Tricorona and Climate Care. We conclude by asking ourselves how the practices of commensuration and differentiation that have made our carbon contracts the same, and yet different, will affect efforts to reform the CDM beyond 2012. We note that carbon accounting techniques that ensure that 'a tonne is a tonne' across sectors and continents, on the one hand, have been central to the imagining and making of the carbon economy as a governable domain.[1] On the other hand, contemporary efforts to reform the CDM suggest that practices that differentiate sustainable offsets from mere emission reductions will play an increasingly central role in future carbon market dynamics. Hence, we conclude that the future design of the CDM is not just a legal or technical challenge of making things the same, but also a political one.[2]

11.2 Methodology

Studies of global governance are often associated with efforts to extend the analysis of rule and authority beyond narrowly defined political communities. Typically, the changing role of state and non-state actors in a time of hybrid, non-hierarchical and network-like modes of governing are the focus of attention (Pattberg 2007; Pattberg, this volume, Chapter 10). While this chapter shares the interest in decentralized forms of rule, it does not ask who has power and authority in contemporary carbon market governance. Rather, following the broad field of governmentality studies (Rose *et al.* 2006; Miller and Rose 2008), we draw attention to the processes that make carbon markets possible in the first place. Hence, we focus on what Dean (2004: 2) calls 'the how of governing', the question of what processes enable or constrain agency.

By moving from *who* to *how* questions, we are able to take a step back and analyse the constituting ways of thinking and acting that 'makes up' the carbon economy and renders it practicable and amendable to intervention. As argued by Miller and Rose (2008: 62):

> In a very real sense, 'the economy' is brought into being by economic theories themselves, which define and individuate a set of characteristics, laws and processes designated economic rather than, say, political or natural. This enables 'the economy' to become something which politicians, academics and industrialists and others think can be governed, managed, evaluated and programmed, in order to increase wealth, profit, and the like.

The claim that economic theories do not only describe an existing external economy but also bring that economy into being, is reiterated by MacKenzie and Millo (2003). Drawing upon what Callon (1998: 2) has called 'the performativity of economics',

[1] See for example the International Carbon Action Partnership's conference invitation for the 1st Global Carbon Market Forum on Monitoring, Reporting, Verification, Compliance and Enforcement, 19 and 20 May 2008 – Management Centre Europe. The invitation frames these practices as the 'backbone of a robust carbon market'. See more at www.icapcarbonaction.com/.

[2] This point was articulated by Peter Zapfel (EU Commission Directorate-General Environment) at the ADAM Post-2012 workshop in Brussels, October 2008. From the perspective of the EU Emissions Trading Scheme, Zapfel argued, linking is a very simple binary decision. Either we recognize other carbon units or we do not. The question of linking is therefore a good example of the tension between commensuration and differentiation and, in general, the performative character of economics (Callon 1998).

they argue that economics create the phenomena it describes. This performativity is not confined in words, but also includes methods, calculation tools and technical instruments (Callon *et al.* 2007: 334). As suggested by McKenzie *et al.* (2007), it is far too narrow to view the economy as a body of ideas alone because 'economics also consists of people, skills, datasets, techniques, procedures, tools, and so on' (McKenzie *et al.* 2007: 5). Hence, Miller and Rose (2008: 32) have suggested that we need to extend our analysis of economic life to the mundane and humble mechanisms (for example computation, notation, surveys, assessments and evaluations) that make it possible to govern. Following this analytical tradition, our chapter highlights the particular ways of thinking and acting that perform the carbon economy.

When studying carbon markets through the multiple ideas and practices that constitute them, the contemporary trade in emission reductions appears far from natural or given. Carbon markets instead emerge as fragile and complicated socio-technical artefacts, or what Callon (2009: 4) calls 'a network of experimentation', wherein all aspects and components are constantly tested, reflected on and critically evaluated. While this fragility was widely recognized when the price for 'European Union allowances' fell in early April 2006 due to a too generous national allocation, we are in this chapter particularly interested in the knowledge and techniques that seek to stabilize the carbon economy after such disturbance and hereby establish it as a fertile ground for government intervention.

11.3 Analysis

11.3.1 Making up the carbon economy

The past years have seen a significant growth in carbon markets and the establishment of a range of new offset currencies (for an overview see Capoor and Ambrosi 2008). The 'European Union allowances', the 'certified emission reductions' and 'verified emission reductions' are just some of the many labels currently used to conceptualize the transfer of 1 tonne of avoided or sequestered carbon dioxide emissions from one part of the world to another. Central to the imagining of such carbon transactions is the idea that reductions of greenhouse gas emissions (translated into carbon dioxide-equivalents) have the same atmospheric effect wherever they are carried out. Although the activities underpinning the avoided emissions may involve different people under very different circumstances, climate science tells us that 1 tonne of carbon dioxide emissions is the same regardless of its location on Earth. The geography of carbon mitigation practices does not matter. This imagining of an atmospheric mixing of carbon can be traced back to a long series of developments within meteorology and biogeochemistry during the past 100 years.[3]

3 A vast array of literature substantiates this claim. Examples from the field of meteorology include Clayton (1927), Chapman (1930) and Haurwitz (1948). For a recent review, see Weart (2003).

The Swedish chemist Svante Arrhenius pioneered the understanding of the human-induced greenhouse effect in 1896. Since then, the exchange of carbon between the atmosphere, oceans and terrestrial biosphere has been studied in further detail, offering empirical support to Arrhenius' theory. When the climate convention was negotiated in the early 1990s, the scientific interest in global carbon flows was prevalent. However, as a result of the interstate negotiations, the notion of 'national' sinks and sources of greenhouse gases gained ground (Lövbrand and Stripple 2006). The idea that carbon flows should be measured and monitored within national borders was institutionalized in the climate convention, and has since then resulted in regular national carbon accounting and reporting practices. While these calculative practices have enabled the imagining of a national carbon space, they have also been central for the construction of the global carbon economy. So has the notion that emission reductions can be spatially fixed (Bumpus and Liverman 2008: 134) and thus treated as a tradable commodity.

Through the workings of the globalized carbon economy, specific ways of reducing emissions are today abstracted from their local context, objectified through a range of standardized techniques and practices, and made available for sale in different carbon markets around the world. The very idea to *trade* emission reductions can be traced back to the resource economics literature and thinkers such as Pigou in the 1920s and later Dales (1968a, b) and Coase (1960). As Callon (2009: 4) puts it, 'without this contribution from economic theory, carbon markets would have been literally unthinkable'. Coase (1960: 44) showed that if polluters are given the right to pollute, and to settle the level of compensation for sufferers through the market, a social optimum level of pollution will arise. If the market is fully competitive, such private bargaining will offer a more efficient pollution control than government regulation. While Coase based his theory on examples from local resource and pollution controversies in the United Kingdom and the United States, carbon markets have extended the theory's geographical reach. Since the marginal costs of reducing greenhouse gas emissions differ within an industry or nation, as well as among nations, emissions trading offers a way to reach a given reduction target at lowest total cost.

Voss (2007: 340) argues that emission trading contains a specific grammar of governance that is not only embedded in economic theory, but also in the skills of a range of market actors (for example brokers, consultants, legal advisors, public administrators and specific interest groups) operating in different geographical sites. Once established, carbon markets are therefore not just cognitive constructs, but have a social life and an institutional trajectory in their own right. According to Larner and Heron (2004), these actors are connected through global supply chains. Supply chains represent global production networks through which individuals and groups create value through mutual association. Global supply chains are 'a relatively new spatialization of the economic, a distinct imaginary that has the governmental effect of institutionalizing conversations between vertically and horizontally linked actors around the expectations of distant consumers' (Larner and Heron 2004: 220). In the carbon economy, the supply chain for carbon offsets rests upon a set of rules for participation, allocation and credit generation, including extensive monitoring, reporting and verification procedures. However, the procedures by which $1tCO_2$ is functionally and spatially abstracted into an individualized commodity differ depending on the type of offset (Bumpus and Liverman 2008: 136).

Currently, there are two market procedures, or systems, by which $1tCO_2$ is transformed into a tangible good: first, cap-and-trade systems; and second, baseline-and-credit systems. A cap-and-trade market is created when a collective of emitters within a certain jurisdiction receives a 'cap' on their emissions. Within that cap, emitters are allowed to trade the allowances among each other. In a baseline-and-credit market, on the other hand, the carbon credits are equal to the difference between emissions after mitigation measures vis-à-vis a baseline (that is, the emission trajectory under a business-as-usual scenario). The cap and the baseline, respectively, are thus essential to turn $1tCO_2$ into a tradable good. We can further distinguish between compliance markets, which result from public regulation (for example the Kyoto Protocol or national emission trading schemes), and voluntary markets, which do not rely on public regulation to generate demand. Compliance markets as well as voluntary markets can be designed as cap-and-trade systems or as baseline-and-credit systems. These distinctions result in the following typology (see Table 11.1).

The carbon markets represented in Table 11.1 all rest upon a range of calculative practices that extend beyond the public–private divide. Whereas the Kyoto

Table 11.1 *Compliance and voluntary carbon markets*

Carbon markets	Compliance market	Voluntary market
Cap and trade	Emissions trading under the Kyoto Protocol, the EU emissions trading scheme, New Zealand emissions trading system, Australian emissions trading system, Regional Greenhouse Gas Initiative, California Climate Registry	Chicago Climate Exchange Japanese voluntary emissions trading system
Baseline and credit	Clean Development Mechanism, Joint Implementation	Voluntary carbon offsetting

Protocol's baseline-and-credit mechanisms (the CDM and Joint Implementation) and cap-and-trade system (emissions trading) build upon a set of intergovernmental decisions negotiated since 1997, they have engaged a broad network of public agencies, private firms and civic groups in a wide range of governance functions (Lövbrand *et al.* 2009). Similarly, the rules and practices that currently regulate the EU Emissions Trading Scheme have evolved among a vast assemblage of state and non-state actors, theories, experiments and techniques beyond the control of any single Member State. While there are notable differences between baseline and crediting in the compliance market and in the voluntary market, we will suggest in the following section that both represent 'regimes of value' (Appadurai 1986: 15)[4] that employ similar calculative practices when establishing 1tCO_2 as a tradable commodity with distinct value and meaning.

11.3.2 The making of 'certified emission reductions'

In December 2007, one of the authors of this chapter purchased a climate neutralization certificate from the Swedish carbon broker Tricorona for USD 288. The offset was bought to neutralize a research trip to the thirteenth conference of the parties to the climate convention in Bali, Indonesia. According to the certificate, the 6.15 tonnes of carbon dioxide generated by the trip are compensated for by the Sri Balaji biomass power plant in Andhra Pradesh, India. The plant's mitigation activities are registered under the CDM of the Kyoto Protocol, and therefore produce certified emission reductions later brokered by Tricorona.

According to decision 3/CMP.1 in the Marrakech Accords under the Kyoto Protocol,[5] a certified emission reduction is a unit that equals 1 metric tonne of carbon dioxide-equivalents, calculated by using the global warming potential developed by the Intergovernmental Panel on Climate Change and defined in article 5 of the Kyoto Protocol. The certified emission reduction travels a long way, moulded by many actors and practices before it reaches the store of carbon brokers such as Tricorona. It is produced by project developers in a developing country, often in collaboration with Northern investors, approved and authorized by a designated national authority in the host country, validated, verified and certified by a third-party auditing firm, and finally given a legal status by the CDM Executive Board, that is, the supervisory body incepted by the conference of the parties to the climate convention in 2001.

[4] Appadurai (1986: 15) talks in *The Social Life of Things* about the creation of value as a politically mediated process. While a regime of value requires a certain cultural sharing of assumptions by the parties subject to a particular commodity exchange, the degree of value coherence may be highly variable from situation to situation, and from commodity to commodity.

[5] See 'CDM rules' in FCCC/KP/CMP/2005/8/Add.1, 30 March 2006.

The global supply chain of the certified emission reductions sold in Tricorona's web store in December 2007 began in February 2003 when the private power company Indur Green Power Private Ltd decided to displace fossil-fuel-based electricity in the Andhra Pradesh state grid with electricity generated from renewable biomass. To fund the power plant and its condensing type steam generator, the company submitted a project design document to the national CDM authority in India for approval and to the auditing firm Det Norske Veritas for validation. According to the project design document, the biomass used in the plant is carbon dioxide neutral.[6] Hence, over a period of 7 years the operation will replace around 296 million kWh of conventional energy and avoid 245 813 tonnes of carbon dioxide emissions. This claim was justified by a simplified baseline methodology for small-scale CDM projects accepted by the CDM Executive Board's methodology panel in 2005.[7] By referring to financial and policy barriers for investments in biomass energy production (for example fuel availability, fluctuating prices and tariffs), these calculative practices have sought to establish the estimated emission reductions as real, measurable and additional to any that would occur in the absence of the project activity, and hereby satisfying the eligibility criteria for CDM projects specified in article 12 of the 1997 Kyoto Protocol.

In September 2005, the Indian Ministry of Forestry and Environment officially approved the 7.5 MW Biomass Power Project of Indur Green Power Pvt Ltd as an activity that would contribute to sustainable development in the Andhra Pradesh region.[8] One month after this second step in the global supply chain of the certified emission reduction, Det Norske Veritas submitted its validation report to the CDM Executive Board (Det Norske Veritas 2005). By measuring the project documentation against the multilateral criteria for CDM projects in the Kyoto Protocol and the Marrakech Accords, the Det Norske Veritas validation team established the estimated emission reductions as measurable (that is, thinkable) and additional (that is, legitimate). In line with Power (2003: 385), we approach these auditing procedures as ritualistic practices that bring order to a previously unordered world. When filtering the emission reduction activities of the Indian power plant through the lens of standardized techniques, auditing firms such as the Det Norske Veritas make it possible to think of organizationally discrete and spatially disparate carbon practices as comparable, or even the same. It is this ability to make the incommensurable seem commensurable (Larner and Heron 2004: 214) that allows a

[6] Available for download at the climate convention webpage: http://cdm.unfccc.int/Projects/DB/DNV-CUK 1144412972.08.

[7] See 'Indicative simplified baseline and monitoring methodologies for selected small-scale CDM project activity categories': http://cdm.unfccc.int/Projects/pac/ssclistmeth.pdf.

[8] The signed host country approval of the 7.5 MW Biomass Power Project of Indur Green Power Pvt Ltd is available for download on the Climate Convention webpage: http://cdm.unfccc.int/Projects/DB/DNV-CUK1144412972.08.

certified emission reduction to be represented as a standardized commodity with the same atmospheric effect regardless of its origin.

Two months after the official registration of Sri Balaji in May 2006, the Indur Green Power Pvt Ltd submitted a monitoring report to the CDM Executive Board. According to the report the power plant had avoided 103 763 tonne carbon dioxide-equivalent between February 2003 and March 2006 (Indur Green Power Pvt Ltd 2006). The accuracy of this claim was later verified by Det Norske Veritas after a desk review of the data in the monitoring report and two on-site inspections in September 2006 (Det Norske Veritas 2006). Based on the Det Norske Veritas's verification/certification report, the CDM Executive Board issued 81 295 certified emission reductions in May 2007.[9] In December the same year, six of these carbon offsets were sold to one of the authors of this chapter through the Tricorona webpage, thus completing the certified emission reductions supply chain.

11.3.3 The making of 'verified emission reductions'

The making of a 'verified emission reduction' on the voluntary carbon market rests upon more heterogeneous and less transparent calculative practices than those used to produce a 'certified emission reduction'. In contrast to the regulated CDM market, the various actors involved in the voluntary market are not necessarily *accredited*, a project is not necessarily *validated*, emission reductions are not necessarily *verified* and there are many different standards of verification working alongside each other. In the final stage of the global supply chain, the emission reductions are not necessarily *certified* and kept in a registry to prevent them from being sold twice. This lack of standardization suggests that the verified emission reduction is more heterogeneous in character, and often seen as incommensurable with the certified emission reduction. However, through the introduction of voluntary offset standards, the global supply chain of the 'verified emission reductions' is gradually conforming to that of the 'certified emission reductions'.

For the sake of illustration, we briefly outline three different types of voluntary offsets that one author of this chapter considered when compensating for a conference trip from Copenhagen to San Francisco in March 2008.

The first offset was for sale on eBay in February 2008. The auction started at USD 0.99, although the total amount of avoided emissions was unclear. According to the eBay seller, the offset was generated by himself. 'I work at home and drive very little. I walk a lot and have a car that gets 36 miles per gallon. I haven't travelled by air for at least 10 months. If you are commuting a lot and flying around the country

[9] See the following webpage for full project documentation: http://cdm.unfccc.int/Projects/DB/DNV-CUK 1144412972.08.

you may need my offset.'[10] Whether this was a sincere attempt to sell emission reductions or not, the example is principally interesting since it departs from the requirement for additionality, central to the production of a 'certified emission reduction' under the Kyoto Protocol. As suggested by the quote, the seller's emission reductions are not calculated vis-à-vis a business-as-usual baseline, but in relation to a generalized idea of an 'equal share' of emissions. Although this way of thinking about climate governance is far from new (see for instance Agarwal and Narain 1991), it has not yet become an operational part of carbon market dynamics.

The second offset under consideration by the author was also sold on eBay (but in a specific store) by the International Small Group and Tree Planting Program (TIST). TIST is a community-based organization that helps subsistence farmers in Tanzania, Kenya, Uganda and India to produce commercial carbon offsets through tree plantation programmes and modern satellite technology, such as GPS. The farmers use the satellite monitoring system to measure and report the amount of atmospheric carbon sequestered within their plantations. Subsequently, they upload the coordinates and measurements to the TIST webpage for increased transparency. According to Hawn (2006: 2), the supply of this virtual cash crop has certain advantages:

> TIST farmers have found that modern technology lets them leapfrog some of the hurdles that in the past had plagued their access to markets. They no longer need to transport products like mangos or tea across long distances on unpaved roads; they simply transmit information across oceans.

By 2006, TIST had sold some 180 verified emission reductions through its eBay store and another 5000 directly to customers, including the World Bank (Hawn 2006). However, none of these offsets were calculated in a manner that satisfied the eligibility criteria under the Kyoto baseline-and-credit regulations.

The third type of voluntary offset, the one finally chosen by the author, was, however, generated by similar calculative practices as those underpinning the CDM market. The 2.53 tonnes of carbon dioxide generated by the author's return trip from Copenhagen to San Francisco were offset by the United Kingdom-based firm ClimateCare at a total cost of USD 38. The offset contract, purchased through the broker's web store, does not specify where or how the verified emission reductions were produced. Instead, the return flight was compensated for by a portfolio of projects located in different parts of the world (see Table 11.2).

However, since ClimateCare employs a number of voluntary offset standards for their projects, it is possible to reconstruct the making of the verified emission

[10] The carbon offset was offered on eBay on 2 February 2008 under the item number 250218013349. The page is now removed, but a copy of the webpage from that day can be found at www.svet.lu.se/documents/JST_Carbonoffsets.pdf.

Table 11.2 *ClimateCare General Portfolio 2007/2008*

Project	Country	Type	Standard
Improved Lao stoves	Cambodia	Efficient stoves	VCS
Efficient cook stoves	Uganda	Efficient stoves	GS VER
Qingdao wind	China	Wind power	VER+
Bagasse co-generation	India	Biomass power	VCS
Treadle pumps	India	Human power	VCS
Mulan wind	China	Wind power	VCS

Source: The portfolio can be found at ClimateCare's home page (www.jpmorganclimatecare.com) or through their direct link (www.jpmorganclimatecare.com/projects/portfolios/portfolio-2007-08, accessed on 29 January 2009).

reductions purchased by the author. As outlined by the official manuals and guidelines for the Voluntary Carbon Standard and the VER+, a standard developed by TÜV SÜD, offset projects following these standards are adjusted to a number of baseline and monitoring methodologies developed to ensure project additionality. In order to avoid double-accounting, both standards also coordinate a registry that accounts for the issued, held and retired verified emission reductions.[11] While these standardized carbon accounting procedures are designed to secure the commensurability of the resulting verified emission reductions, the Gold Standard verified emission reductions is, instead, a standard that seeks to differentiate between verified emission reductions. Using a set of criteria – for example a positive list of project types, sustainable development and civic involvement – the Gold Standard seeks to produce high-quality verified emission reductions.[12]

Although these standards govern the voluntary market according to different logics (making things the same versus making things different), they have, interestingly, made the supply chain of the verified emission reductions very similar to that of the certified emission reductions. The baseline-and-credit methodologies used to secure additionality by the Voluntary Carbon Standard and VER Plus, or the sustainability indicators developed by the Gold Standard, all draw upon the multilateral rules developed for the CDM. Hence, when analysing carbon markets through the practices that constitute or perform them, they can no longer be divided into one *regulated* and one *unregulated* zone as often-implicated in writings on the carbon economy. Rules for governing carbon markets are evolving with the markets, and accounting practices specified for the Kyoto Protocol are now copied and

[11] On the Voluntary Carbon Standard, see http://www.v-c-s.org/about.html, and VER+ at http://www.tuev-sued.de/uploads/images/1179142340972697520616/Standard_VER_e.pdf.

[12] Read more about the Gold Standard at http://www.cdmgoldstandard.org.

transformed beyond their initial reach. The current search for common standards, registries and accounting procedures for 1tCO_2 thus indicates a trend towards an institutionalization of carbon market governance that extends beyond the public–private divide (see Pattberg and Stripple, this volume, Chapter 9).

11.3.4 Governing by practices of commensuration and differentiation

An analysis of the 'how' of carbon markets governance inevitably draws attention the procedures by which carbon emissions produced in one place (for example through air travel) today are made exchangeable with emission reductions generated in another place (for example fossil-fuel substitution, forest plantations or efficient stoves). As illustrated above, the contemporary trade in carbon offsets hinges on making diverse and disparate activities part of the same balance sheet. Numbers transform qualitative distinctions into quantitative differences and allow seemingly incommensurable practices to appear commensurable (Samuel 2008). Through a wide range of measurement, monitoring, verification and certification techniques, diverse practices such as driving a SUV in Sweden, running a refrigerant plant in China or managing a forest in Kenya are today made 'the same' in the carbon accountant's book. Hence, carbon market governance relies on managing flows of carbon in a particular calculative sense, which translates into measuring, balancing and exchanging carbon in accordance with universal currency systems.

Interestingly, these practices of commensuration are not confined to carbon market governance alone, but are integral to the history of sustainability.[13] To quantify and statistically aggregate nature according to accounting ideals that associate natural objects with the objects of market economies, has been crucial to sectors such as forestry, fishery and agriculture. Such 'environmental accounting' allows nature to be allocated and exchanged as stocks and shares, as profitable or social liabilities. To reconfigure the 'natural' as 'environment' – a calculate object – is the first step towards its 'resourcification' and 'commodification'. While resting upon a long tradition in environmental management, the carbon economy has, however, given these accounting practices a new global reach. According to Hulme (2007: 6) the climate is today 'defined in purely physical terms, constructed from meteorological observations, predicted inside the software of Earth system science models and governed (or not) through multi-lateral agreements and institutions'. As a result, reduced emissions have been 'made up'

[13] See, for example, the papers presented at the workshop Nature's Accountability: Aggregation and Governmentality in the History of Sustainability organized by German Historical Institute, Washington, DC, 9–11 October 2008. See more at www.ghi-dc.org.

as a global public good that can be commensurated, commodified and traded in the global marketplace.

Although these practices of commensuration are essential for the imagining of a global carbon economy where certified emission reductions and 'verified emission reductions' are abstracted from their local context and bought and sold as regular commodities, carbon markets are, interestingly, also made by practices of differentiation. Already at the time when the climate convention was negotiated in the early 1990s, Agarwal and Narain (1991) questioned the ways in which the share of atmospheric carbon dioxide of developing countries was calculated. In their view, emissions cannot easily be compared and should therefore not be added together. Their claim that we need to differentiate between 'subsistence' and 'luxury' emissions (Agarwal and Narain 1991) gained ground in the negotiations and is institutionalized as the principle of 'common but differentiated responsibilities' in article 3 of the climate convention. This early effort to make global climate governance account for inequalities in terms of material, social and economic situations resonates with contemporary debates over the future design of the CDM. Today, an increasing number of market actors and scholars alike question the idea that all emission reductions can or should be treated as the same. As noted in a recent literature review (Paulsson 2009: 14),

> most evaluations are rather moderate in their judgment, acknowledging that the CDM is working quite well in terms of its ability to provide CERs [certified emission reductions], but also highlighting problems such as the inequitable geographical distribution of projects and the lack of sustainable development benefits from many projects.

To address the unequal allocation of CDM investments among continents and regions, and to encourage the production of certified emission reductions in least developed countries, a range of techniques to make emission reductions *different* have been proposed. Among the various suggestions are differentiated levies on CDM projects, discounting credits from non-least-developed countries, or the use of quota systems (for example suggested by Bolivia). Although politically contested, such practices of differentiation are expected to reflect the unequal economic conditions under which a certified emission reduction is made. An increased differentiation between project types has also been suggested as a way to enhance the sustainable development benefits of CDM projects (Schneider 2008: 23). While a global standard for sustainability seems unfeasible (defining sustainability top–down has always been met with opposition from developing countries), voluntary performance standards such as the Gold Standard already offer certification procedures that differentiate carbon offsets according to their sustainability performance.

Another way to make carbon offsets different would be to separate the two objectives of the CDM (cost-effective emission reductions and sustainable development), effectively leaving the achievement of sustainable development in developing states to other mechanisms (Falaleeva and Stripple 2008: 12). According to Schneider (2008), such changes in the multilateral rule system would turn the CDM into a mere instrument for offset trading. The current requirement for additional sustainability benefits would, according to this model, be taken care of by a new multilateral fund aimed at projects with high costs and questionable additionality, such as some renewable energy and energy efficiency projects (van Asselt and Gupta 2009: 46–47). Whether any of these suggestions for CDM reform will gain ground in the future regime is yet too early to say. However, they all point to a growing tension between the many calculative practices that currently make emission reductions the same in the carbon economy, and political efforts to make them different. We suggest that this tension is likely to characterize carbon market dynamics after 2012. Future efforts to govern this domain will thus balance between the calculative practices (for example monitoring, validation, verification or certification) that still seek to stabilize $1tCO_2$ as a solid good (and the carbon economy as a credible sector of reality), and those practices that shed light on the different locations and socio-economic circumstances under which efforts to reduce greenhouse gas emissions take place.

11.4 Conclusions and policy recommendations

In this chapter, we have advanced an understanding of carbon markets that is based on the calculative practices that constitute or perform them. Rather than asking *who* or *which* entities govern this domain, we have asked *how* or by which procedures that carbon markets are made thinkable and operational in the first place. To that end, we have analysed baseline-and-credit markets in particular, where a complex measurement of counterfactuals (current emissions vis-à-vis a business-as-usual scenario) enables reductions of carbon dioxide-equivalents to be assigned market value and transformed into various offset currencies. By examining the global supply chain of two concrete carbon offset contracts, this chapter has come in contact with a wide range of actors beyond the state such as investors, developers, managers, auditors, brokers, retailers and buyers, not to mention individuals as geographically dispersed (but functionally linked) producers and consumers of carbon dioxide emissions.

Our analysis suggests that these actors perform a range of governance functions, including enhancing the credibility of offsets (that is, auditing, verification and creating standards), providing information (technical advice, capacity-building), enabling aggregation (that is, project finance, facilitate investments and structuring deals), facilitating transactions, influencing regulation (lobbying and building market infrastructure) and finally adjudicating conflicts (interpreting laws to resolve conflicts).[14] While this global production network signifies a shift from hierarchical forms of government to more decentralized forms of rule, this chapter has not interpreted carbon market governance as a retreat of politics or the state in favour of the market. Instead, *agency beyond the state* is here understood as a different kind of statehood that governs social behaviour 'at a distance' (Rose and Miller 1992: 180). Hence, of interest is how the introduction of new actors on the scene of government has helped to create an economic domain where people interact to attach meaning and value to 1tCO_2.

Although the analytical approach advanced in this chapter does not allow us to make a normative evaluation of the various options for future carbon market governance, it does help us to interpret the ways of thinking and acting that underpin such options. By analysing carbon markets through the procedures that constitute them, we have found that effective governance of carbon markets does not only involve finding the right institutional environment, involving certain types of actors or choosing between hierarchical versus bottom–up governance arrangements. The success of carbon market governance also hinges on the ability to commensurate diverse and dispersed activities in time and space. We have suggested that climate science and resource economics offer the underlying ways of thinking that make it possible to temporarily freeze people and places into an imaginary space so that the necessary calculative work can be performed (Larner and Heron 2004: 226). While conceptualizing emission reductions as the same is crucial for the functioning of carbon markets, practices of commensuration obscure the complex political geography of emissions including the various meanings invested in them. Consequently, recent debates over carbon market governance are increasingly dominated by efforts to make emission reductions different.

The introduction of offset quality standards such as the Gold Standard, or debates over the future design of the Kyoto cap-and-trade regime, highlight that carbon offsets often are made up by different people, under very different circumstances. Whether such debates will result in more diversified offset currencies after 2012 or a more profound critique of efforts to stabilize the carbon economy as a homogeneous, coherent space, is yet too early to say. Given the experimental stage of carbon markets, we know that there is no fixed trajectory. We also know that there is no fixed point, no firm ground, from where the carbon economy can be governed. However, when focusing on the *how* of governing, we do know that understanding a phenomenon in such a way that it *can* be governed is more than a speculative activity (Miller and Rose 2008: 30). It requires the invention of a wide range of

[14] This indicative list builds on Marc Ventrasca's presentation: 'How markets get built: Institutional intermediation and commensuration in emerging carbon markets' at the conference *Energizing Markets – Making and Breaking Boundaries for the Regimes of Value*, Copenhagen Business School, 30 October – 1 November 2008.

calculative practices that make the carbon economy amendable to intervention. Hence, we conclude that any effort to reform carbon markets in the post-2012 era will have to pay close attention to the various practices that constitute them as operable and legitimate domains of social interaction.

Acknowledgements

The authors acknowledge the financial support provided by the research programme *Adaptation and Mitigation Strategies: Supporting European Climate Policy*, funded by the European Union under its sixth framework programme; the ClimateColl project funded by the Swedish Energy Agency; and the GreenGovern project, funded by the Swedish Research Council (FORMAS).

References

Agarwal, A. and S. Narain 1991. *Global Warming in an Unequal World: A Case of Environmental Colonialism*. New Delhi: Centre for Science and Environment.

Appadurai, A. 1986. 'Introduction: commodities and the politics of value', in A. Appadurai (ed.), *The Social Life of Things*. Cambridge, UK: Cambridge University Press, pp. 3–63.

Asselt, H. van and J. Gupta 2009 (in press). 'Stretching too far: developing countries and the role of flexibility mechanisms beyond Kyoto', *Stanford Environmental Law Journal* **28**.

Biersteker, T. and R. H. Hall (eds.) 2002. *The Emergence of Private Authority in International Affairs*. Cambridge, UK: Cambridge University Press.

Bumpus, A. and D. Liverman 2008. 'Accumulation by decarbonisation and the governance of carbon', *Economic Geography* **84**: 127–155.

Callon, M. 1998. *The Laws of the Markets*. Malden, MA: Blackwell Publishers/Sociological Review.

Callon, M. 2009. 'Civilizing markets: carbon trading between in vitro and in vivo experiments', *Accounting, Organizations and Society* **20**: 535–548.

Callon, M., Y. Millo and F. Muniesa 2007. *Market Devices*. Malden, MA: Blackwell Publishers.

Capoor, K. and P. Ambrosi 2008. *State and Trends of the Carbon Market 2008*. Washington, DC: World Bank.

Chapman, S. 1930. 'A theory of upper atmospheric ozone', *Memoirs of the Royal Meteorological Society* **3**: 103–125.

Clayton, H. H. 1927. *World Weather Records*, Smithsonian Miscellaneous Collections No. 79. Washington, DC: Smithsonian Institution.

Coase, R. 1960. 'The problem of social cost', *Journal of Law and Economics* **3**: 1–44.

Dales, J. H. 1968a. 'Land, water and ownership', *Canadian Journal of Economics* **1**: 791–804.

Dales, J. H. 1968b. *Pollution Property and Prices*. Toronto: University of Toronto Press.

Dean, M. 2004. *Governmentality: Power and Rule in Modern Society*. London: Sage Publications.

Det Norske Veritas 2005. *Validation Report: Indur 7.5 Non-Conventional Renewable Sources Biomass Power Project in India*, Report No. 2005–9008. Available at http://cdm.unfccc.int/Projects/DB/DNV-CUK1144412972.08.

Det Norske Veritas 2006. *Verification/Certification Report: Indur 7.5 Non-Conventional Renewable Sources Biomass Power Project in India*, Report No. 2006–9116. CDM Registration No. 0391. Availabel at http://cdm.unfccc.int/Projects/DB/DNV-CUK1144412972.08.

Falaleeva, M. and J. Stripple 2008. *CDM Post-2012: Practices, Possibilities, Politics*, workshop report, Lund University, Sweden. Available at Http://www.svet.lu.se/documents/ht2008_CDM_workshop_report.pdf.

Haurwitz, B. 1948. 'Insolation in relation to cloud type', *Journal of Meteorology* **5**: 110–113.

Hawn, A. 2006. 'Carbon credits on eBay: subsistence farmers sell ecosystem services in a virtual marketplace', *Conservation Magazine* **7**: 4.

Hulme, M. 2007. 'Geographical work at the boundaries of climate change', *Transactions of the Institute of British Geographers* **33**: 5–11.

Indur Green Power Pvt Ltd 2006. *Monitoring Report Indur 7.5 Non-Conventional Renewable Sources Biomass Power Project in India*, Reference No. UNFCCC 0391. Available at: http://cdm.unfccc.int/Projects/DB/DNV-CUK1144412972.08

Larner, W. and R. L. Heron 2004. 'Globalizing benchmarking: participating at a distance in the globalizing economy', in W. Larner and W. Walters (eds.), *Global Governmentality: Governing International Spaces*. New York: Routledge, pp. 212–232.

Lemke, T. 2002. 'Foucault, governmentality and critique', *Rethinking Marxism* **14**: 49–64.

Lövbrand, E. and J. Stripple 2006. 'The climate as political space: on the territorialization of the global carbon cycle', *Review of International Studies* **32**: 1–19.

Lövbrand, E., J. Nordqvist and T. Rindefjäll 2009 (in press). 'Closing the legitimacy gap in global environmental governance: examples from the emerging CDM market', *Global Environmental Politics* **9**: 74–100.

MacKenzie, D. 2009. 'Making things the same: gases, emission rights and the politics of carbon markets', *Accounting, Organizations and Society* **34**: 440–455.

MacKenzie, D. and Y. Millo 2003. 'Constructing a market, performing theory: the historical sociology of a financial derivatives exchange', *American Journal of Sociology* **109**: 107–145.

MacKenzie, D. A., F. Muniesa and L. Siu 2007. *Do Economists Make Markets? On the Performativity of Economics*. Princeton, NJ: Princeton University Press.

Miller, P. and N. Rose 2008. *Governing the Present*. Cambridge, UK: Polity Press.

Paulsson, E. 2009. 'A review of the CDM literature: from fine-tuning to critical scrutiny?', *International Environmental Agreements: Politics, Law and Economics* **9**: 63–80.

Pattberg, P. 2007. *Private Institutions and Global Governance: The New Politics of Environmental Sustainability*. Cheltenham, UK: Edward Elgar.

Power, M. K. 2003. 'Auditing and the production of legitimacy', *Accounting, Organizations and Society* **28**: 379–394.

Rose, N. and P. Miller 1992. 'Political power beyond the state: problematics of government', *British Journal of Sociology* **43**: 173–205.

Rose, N., P. O'Malley and M. Valverde 2006. 'Governmentality', *Annual Review of Law and Social Science* **2**: 83–104.

Rosenau, J. N. 1999. 'Toward an ontology for global governance', in M. Hewson and T. Sinclair (eds.), *Global Governance Theory*. Albany, NY: State University of New York Press, pp. 287–302.

Samuel, S. 2008. 'The fishy business of eco-economics', presented at the workshop *Nature's Accountability: Aggregation and Governmentality in the History of Sustainability* organized by German Historical Institute, Washington DC, 9–11 October 2008. Available at www.ghi-dc.org

Schneider, L. 2008. *A Clean Development Mechanism (CDM) with Atmospheric Benefits for a Post-2012 Climate Regime*'. Freiburg, Germany: Oeko Institut. Available at www.oeko.de/oekodoc/779/2008-227-en.pdf

Voss, J.-P. 2007. 'Innovation processes in governance: the development of "emissions trading" as a new policy instrument'. *Science and Public Policy* **34**: 329–343.

Weart, Spencer R. 2003. *The Discovery of Global Warming*. Cambridge, MA: Harvard University Press.

12

A staged sectoral approach for climate mitigation

MICHEL DEN ELZEN, ANDRIES HOF, JASPER VAN VLIET AND PAUL LUCAS

12.1 Introduction

A major question in negotiations on international climate mitigation is how to allocate future greenhouse gas emission reduction targets among countries. For this reason, many proposals for allocating emission reductions among countries have been developed (Aldy *et al.* 2003; Bodansky 2004; Kameyama 2004; Torvanger and Godal 2004; Blok *et al.* 2005; Gupta *et al.* 2007; Hof *et al.*, this volume, Chapter 4). Most of these proposals are based on one or more equity principles. According to these principles, emission reductions are allocated on the basis of current emissions (sovereignty principle), population (egalitarian principle), gross domestic product (GDP) (ability to pay principle) or their share of responsibility for climate change (polluter-pays principle) (Rose *et al.* 1998; Hof *et al.*, this volume, Chapter 4). Another type of proposal takes specific national circumstances better into account by basing emission allocations on sectoral targets. This could potentially help improve the involvement of private actors, since targets are set for market-based sectors instead of for the national government.

The basic idea of this sectoral target approach is that sectors need to improve their efficiency to the same international level over time. The advantages are equal treatment of international competitive sectors in all countries, detailed consideration of mitigation potential and increased technological transfer. However, it also involves some disadvantages, one of the most important being the need for detailed information about efficiencies and emissions for a large number of subsectors for all countries (Baron *et al.* 2007). Höhne *et al.* (2008) conclude that such detailed information will not be available in the short term.

Therefore, this chapter focuses on a sectoral approach with only a few major sectors. This approach, called Triptych, has originally been designed to combine the reductions that can be achieved in individual sectors to produce an economy-wide target for a country. This chapter, however, will also focus on specific sectoral targets

Global Climate Governance Beyond 2012: Architecture, Agency and Adaptation, eds. F. Biermann, P. Pattberg and F. Zelli. Published by Cambridge University Press.

resulting from Triptych. The methodology was originally developed at the University of Utrecht and has been used for supporting decision-making when differentiating the European Union's internal Kyoto target among its member states both before and after the third Conference of the Parties in 1997 in Kyoto (Blok *et al.* 1997; Phylipsen *et al.* 1998; Ringius *et al.* 1998).

One factor influencing the future emission allowances that result from the Triptych approach are sectoral historical emissions data from 1990 until present. These data determine the initial and projected baseline emissions for the Triptych calculations, on which the allowances are (partly) based. The political feasibility of implementing Triptych or other more sophisticated sectoral approaches strongly depends on the consistency in the data. This chapter compares the emission allowances under the Triptych approach resulting from two main datasets used before in model studies of Triptych. (1) Most studies (Blok *et al.* 1997; Phylipsen *et al.* 1998; Groenenberg *et al.* 2004; den Elzen *et al.* 2008) use international carbon dioxide emissions datasets of the International Energy Agency (2005), combined with the non-carbon dioxide emissions dataset of EDGAR (Olivier *et al.* 2005). (2) Höhne *et al.* (2005), however, use emissions datasets of the national submissions to the climate convention for all greenhouse gases, and where not otherwise available, carbon dioxide emissions from International Energy Agency and non-carbon dioxide from the US Environmental Protection Agency (EPA 2006).

Our main conclusion is that the two datasets result in substantial differences in allowances for a number of sectors and countries, even for such a very simple sectoral approach. For more sophisticated sectoral approaches this data problem will be even more pronounced. Increasing the involvement of private sectors in emission reductions by allocating emissions allowances to sectors instead of whole economies might thus be difficult to implement due to this inconsistency of sectoral emissions data. (Note that the current analysis does not focus on explaining differences in emissions between the datasets. See here Reis *et al.* (2008) for an analysis of different methodologies used on a European level, and Olivier and Van Aardenne (2007) for a comparison of the two datasets.)

12.2 Methodology

The country model Framework Assess International Regimes for differentiations of commitments (FAIR 2.2) is used here for the analysis of the Triptych approach (available at www.mnp.nl\fair). FAIR is a policy-decision support tool for analysing regional emission allowances and abatement costs for various regimes for future commitments. The Triptych approach assigns emission allowances according to common rules using country-specific sector and technology information. These common rules allow for growth in economic activities (more for developing countries and less for industrialized countries) and require an improvement in energy efficiency or emission intensity over time. The Triptych approach attempts to incorporate some of the major principles of international climate policy as set out in article 3 of the climate convention (see Sections 12.2.1 and 12.2.2): first, the notion of equity and the principle of 'common but differentiated responsibilities', that is, that industrialized countries should take the lead in combating climate change (see Section 12.2.1); second, policies and measures to deal with climate change should take into account all economic sectors; third, harmonization of climate change mitigation measures and sustainable development.

12.2.1 Common but differentiated responsibilities

First of all, the principle of 'common but differentiated responsibilities' is implemented by convergence trajectories based on a 'common convergence', but 'differentiated' in time in that developing countries have the same obligation as industrialized countries to reduce emissions, but this obligation is delayed and conditional on the actions of the industrialized countries (Höhne *et al.* 2006). More specifically for the Triptych approach, the per capita domestic emissions of industrialized countries or efficiency indicators in the industry and electricity sector converge linearly from their levels in 2010 to a common level in 2030. Per capita emissions of individual developing countries or efficiency indicators also converge to the same level within the same time period, but starting (and therefore converging) later. The differentiation of all countries into three convergence groups draws on the criteria of responsibility, capability and the potential to mitigate, following the South–North dialogue proposal (Ott *et al.* 2004). Convergence groups are shown in Figure 12.1.

Newly industrialized countries start to converge at the same year as industrialized countries. Advanced developing countries start to converge five years after the industrialized countries and newly industrialized countries start to converge, while the other developing countries start to converge ten years after industrialized countries and newly industrialized countries start to converge.

Until the start of the convergence is reached, advanced developing countries commit to no-lose targets or policies and measures that focus on their sustainable development objectives, in line with the climate convention principle of harmonization of climate change measures and sustainable development. These so-called 'sustainable development policies and measures' are government actions that have both development and greenhouse gas emissions benefits (Winkler, this volume, Chapter 7; also Winkler *et al.* 2002; Baumert and Winkler 2005). Emissions are reduced only in relative terms, as energy use and emissions in developing countries will need to grow to meet the requirement of sustainable

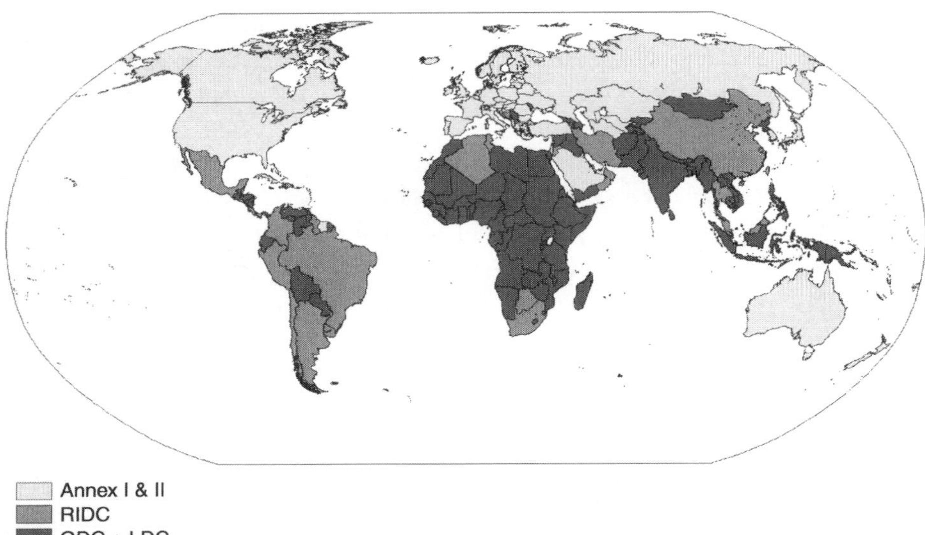

Figure 12.1 Differentiation of convergence groups. RIDC, rapidly industrializing or advanced developing countries; ODC, other developing countries; LDC, least developed countries. *Source:* den Elzen *et al.* 2008.

economic development. No-lose targets set a limit for total emissions in a target year. If this target is exceeded (real emissions are below the target), the additional emission allowances can be sold on the carbon market. If the target is not met (real emissions are above the target), no additional rights have to be bought. As such, no-lose targets could be an incentive for developing countries to participate in the system, but it is an option that requires better abilities to quantify emissions and its reductions. It is assumed that the emission reductions of these sustainable development policies and no-lose targets could reduce emissions by roughly 10 per cent in 2020 compared to their baseline emissions, and stay at a constant level afterwards (Höhne and Moltmann 2007).

12.2.2 Sectoral targets

The climate convention principle of taking into account all economic sectors is incorporated in Triptych by distinguishing the original three energy-using sectors (industry, domestic, power generation) and non-combustion emissions from fossil-fuel production, agriculture and waste. Note that land-use-related carbon dioxide emissions are not included in Triptych, as data on land-use change and forestry are difficult to estimate and to assess (Grassl *et al.* 2003). Table 12.1 summarizes all sectoral targets and convergence trajectories. First, each sector in every country has

Table 12.1 *Sectoral targets applied for the analysis*

Sector	Type of target	Convergence level[a]	Target after convergence[b]
Industry	Energy Efficiency Indicator (EEI)	1.0	0.5
Domestic	Per capita emissions in tCO_2/year	1.25	
	Per capita reduction rate (%/year)		2
Power generation	Emissions from coal (gCO_2/kWh)	600	400
	Emissions from oil (gCO_2/kWh)	450	300
	Emissions from gas (gCO_2/kWh)	300	250
	Reduction in the share of coal and oil (%)	60	90
	Annual efficiency improvement in consumption (%)	2	
Fossil fuel production	Reduction compared to baseline emissions (%)	90	95
Agriculture	Reduction compared to baseline emissions:		
	All developing countries (%)	30	50
	Industrialized and newly industrialized countries (%)	40	50
Waste	Reduction below base-year per capita emissions (%)	90	

[a] Convergence year is 2030 for industrialized countries and newly industrialized countries, 2035 for advanced developing countries and 2040 for other developing countries.
[b] Targets are for 20 years after convergence year.
Source: den Elzen *et al.* (2008).

to converge to the present level of the best-performing country for that sector (such as carbon dioxide emissions per kilowatt hour per fuel type). These convergence trajectories start in 2010 and should be completed by 2030 for industrialized countries and newly industrialized countries. For the advanced developing countries the convergence trajectory starts five years later and for the other developing countries ten years later. After these first convergence trajectories, a common convergence to the level possible by implementing best available technologies is implemented. The convergence year for the second trajectory is 2050 for industrialized countries and newly industrialized countries and again five and ten years later for advanced developing countries and other developing countries, respectively.

The industry sector includes manufacturing and construction. The targets for the industry sector consist of energy efficiency improvements, measured by the energy efficiency index (Phylipsen *et al.* 1998). This index is defined as the ratio between the specific energy consumption (energy consumption per tonne of product) for each region divided by the theoretical specific energy consumption using current best practices or best available technologies (Phylipsen *et al.* 1998). Hence, an energy efficiency index of 1 in a region means that best practices are currently in place, while an energy efficiency index of 1.05, for instance, means that 5 per cent of energy could be saved in the given sector by implementing the best practice. In this way, the energy efficiency index indicates by how much the energy consumption can be improved by using current best practices or best available technologies. A worldwide differentiated convergence in the energy efficiency index has to occur. Subsequently, the energy efficiency index has to improve further. Table 12.1 shows the convergence level and subsequent targets for the energy efficiency index in the industry sector. Industries in countries with current low energy efficiencies (such as South Africa, Russia, China, New Zealand and several Eastern European countries) will face the biggest challenge to reach these targets.

The domestic sector includes fossil fuel combustion from the residential, commercial, agriculture and (domestic) transport sectors and F-gases emissions (HFCs, PFCs and SF_6) from a range of sources (refrigeration, air conditioning equipment, fire extinguishers and aerosol applications). The target indicator is greenhouse gas emissions per capita, which will have to converge (differentiated, as with all sectors) to a common level and decrease subsequently according to Table 12.1, based on Groenenberg *et al.* (2004). Note that as the United States currently have the highest emissions per capita, they will have to make the greatest effort of all countries in this sector.

In the power generation sector, carbon dioxide emissions differ greatly from country to country due to large differences in the shares of nuclear and renewables generation capacity as well as in the fuel mix in fossil-fuel-fired power plants. Furthermore, the potential for renewable energy is different for each country. The fuel mix in power generation is an important national characteristic to take into account in the differentiation of commitments. This is done in Triptych by distinguishing three types of targets: convergence and reduction of emissions per kWh per fuel; decrease in the share of coal and oil in the fuel mix; and annual improvements in the efficiency of electric consumption.

Emissions from fossil-fuel production are treated as a separate sector, as emissions from this sector can be reduced drastically (Delhotal *et al.* 2006). Therefore, strong emission reduction targets are set for this sector, as depicted in Table 12.1. Targets for the non-carbon dioxide emissions in the agricultural sector are also simply expressed as emission reductions compared to the baseline. Two groups of countries are distinguished: industrialized countries and newly industrialized countries have stricter convergence targets than developing countries (based on studies by Graus *et al.* 2004 and the United States Environmental Protection Agency, EPA

2006). Targets for the waste sector are expressed as per capita reduction levels below the base year of 2010. These emission reduction targets are substantial, because many reduction options exist, such as landfills covering or capture and usage of methane (Höhne 2005; Lucas *et al.* 2007).

12.2.3 Data and downscaling

One important factor influencing the emission reduction target in the Triptych approach is the historical emissions dataset, as it highly influences the initial 2010 emissions (starting point of Triptych calculations) and the baseline emission projection, particularly in the short term. Both determine the emission level of the starting year of convergence (2010 for industrialized countries and newly industrialized countries, 2015 for advanced developing countries and 2020 for 'other developing countries'), which influences the necessary emission reductions in the domestic, industry and power sector. The baseline emissions directly influence the future emission for the fossil fuel production, waste and agricultural sector, as these are calculated as a fraction of the baseline. Two different historical (1990–2005) emissions datasets are used here.

The first is the dataset that has most often been used in earlier Triptych analyses (except those by Höhne): historical sectoral carbon dioxide emissions from fossil fuel combustion are based on data by the International Energy Agency (2005), while carbon dioxide emissions other than from fossil fuel combustion and non-carbon dioxide greenhouse gas emissions are based on the EDGAR database (Olivier *et al.* 2005) (hereafter this dataset is referred to as IEA/EDGAR).

The other dataset is based on Höhne *et al.* (2007). The historical emissions in this dataset are based on national emission inventories, submitted to the climate convention (hereafter, this dataset is referred to as the climate convention dataset). Where national inventories are not available, other sources such as the International Energy Agency (carbon dioxide) and United States Environmental Protection Agency (EPA) (non-carbon dioxide) are used.

More specifically, historical emissions are gathered from the following sources and in the following hierarchy:

(1) National submissions to the climate convention as collected by the climate convention secretariat and published in the greenhouse gas emission database available at their website.[1] For industrialized countries, the latest available year is usually 2004. Most developing countries report only or until 1994.

(2) Carbon dioxide emissions from fuel combustion as published by the International Energy Agency (usually covering 1970 to 2004). This dataset was

[1] Data and sources available from their website: http://unfccc.int/ghg_data/items/3800.php.

supplemented by process emissions from cement production from Marland *et al.* (2003) to cover all industrial carbon dioxide emissions.

(3) Emissions from CH_4 and N_2O as estimated by the EPA (2006), covering the years 1990 to 2005.

We estimated emissions up to 2010 for both datasets by assuming that industrialized countries, excluding the United States, implement their Kyoto targets by 2010. We also assumed that emissions are reduced equally in all sectors. The years from the last available year up to 2010 are linearly interpolated. For developing countries and the United States, emissions up to 2010 are taken from the ADAM baseline (see next section), as are future baseline emissions, income levels and population.

To project baseline emissions from 26 regions to the country level, we applied the downscaling methodology of van Vuuren *et al.* (2007) to both datasets.[2] First, for downscaling of the population, the relative sizes of countries in the long-range population projections of the United Nations (2004) are used. Second, for downscaling per capita income and partially for emission intensity, convergence to the regional average is assumed, while ensuring that the total of the countries in the region equals the baseline of that region. Third, for the sectors industry, domestic, power generation and fossil-fuel production, relative changes in population, per capita GDP and emission intensity compared to the base year 2000 are used to determine the future sectoral emissions per country. Fourth, for the agriculture and waste sectors, simple linear downscaling is used (all countries within the region follow the same regional emission trend), as these sectors are only loosely linked to consumption and much more closely related to production levels.

12.2.4 Baseline emissions

The ADAM baseline (van Vuuren *et al.* 2010) used in this analysis is a high economic growth scenario based primarily on optimistic growth assumptions in China and India. Outside these regions, growth assumptions are considered to be comparable to other more medium economic growth projections. The economic projections also show that in per capita terms, the current industrialized regions remain the wealthier regions in the world. At the same time, however, in terms of total economic activity, some regions become more important, including China, India, Latin America and the rest of Asia. The population projection used is the UN medium scenario.

The outcomes in terms of energy are broadly similar to those of the World Energy, Technology and Climate Policy Outlook reference scenario. Although the share of oil in total energy demand decreases from about a third now to a quarter in 2050,

[2] Höhne *et al.* (2007) apply a linear downscaling methodology, using the regional trend for the IMAGE 2.2 SRES emission scenarios and the dataset of Höhne. This methodology was criticized in van Vuuren *et al.* (2007), as it may lead to unrealistic baseline emissions for some countries.

absolute oil production is not expected to decrease. A decrease in the production of oil from conventional sources is offset by an increased production from unconventional sources. The oil price shows a more or less constant price level at 2005 levels (about USD 55/barrel) over the period 2005–2050. The share of non-fossil energy sources in the total energy demand slowly increases from 14 per cent now to 18 per cent in 2050. The share of coal in total energy demand remains fairly constant at a third of total demand, while the share of natural gas increases from 20 per cent now to a quarter in 2050.

12.3 Analysis

12.3.1 Global emission reductions and costs

Figure 12.2 shows the global emissions and emission reductions of the different sectors between 2000 and 2050, given the Triptych settings of Table 12.1. Total emission reductions compared to baseline emissions in 2050 are slightly higher for the IEA/EDGAR dataset (49 versus 46 gigatonne carbon dioxide-equivalent), due to the higher projected baseline emissions for the IEA/EDGAR dataset. Emission reductions compared to the 2005 level are very similar for both datasets. For both datasets we see that the power generation sector contributes half of these total emission reductions. The domestic and industry sectors are the next important contributors of emission reductions; together they are responsible for 30 per cent (IEA/EDGAR) to 33 per cent (Climate Convention) of the total emission reductions in 2050. According to the Climate Convention dataset, domestic emissions are slightly higher, resulting in the relatively higher emission reductions for this sector. Despite their small contribution to total emissions, fossil fuel production still contributes 10 per cent to total emission reductions in 2050 for both datasets,

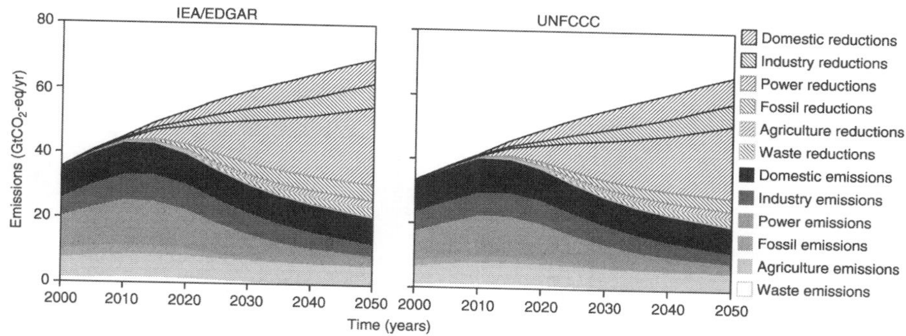

Figure 12.2 Global emission reduction targets by sector, based on datasets of IEA/EDGAR and the Climate Convention dataset.

because of the large potential for reducing emissions in this sector. The share of the waste and agriculture sectors in global emission reduction is more limited.

Only the emission reduction target of the fossil fuel production sector differs substantially between the two datasets. This is due to lower emissions in 2005 for this sector according to the Climate Convention dataset as compared to the IEA/EDGAR dataset. Therefore, absolute emission reduction targets are also lower for this sector. For the other sectors, the absolute global emission reduction targets compared to baseline differ less than 10 per cent between the two datasets.

Total emissions in 2050 are 32 per cent (IEA/EDGAR) or 34 per cent (Climate Convention) below the level of 1990. This is in the range of stabilizing greenhouse gas concentrations at 450 carbon dioxide-equivalent, for which emissions have to be reduced by 23 per cent to 45 per cent below 1990 levels (den Elzen *et al.* 2007).

For both datasets, emission reductions in 2050 correspond with a 70 per cent emission reduction compared to the baseline. This is an ambitious target, but can nevertheless be achieved at relatively modest costs as share of GDP, as shown by Figure 12.3. Annual global mitigation costs increase to 1.8 per cent of global GDP in 2035, after which costs decline again. Note that full carbon trading between countries and sectors is assumed to efficiently reduce emissions globally. In practice, this means that sectors or countries with high internal domestic mitigation costs can buy emission credits abroad to achieve their targets.

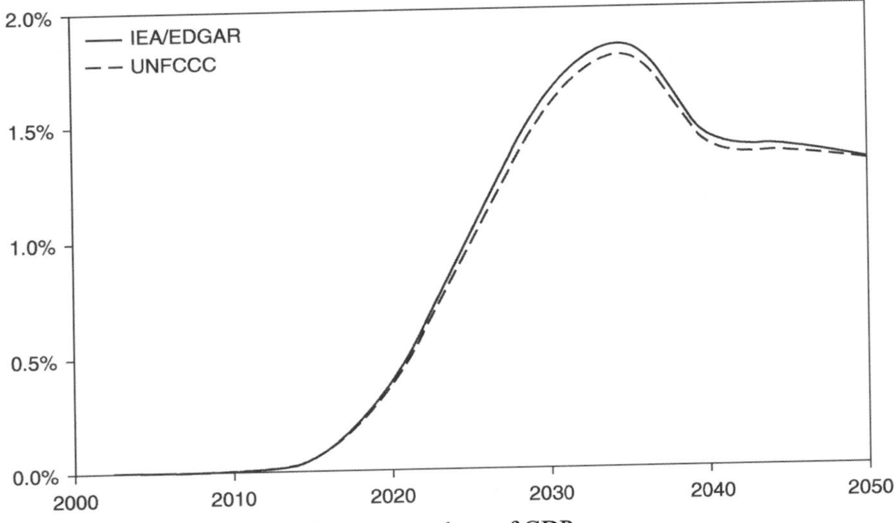

Figure 12.3 Global mitigation costs as share of GDP.

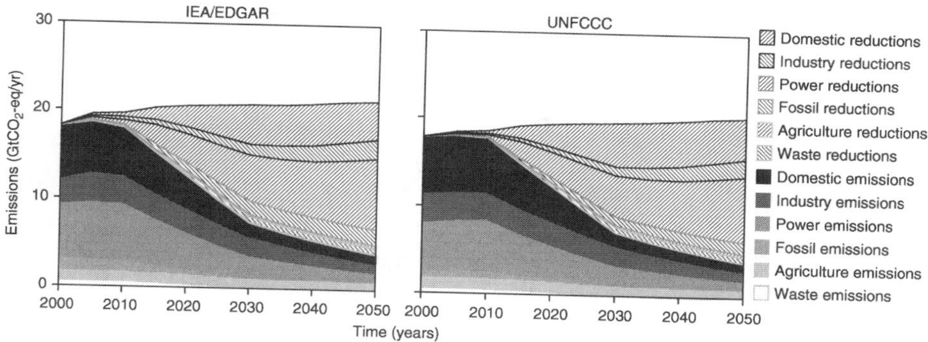

Figure 12.4 Emission reduction targets for all industrialized countries based on datasets from IEA/EDGAR and the Climate Convention dataset.

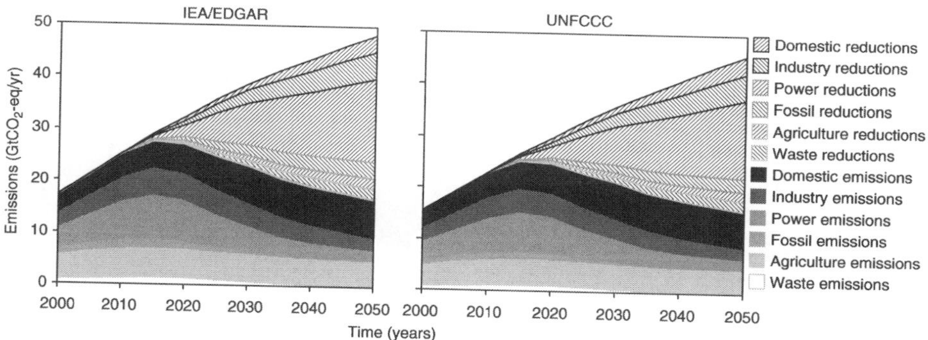

Figure 12.5 Emission reduction targets for all developing countries based on datasets from IEA/EDGAR and the Climate Convention dataset.

12.3.2 Regional emission targets and mitigation costs

The differentiated responsibilities lead to stronger average emission reductions for industrialized countries than for developing countries for both datasets, as shown by Figures 12.4 and 12.5. Total emissions of industrialized countries have to decrease strongly from 2010 onwards, whereas emissions from developing countries have to peak in 2015 and should decrease more gradually afterwards. For developing countries this still implies a doubling in emissions in 2020 compared to their 1990 levels. From 2025 onwards, more than half of the global emission reduction should take place in developing countries.

For both industrialized and developing countries, the strongest emission reductions have to occur in the power generation sector. The main difference in the distribution of emission reductions between industrialized and developing countries is the domestic sector. For industrialized countries, the domestic sector accounts for a quarter of total emission reductions by 2050, whereas for developing countries this is only 10 per cent. This is due to the much lower per capita emissions in developing countries.

The distribution of emission reductions across sectors for developing countries hardly differs between the two different datasets. This is in line with expectations, because for most developing countries recent carbon dioxide emissions data is not available from the national submissions to the climate convention. For these countries, the carbon dioxide data in both the Climate Convention dataset and the IEA/EDGAR dataset are taken from the International Energy Agency. For industrialized countries, the main difference is that for the Climate Convention dataset, higher reductions are necessary for the domestic sector, while the industry and fossil fuel production sector have more relaxed targets. This is explained by higher initial emissions for the domestic sector and lower initial emissions for the other sectors according to the Climate Convention dataset.

Figure 12.6 shows the total emission allowances for all sectors for more disaggregated regions in 2020 and 2050 (see Tables 12.2 and 12.3 for more detail). Note that the actual emission levels may be different from the emission allowances, due to the flexible mechanisms under the Kyoto Protocol (that is, emissions trading, joint implementation and the Clean Development Mechanism).

The largest differences between the datasets in emission targets for 2020 compared to baseline exist in industrialized countries. Overall, the climate convention dataset gives somewhat less stringent emission reduction targets for industrialized countries because of lower reported emissions in 2005. Especially for Russia, Japan and the new EU Member States, the IEA/EDGAR dataset leads to substantially more stringent reduction targets than the Climate Convention dataset. These differences are 21 percentage points for Russia, 11 percentage points for Japan and 8 percentage points for the new EU Member States. Different baseline projections are the main reason for these inconsistencies. In fact, the difference in targets compared to the 2005 level is very small for Japan and the new EU Member States. Section 12.3.3 provides more sectoral detail for Russia and Japan.

For emission targets in 2020 compared to the 2005 level, the largest differences are in developing countries, with the exception of Russia. The difference in South Asia, South Africa and the former Soviet Union is more than 4 percentage points, with the IEA/EDGAR dataset leading to more stringent emission targets. Although the differences for Latin America as a whole are not so large, some individual countries experience large differences in targets between the datasets (see Table 12.2). One example is Brazil, for which the IEA/EDGAR dataset leads to more stringent targets. Therefore, Section 12.3.3 also examines Brazil and South Africa in more detail.

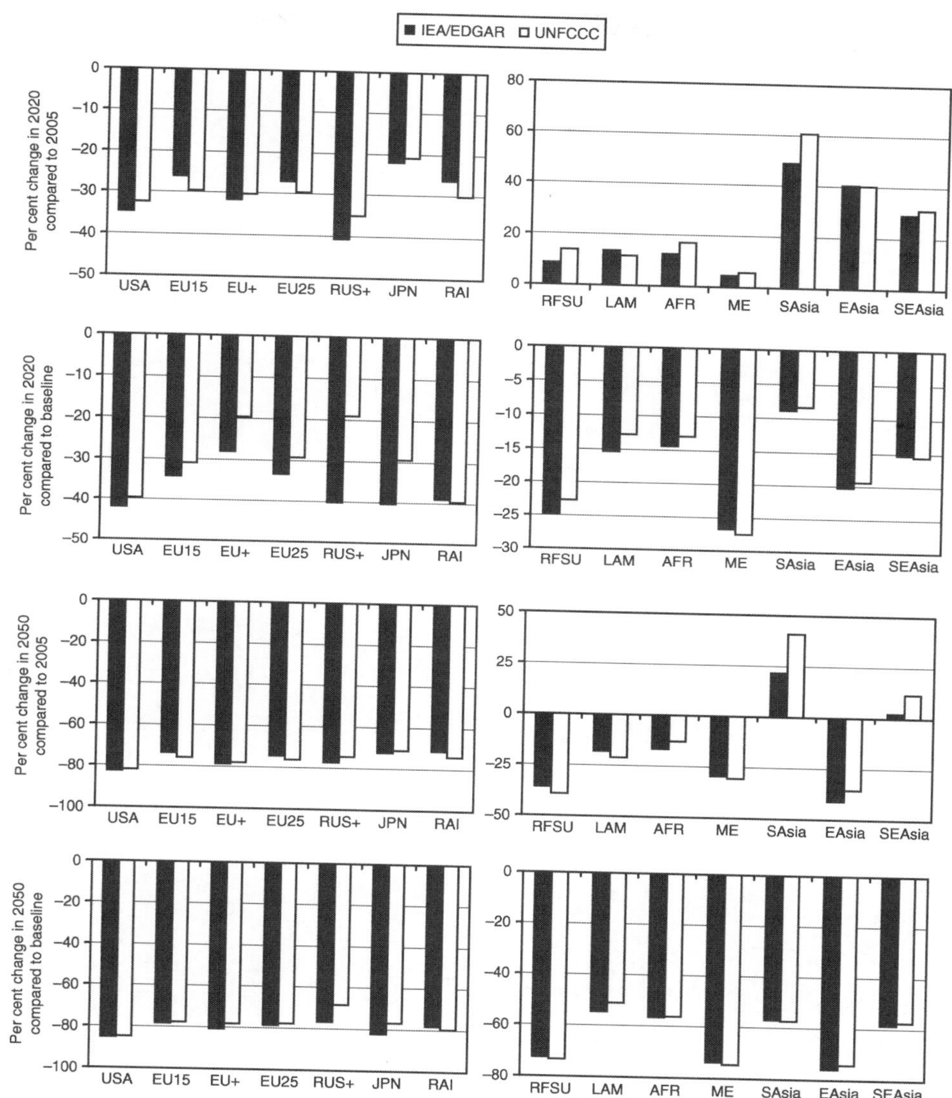

Figure 12.6 Change in emission allowances compared to 2005 and baseline levels in 2020 and 2050 for several regions and countries. EU15, old EU Member States; EU+, new EU Member States; RUS+, Russia and former Soviet Union states part of Annex I; JPN, Japan; RAI, Rest of Annex I; RFSU, Rest of former Soviet Union; LAM, Latin America; AFR, Africa; ME, Middle East; SAsia, South Asia; EAsia, East Asia; SEAsia, South East Asia.

Table 12.2 *Emission allowances compared to 2005 levels and baseline in 2020 under the Triptych settings*

	2005 level		2020 baseline		2020 target			
REGIONS	IEA/EDGAR $MtCo_2$-eq	UNFCC	IEA/EDGAR relative to 2005 (%)	UNFCCC	IEA/EDGAR relative to 2005 (%)	UNFCCC	IEA/EDGAR relative to baseline(%)	UNFCCC
Industrialized countries	19 651	18 405	+6	+6	−36	−33	−39	−36
Developing countries	21 351	19 158	+54	+56	+27	+30	−17	−17
World	41 002	37 563	+31	+31	−3	−1	−26	−24
USA	7 918	7 436	+12	+12	−35	−32	−42	−40
EU-15	4 319	4 279	+4	+2	−31	−30	−33	−33
EU-10	781	764	+1	−1	−33	−30	−36	−28
Other Western Europe	120	119	+2	−5	−34	−30	−36	−28
Russia	2 611	2 161	−2	−3	−43	−35	−42	−33
Other Annex I Eastern Europe	863	802	−11	−11	−49	−34	−42	−26
Japan	1 606	1 455	+5	+6	−34	−21	−37	−32
Oceania and Canada	1 445	1 404	+11	+12	−35	−30	−41	−44
Turkey	358	345	+57	+63	+12	+20	−29	−27
Eastern Europe and Central Asia	875	898	+31	+35	−0	+8	−24	−20
Argentina	330	327	+23	+21	+7	+7	−13	−11
Brazil	1 132	1 022	+33	+32	+10	+16	−18	−13
Mexico	610	522	+25	+28	+4	+6	−17	−17
Venezuela	241	212	+36	+31	+14	+22	−16	−7
Other Latin America	816	1 206	+36	+26	+19	+10	−12	−13
Egypt	212	187	+53	+54	+36	+41	−11	−8
South Africa	544	461	+26	+21	−27	−19	−42	−33
Nigeria	223	387	+45	+40	+40	+32	−4	−6
Other North Africa	280	300	+42	+43	+26	+21	−12	−16
Other Africa	1 361	999	+25	+32	+16	+22	−7	−8

Saudi Arabia	452	374	+46	+56	−22	−11	−47	−43
United Arab Emirates	140	173	+10	+15	−21	−23	−28	−33
Other Middle East	1 082	1 059	+44	+48	+18	+17	−18	−21
China	6 728	5 879	+78	+78	+40	+41	−21	−20
India	2 503	1 891	+63	+75	+48	+60	−9	−9
Indonesia	686	532	+45	−57	+36	+44	−6	−8
South Korea	130	131	+100	+88	−7	−7	−54	−51
Malaysia	188	116	+73	+63	+35	+12	−22	−31
Phillippines	163	138	+82	+95	+41	+51	−23	−23
Singapore	71	53	+33	+36	−18	+8	−38	−21
Thailand	316	301	+51	+52	+29	+31	−15	−14
Other Asia	2 006	1 720	+45	+48	+33	+38	−9	−7

Table 12.3 *Emission allowances compared to 2005 levels and baseline in 2050 under the Triptych settings*

	Baseline				Target			
REGIONS	IEA/EDGAR $MtCO_2$-eq	UNFCCC	IEA/EDGAR relative to 2005 (%)	UNFCCC	IEA/EDGAR relative to 2005 (%)	UNFCCC	IEA/EDGAR relative to baseline (%)	UNFCCC
Industrialized countries	21 672	20 456	+10	+11	−79	−78	−81	−80
Developing countries	49 707	45 892	+133	+140	−21	−16	−66	−65
World	71 379	66 347	+74	+77	−49	−47	−71	−70
USA	9 245	8 861	+17	+19	−83	−82	−85	−85
EU-15	4 696	4 580	+9	+7	−75	−76	−77	−78
EU-10	918	872	+17	+14	−79	−78	−82	−81
Other Western Europe	125	104	+4	−12	−73	−74	−75	−70
Russia	2 383	1 918	−9	−11	−79	−75	−77	−72
Other Annex I Eastern Europe	838	799	−3	−0	−77	−73	−79	−77
Japan	1 805	1 677	+12	+15	−76	−73	−79	−77
Oceania and Canada	1 681	1 665	+16	+19	−74	−77	−78	−80
Turkey	949	942	+165	+173	−22	−20	−71	−71
Eastern Europe and Central Asia	1 760	1 767	+101	+97	−46	−49	−73	−74
Argentina	498	464	+51	+42	−26	−31	−51	−51
Brazil	2 065	1 779	+82	+74	−19	−13	−56	−50
Mexico	952	828	+56	+59	−34	−32	−58	−57
Venezuela	451	360	+87	+70	−30	−30	−63	−59
Other Latin America	1 556	1 861	+91	+54	−4	−19	−50	−48
Egypt	529	471	+149	+151	+26	+39	−50	−45
South Africa	979	785	+80	+70	−76	−72	−87	−83
Nigeria	460	696	+106	+80	−18	−39	−60	−66
Other North Africa	553	580	+97	+93	−4	−11	−51	−54

Other Africa	2 469	2 106	+81	+111	−3	+15	−47	−46
Saudi Arabia	1 229	1 115	+172	+198	−66	−63	−88	−88
United Arab Emirates	151	185	+8	+7	−59	−73	−62	−74
Other Middle East	3 054	3 076	+182	+190	−12	−12	−69	−70
China	16 847	15 275	+150	+160	−41	−37	−77	−76
India	6 864	6 093	+174	+222	+15	+32	−58	−59
Indonesia	1 574	1 435	+130	+170	+12	+28	−51	−53
South Korea	426	365	+228	+179	−61	−64	−88	−87
Malaysia	593	371	+216	+219	−21	−13	−75	−73
Phillippines	653	628	+301	+356	+45	+64	−64	−64
Singapore	100	84	+43	+57	−60	−56	−72	−72
Thailand	775	749	+145	+149	−8	−4	−62	−61
Other Asia	4 295	3 943	+114	+129	+2	+18	−52	−49

By 2050, the total emission reduction target of industrialized countries and newly industrialized countries is about 80 per cent compared to their baseline emissions. The differences in targets between the datasets have diminished by now. Developing countries have to reduce their emissions substantially as well by 2050: even Africa has to reduce emissions by more than 50 per cent compared to baseline by 2050. Still, there is room for absolute growth in emissions especially in the short term, as shown by the emissions allowances for 2020 compared to the 2005 level.

Figure 12.6 also shows that the emission reduction targets in 2020 compared to baseline levels are much higher compared to the 2005 level for Japan and the United States. This is due to an increasing usage of coal in the electricity production in the United States and Japan, causing their baseline emissions to grow faster compared to for example the European Union. Japan is projected to use more coal because of the relatively high gas prices; for the United States, it is a combination of low coal prices combined with projected higher gas prices. Because of easier access to large gas fields in the European Union, gas prices are lower here, causing a decline in the usage of coal in power generation. Therefore, emissions in Europe do not increase as much as in Japan and the United States.

12.3.3 Discussion of country results

This section provides more detailed results for several countries for which the emission reduction targets differ substantially between the two datasets, namely Brazil, Russia, Japan and South Africa.

Brazil

Brazil has a particularly unusual emission profile with relatively low emissions in power generation due to the high availability of hydropower and relatively high emissions from agriculture (Figure 12.7). Therefore, the relative emission reductions for power generation are much lower than the world average. Since many emission reduction options in the agriculture sector are not readily available due to implementation barriers, the total emission reduction target for Brazil is lower compared to most other advanced developing countries. According to the Climate Convention dataset, emissions in 2050 have to be 13 per cent below the 2005 level. According to the IEA/EDGAR dataset, emissions in 2050 have to be 17 per cent below the 2005 level. This difference is caused by less stringent targets for the waste, fossil fuel production, power generation and industry sectors according to the Climate Convention dataset. For all these sectors, IEA/EDGAR reports both higher 2005 emissions and a higher growth in emissions than the Climate Convention dataset. For the agricultural and domestic sector, emissions are similar for the two datasets.

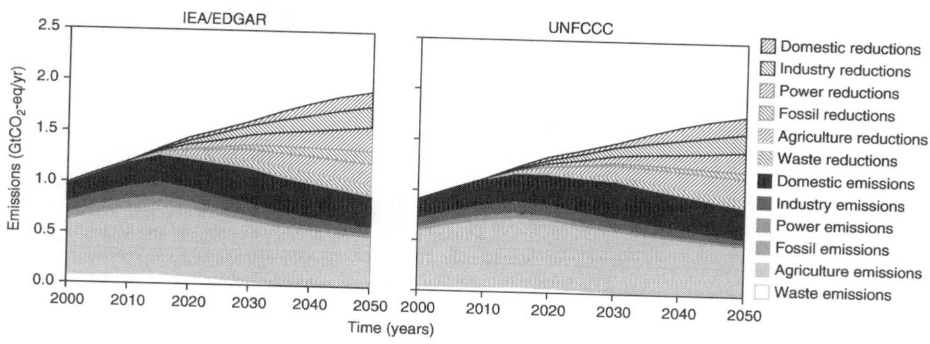

Figure 12.7 Emission reduction targets by sector for Brazil, based on datasets of the IEA/EDGAR and the Climate Convention dataset.

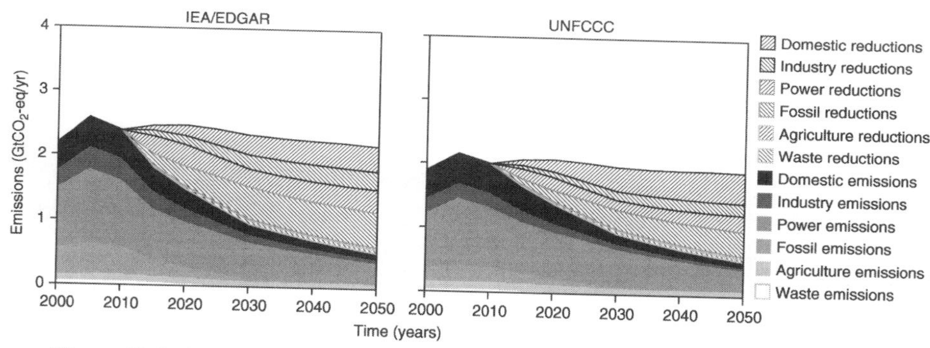

Figure 12.8 Emission reduction targets by sector for Russia, based on datasets of the IEA/EDGAR and the Climate Convention dataset.

Russia

By far the largest share of Russia's emissions comes from the power generation sector, which is based mainly on gas and coal (Figure 12.8). Similar to other industrialized countries and high-emission countries, the Triptych approach leads to a significant decrease in the share of fossil sources in the electricity mix by 2050 (see Table 12.1). While the share of gas can stay constant, the use of oil and coal will have to decrease sharply. Emissions from the power generation sector have to decrease to 300 million tonnes carbon dioxide-equivalent in 2050 for both datasets. Since the IEA/EDGAR dataset leads to higher projected baseline emissions in 2050 for this sector, the emission reduction target compared to baseline is higher for the IEA/EDGAR dataset.

Emissions from fossil fuel production are particularly high as well, with one reason being leaking gas pipelines. For 2005, emissions for this sector as reported by IEA/EDGAR are twice the level of emissions as reported by the climate convention (which is, in this case, the national submission). This results in much higher baseline emissions and consequently in more stringent emission reduction targets according to the IEA/EDGAR dataset, compared both to the 2005 level and to baseline emissions.

Reported emissions from the domestic sector in 2005 and emission reductions compared to 2005 for this sector are the same for both datasets. Their baseline projections differ, however, due to different emission trends in the period 1990–2005. In this case, the Climate Convention dataset leads to higher baseline emissions and therefore to more stringent emission reductions compared to baseline.

Differences in emissions from the industry sector between the two datasets are also substantial. The IEA/EDGAR dataset reports much higher emissions in 2005, resulting in more stringent emission reduction targets compared to baseline according to this dataset. Note, however, that emission reduction targets compared to 2005 are similar between the datasets, because the projected growth in baseline emissions is similar.

Finally, emissions in 2005 and baseline emissions for the waste and agricultural sectors are similar for both datasets. Overall, the higher emission reduction targets for the power generation, fossil fuel production and industry sector results in more stringent emission reduction targets for Russia according to the IEA/EDGAR dataset.

Japan

The IEA/EDGAR dataset reports similar emissions as the Climate Convention dataset in 2000, but much higher emissions in 2005. This is caused by different emission data for the industry sector: according to the Climate Convention dataset, industry emissions remain constant between 2000 and 2005, while IEA/EDGAR reports a substantial increase in emissions in this period (Figure 12.9). For the fossil fuel production and power generation sector, the IEA/EDGAR dataset reports much higher emissions both in 2005 and in the baseline. The overall result is that baseline emissions increase more sharply according to the IEA/EDGAR dataset. This results in higher emission reduction targets according to this dataset.

South Africa

South Africa's emissions are dominated by the coal-intensive and fast-growing power generation sector, followed by the coal production, industrial and domestic sectors (Figure 12.10). Consequently, the Triptych approach requires significant reductions in the power generation sector, not only in terms of slowing growth but also of reducing emissions in absolute terms. The reduction targets for this sector are almost equal for the two datasets. The same is true for all other sectors with the

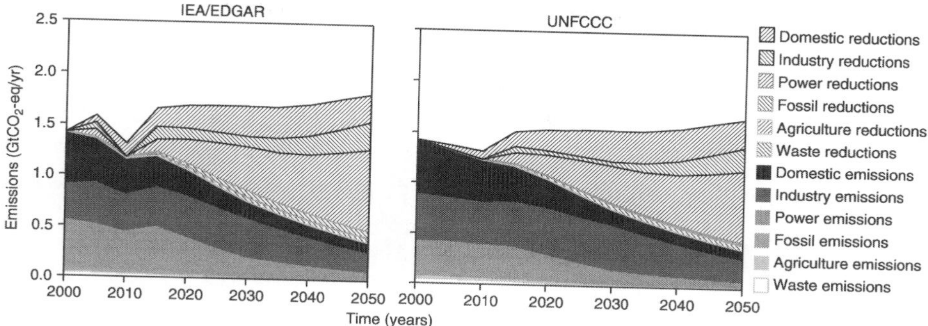

Figure 12.9 Emission reduction targets by sector for Japan, based on datasets of the IEA/EDGAR and the Climate Convention dataset.

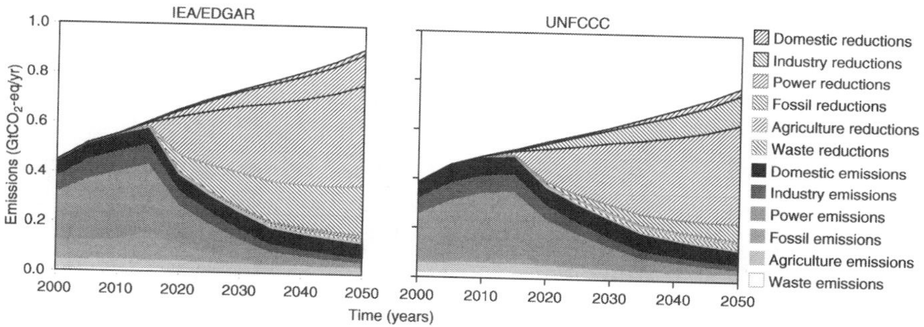

Figure 12.10 Emission reduction targets by sector for South Africa, based on datasets of the IEA/EDGAR and the Climate Convention dataset.

exception of the fossil fuel production sector. According to the IEA/EDGAR dataset, the fossil fuel production sector is the second most important contributor in emissions in 2005. According to the Climate Convention dataset, however, emissions from the fossil fuel production sector contribute the least to total emissions in 2005. The much lower emissions according to the Climate Convention dataset in both 2005 and the baseline are therefore solely the consequence of lower emissions in the fossil fuel production sector. This leads to substantially higher emission reduction targets according to the IEA/EDGAR dataset.

12.4 Conclusions and policy recommendations

The Triptych approach differentiates emission reduction requirements under a future international climate agreement based on technological considerations at the sector level. The advantage of decomposing targets according to sectors is that it enables non-state actor involvement in emission reduction targets more directly. Its framework allows for discussions on sectors that compete worldwide. The disadvantage is that it requires projections of sectoral growth rates for each country. These projected growth rates are based on historical sectoral emissions data, for which different datasets are available. We compared the emission allowances resulting from the Triptych approach for two of these datasets: IEA/EDGAR and submissions to the climate convention. We have used the FAIR policy support tool to analyse how large the differences in future emission allowances are between these two different historical emissions datasets. The main goal was to identify the robustness of the outcomes of a simple sectoral approach in reducing greenhouse gas emissions.

Globally, the two different datasets come to similar reduction targets. The Climate Convention dataset has slightly less stringent global targets due to their lower reported emissions in 2005. Total emissions have to reach the 2005 level in 2020, which is equal to 25 per cent below baseline emissions. The emission reduction target of all developing countries together are 17 per cent below baseline for both datasets. For all industrialized countries together, the IEA/EDGAR dataset comes to a slightly more stringent reduction target than the Climate Convention dataset: 39 per cent versus 36 per cent below baseline in 2020.

At country level, the analysis shows that the emission reduction targets, both compared to the 2005 level and baseline, differ substantially for quite some countries between the two datasets. More specifically, Table 12.1 shows that for 18 of the 32 countries and regions considered, the difference in emission reduction targets in 2020 relative to 2005 is more than 5 percentage points. The differences in emission reduction targets in 2020 compared to baseline are less: the differences for only eight of the 32 countries and regions considered are more than 5 percentage points. In most of these cases, the IEA/EDGAR dataset leads to more stringent targets than the Climate Convention dataset.

Some country case studies show that these differences are caused by either a difference in emission levels in the starting year of convergence or by a difference in projections of baseline emissions. These projections are derived from the trend in emissions between 1990 and 2005.

Our main conclusion is that the differences in sectoral emission allowances for a large number of countries between the two datasets currently hinder involving non-state actors by implementation of a sectoral approach. These differences are particularly present in the short-term emission allowances. As the Triptych approach is relatively simple in that it only considers a very small number of sectors, this data problem will be even more important for the more sophisticated sectoral approaches proposed.

This conclusion highlights the need for reliable, uniform sectoral emissions registrations on country level in order to allocate emissions to sectors. The current discrepancies in sectoral targets resulting from the IEA/EDGAR emissions dataset and the national emission inventories submitted to the climate convention could represent a major obstacle to negotiations. How such a uniform emission registration should be structured (that is, the international institutional environment) is another question and beyond the scope of this chapter. It seems obvious, however, that an international institution should supervise the emission registration in a way that involves governments and the private sector.

References

Aldy, J. E., J. Ashton, R. Baron, D. Bodansky, S. Charnovitz, E. Diringer, T. C. Heller, J. Pershing, P. R. Shukla, L. Tubiana, F. Tudela and X. Wang 2003. *Beyond Kyoto: Advancing the International Effort against Climate Change*. Arlington, VA: Pew Center on Global Climate Change.

Baron, R., J. Reinaud, M. Genasci and C. Philibert 2007. *Sectoral Approaches to Greenhouse Gas Mitigation: Exploring Issues for Heavy Industry*. Paris: Organization for Economic Cooperation and Development/International Energy Agency.

Baumert, K. and H. Winkler 2005. 'Sustainable development: policies and measures and international climate agreements', in R. Bradley and K. Baumert (eds.), *Growing in the Greenhouse: Protecting the Climate by Putting Development First*. Washington, DC: World Resource Institute, pp. 15–23.

Blok, K., N. Höhne, A. Torvanger and R. Janzic 2005. *Towards a Post-2012 Climate Change Regime*. Brussels: 3E.

Blok, K., G. J. M. Phylipsen and J. W. Bode 1997. *The Triptique Approach: Burden Differentiation of CO_2 Emission Reduction among European Union Member States*, discussion paper, informal workshop for the European Union ad hoc group on climate. Zeist: Utrecht University, Department of Science, Technology and Society.

Bodansky, D. 2004. *International Climate Efforts beyond 2012: A Survey of Approaches*. Arlington, VA: Pew Center on Global Climate Change.

Delhotal, K. C., F. C. DelaChesnaye, A. Gardiner, J. Bates and A. Sankovski 2006. 'Mitigation of methane and nitrous oxide emissions from waste, energy and industry', *The Energy Journal, Multi-Greenhouse Gas Mitigation and Climate Policy* (Special Issue No. 3): 89–103.

den Elzen, M. G. J., N. Höhne and S. Moltman 2008. 'The Triptych approach revisited: a staged sectoral approach for climate mitigation', *Energy Policy* **36**: 1107–1124.

den Elzen, M. G. J., M. Meinshausen and D. P. van Vuuren 2007. 'Multi-gas emission envelopes to meet greenhouse gas concentration targets: costs versus certainty of limiting temperature increase', *Global Environmental Change* **17**: 260–280.

EPA 2006. *Global Mitigation of Non-CO_2 Greenhouse Gases*. Washington, DC: United States Environmental Protection Agency.

Grassl, H., J. Kokott, M. Kulessa, J. Luther, F. Nuscheler, R. Sauerborn, H.-J. Schellnhuber, R. Schubert and E.-D. Schulze 2003. *Climate Protection Strategies for the 21st Century: Kyoto and Beyond*. Berlin: German Advisory Council on Global Change (WBGU).

Graus, W., M. Harmelink and C. Hendriks 2004. *Marginal GHG-Abatement Curves for Agriculture*. Utrecht: Ecofys.

Groenenberg, H., K. Blok and J. P. van der Sluijs 2004. 'Global triptych: a bottom–up approach for the differentiation of commitments under the Climate Convention', *Climate Policy* **4**: 153–175.

Gupta, S., D. A. Tirpak, N. Burger, J. Gupta, N. Höhne, A. I. Boncheva, G. M. Kanoan, C. Kolstad, J. A. Kruger, A. Michaelowa, S. Murase, J. Pershing, T. Saijo and A. Sari 2007. 'Policies, instruments and co-operative arrangements', in B. Metz, O. R. Davidson, P. R. Bosch, R. Dave and L. A. Meyer (eds.), *Climate Change 2007: Mitigation. Contribution of Working Group III to the Fourth Assessment Report of the Intergovernmental Panel on Climate Change*. Cambridge, UK: Cambridge University Press, pp. 745–807.

Höhne, N. 2005. *What is Next after the Kyoto Protocol: Assessment of Options for International Climate Policy Post 2012*. Utrecht: University of Utrecht.

Höhne, N. and S. Moltmann 2007. *Linking National Climate and Sustainable Development Policies with the Post-2012 Climate Regime: Proposals in the Energy Sector for Brazil, China, India, Indonesia, Mexico, South Africa and South Korea*. Cologne: Ecofys.

Höhne, N., M. G. J. den Elzen and M. Weiss 2006. 'Common but differentiated convergence (CDC), a new conceptual approach to long-term climate policy', *Climate Policy* **6**: 181–199.

Höhne, N., D. Phylipsen and S. Moltman 2007. *Factors Underpinning Future Action: 2007 Update*. Cologne: Ecofys.

Höhne, N., D. Phylipsen, S. Ullrich and K. Blok 2005. *Options for the Second Commitment Period of the Kyoto Protocol*, research report for the German Federal Environmental Agency. Berlin: Ecofys.

Höhne, N., E. Worrell, C. Ellerman, M. Vieweg and M. Hagemann 2008. 'Sectoral approach and development', Input paper for the workshop *Where Development Meets Climate: Development Related Mitigation Options for a Global Climate Change Agreement*. Cologne: Ecofys.

International Energy Agency 2005. *CO_2 Emissions from Fuel Combustion 1971–2003*, 2005 edn. Paris: International Energy Agency.

Kameyama, Y. 2004. 'The future climate regime: a regional comparison of proposals', *International Environmental Agreements: Politics, Law and Economics* **4**: 307–326.

Lucas, P., D. P. van Vuuren, J. A. Olivier and M. G. J. den Elzen 2007. 'Long-term reduction potential of non-CO_2 greenhouse gases', *Environmental Science and Policy* **10**: 85–103.

Marland, G., T. A. Boden and R. J. Andres 2003. 'Global, regional, and national fossil fuel CO_2 emissions', in *Trends: A Compendium of Data on Global Change*. Oak Ridge, TN: Carbon Dioxide Information Analysis Center.

Olivier, J. G. J. and J. A. van Aardenne 2007. 'EDGAR and UNFCCC greenhouse gas datasets: comparisons as indicator of accuracy', in P. Bergamaschi (ed.), *Atmospheric Monitoring and Inverse Modelling for Verification of National and EU and National and EU Bottom-Up GHG Inventories*. Brussels: European Commission Joint Research Centre. pp. 87–90.

Olivier, J. G. J., J. A. van Aardenne, F. Dentener, V. Pagliari, L. N. Ganzeveld and J. A. H. W. Peters 2005. 'Recent trends in global greenhouse gas emissions: regional trends and spatial distribution of key sources', *Environmental Sciences* **2**: 81–100.

Ott, H. E., H. Winkler, B. Brouns, S. Kartha, M. Mace, S. Huq, Y. Kameyama, A. P. Sari, J. Pan, Y. Sokona, P. M. Bhandari, A. Kassenberg, E. L. La Rovere and A. Rahman 2004. *South–North Dialogue on Equity in the Greenhouse: A Proposal for an*

Adequate and Equitable Global Climate Agreement. Eschborn: Deutsche Gesellschaft für Technische Zusammenarbeit GmbH.

Phylipsen, G. J. M., J. W. Bode, K. Blok, H. Merkus and B. Metz 1998. 'A Triptych sectoral approach to burden differentiation: GHG emissions in the European bubble', *Energy Policy* **26**: 929–943.

Reis, S., H. Pfeiffer, J. Theloke and Y. Scholz 2008. 'Temporal and spatial distribution of carbon emissions', in A. J. Dolman, R. Valentini and A. Freibauer (eds.), *The Continental-Scale Greenhouse Gas Balance of Europe*. New York: Springer, pp. 73–90.

Ringius, L., A. Torvanger and B. Holtsmark 1998. 'Can multi-criteria rules fairly distribute climate burdens? OECD results from three burden sharing rules', *Energy Policy* **26**: 777–793.

Rose, A., B. Stevens, J. Edmonds and M. Wise 1998. 'International equity and differentiation in global warming policy: an application to tradeable emission permits', *Environmental and Resource Economics* **12**: 25–51.

Torvanger, A. and O. Godal 2004. 'An evaluation of pre-Kyoto differentiation proposals for national greenhouse gas abatement targets', *International Environmental Agreements: Politics, Law and Economics* **4**: 65–91.

UN 2004. *World Population Prospects: The 2004 Revision*. New York: United Nations Department for Economic and Social Information and Policy Analysis.

van Vuuren, D. P., M. Isaac, A. F. Hof, R. Mechler, P. Criqui, A. Kitous, T. Barker, S. Scrieciu and Z. Kundzewicz 2010 (in press). 'Scenarios as the basis for assessment of mitigation and adaptation', in M. Hulme and H. Neufeldt (eds.), *Making Climate Change Work for Us: European Perspectives on Adaptation and Mitigation Strategies*. Cambridge, UK: Cambridge University Press.

van Vuuren, D. P., P. L. Lucas and H. Hilderink 2007. 'Downscaling drivers of global environmental change scenarios: Enabling use of the IPCC SRES scenarios at the national and grid level', *Global Environmental Change* **17**: 114–130.

Winkler, H., R. Spalding-Fecher, S. Mwakasonda and O. Davidson 2002. 'Sustainable development policies and measures: starting from development to tackle climate change', in K. A. Baumert, O. Blanchard, L. Llose and J. F. Perkaus (eds.), *Options for Protecting the Climate*. Washington, DC: World Resource Institute, pp. 61–87.

13

Technological change and the role of non-state actors

KNUT H. ALFSEN, GUNNAR S. ESKELAND
AND KRISTIN LINNERUD

13.1 Introduction

This chapter addresses technological change as a key response to the challenge of climate change and focuses on the roles of private and government sectors in providing emissions reductions throughout this century. While standard economic theory recommends that governments set a price on emissions (Pigou 1920), we argue that market imperfections and dynamic inconsistencies may require that in addition, governments support far-reaching technological change by means of publicly funded research, development and demonstration. In fact, public funding of research and development and carbon pricing policies are, at least in theory, mutually supportive and should not be seen as unbridgeable alternatives.

The Kyoto Protocol may fail to meet even its modest goals. This underscores the need for more far-reaching technological change. The political target of avoiding a temperature increase greater than 2 °C – the European Union target – may already be beyond reach. The Stern Review (Stern 2007) estimates that stabilizing the greenhouse gas concentration level at 550 ppm carbon dioxide-equivalents by the middle of this century should avoid the most dangerous climate changes. Figure 13.1, adopted from the Stern Review, illustrates what stabilization targets of 550 ppm carbon dioxide-equivalents and below will imply in terms of emissions reductions by industrialized and developing countries. These targets aim at eliminating greenhouse gas emissions by industrialized nations by the end of the century.

13.2 Methodology

Several contributions on climate policy emphasize technological change (Barrett 2008; Hepburn and Stern 2008). Hepburn and Stern (2008: 275) emphasize that before developing countries can be expected to accept binding commitments, low-carbon growth must be proven feasible. They recommend institutional arrangements

Global Climate Governance Beyond 2012: Architecture, Agency and Adaptation, eds. F. Biermann, P. Pattberg and F. Zelli. Published by Cambridge University Press.

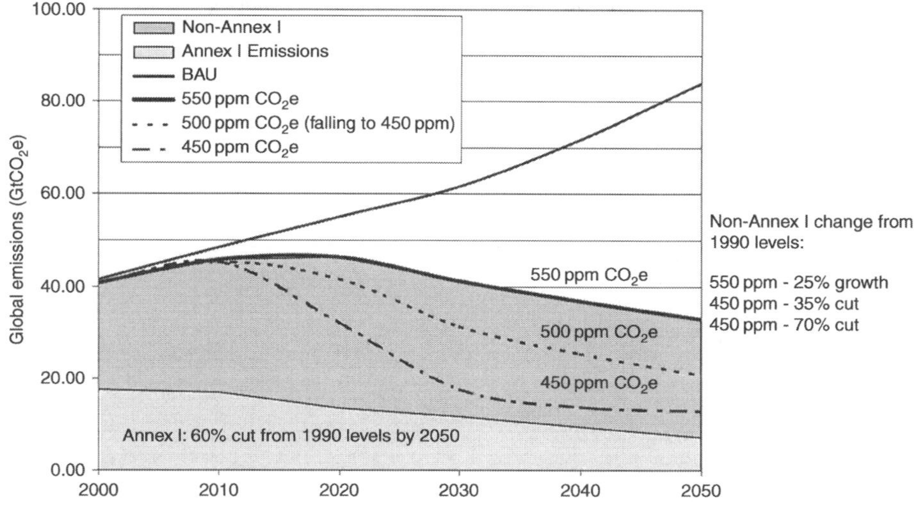

Figure 13.1 Illustrations of emissions paths compatible with different concentration stabilization targets. *Source:* The Stern Review (Stern 2007).

to ensure sufficient funding, for instance through revenues from auctions of quotas. Barrett (2008: 255) notes that the Kyoto Protocol cannot raise the price of emissions significantly given weak incentives for innovation. Helm (1998: 235) points out how present approaches are skewed towards the short term and concludes that long-term technology development may require large-scale research and development programmes. We contribute to this growing literature by making three points.

First, the expected rewards through emissions pricing will be *too low* to induce sufficient far-reaching technological change. This weakness is caused by three factors. First, too few countries participate in cap-and-trade approaches such as the Kyoto Protocol and the European Union emissions trading scheme; second, governments cannot credibly commit to the level of future quotas; and third, governments will choose strategies to assist near-term transition to low-carbon technologies that reduce prices of emissions and emissions-intensive goods and services.

Second, governments should therefore adopt strategies to support far-reaching technological change. In practice, emissions price instruments like the EU emissions trading scheme *are* supplemented by instruments facilitating emissions reductions, particularly through support schemes for renewables and energy efficiency. These support schemes reduce the industry's costs of deploying existing abatement technology. However, they may not produce the more far-reaching technological changes that are required both for Europe beyond 2020 and for the rest of the world.

Third, policies that pursue long-term technological development, including public funding, are needed. We find that such government policies mutually support cap-and-trade approaches. The tightness of future emissions targets will depend on research and development programmes, and vice versa. On the one hand, research and development programmes that reduce the expected costs to governments of meeting future emissions targets will raise expectations that future emissions targets will be even tighter; on the other, research and development programmes will be more successful when a price on emissions is in place and can be expected to increase in the future.

Particularly when highlighting the complementarity between cap-and-trade and research and development support, our conclusions differ from those of a number of critics of the Kyoto Protocol. Prins and Rayner (2007: 974), for instance, recommend putting ‘public investments in research and development on a wartime footing’, but suggest abandoning the Kyoto Protocol ‘to let emissions markets evolve from the bottom up’. Our opinion is that the Kyoto Protocol is essentially sound and that its sequels can be stretched in the right directions, but that research and development must be implemented explicitly and forcefully both in international cooperation and in national policies.

13.3 Analysis

13.3.1 The scale of the problem and the need for new technologies

Solving the climate change problem requires a wholesale change of technology, particularly in the energy and transport sectors. Such change must be implemented with the greatest urgency. Due to expected economic growth in large countries like China and India, more urban infrastructure will be built than ever before. The same will likely hold for power plants and automobiles. The technology we use for these purposes will have enormous impacts on greenhouse gas emissions and the cost of reducing them in the future.

One might argue that rich countries will have modest future needs for more electric power and capacity in energy-intensive sectors because their greatest growth will be in services and other sectors that are not emission or energy intensive. Thus, even if costs of cutting emissions in power and manufacturing sectors prove to be high, as currently with renewables in electric power production, they will be more affordable in high-income countries.

For lower-income developing countries, the opposite holds true. Current costs of cutting emissions are low, as proven through the Clean Development Mechanism. However, these countries’ growth aspirations require massive development in sectors such as power production. In short, only if climate friendly technologies become relatively cheap and affordable will countries like China and India implement them. Making them cheaper and more affordable requires *significant* research and development, not only to find genuinely new solutions, but also to improve the economic efficiency of existing products and thereby making them affordable for developing countries.

Once these solutions become available, an equally daunting task will be to implement climate friendly solutions. No matter how forceful and successful the research and development efforts will be, climate friendly solutions likely will be more expensive than today's fossil-fuel-based solutions. Thus, some sort of market implementation framework (cap-and-trade or greenhouse gas taxes) must be in place to ensure use of the most climate friendly solution.

The Kyoto Protocol is only a small step in this direction. Regarding providing incentives for and financing of the necessary research and development, however, the protocol and similar command-and-control policies are inadequate for the reasons explained below.

13.3.2 Can expected rewards alone drive mitigation efforts?

Expected revenues from selling products that embody commercially feasible technology motivate private investments in research and development. However, the expected revenues from research and development on climate friendly technologies may be too low for the following reasons:

First, because governments control the price of greenhouse gas emissions, through quotas or taxes, there may be inadequate investments in research and development on climate friendly technology. In setting the price, governments continuously balance the expected costs and benefits of greenhouse gas mitigation. They will be under political pressure since some groups benefit from higher and some from lower emissions prices.

Furthermore, even if governments decide to promise high future emissions prices, they will have every incentive to renege on their promises and lower these prices when new technologies are available. Then, governments will set emissions prices to ensure deployment of new technologies, but not so high as to cover industry's research and development investments. These are now sunk costs. We thus have what is sometimes called a dynamic inconsistency between private investment costs and the governmental control of the payback to investors (Montgomery and Smith 2005). Since governments cannot credibly commit to future emission prices, there will be insufficient investments in research and development. In addition, investors will be unsure of emissions pricing policies in countries presently not committed to mitigation.

Finally, a more general problem is that future revenues from current investment in research and development may spill over to competitors, reducing private investors' incentives for acting *now.* Knowledge is a public good in the sense that its value is easily subject to spill-over. Society has many powerful mechanisms to deal with this phenomenon. Most importantly, taxpayers largely finance both schooling and basic research at universities and other research institutes. For more applied research, governments create and protect intellectual property rights (patents) that attempt to preserve a good portion of the returns from new technology for the originator. These solutions are general, but can be adequate also for the development of climate technologies, if two prior problems are addressed. First, insufficient emissions price expectations; and second, limitations to full country participation in mitigation policies.

The solution to this dilemma is rather simple, but unpleasant. Since part of the rewards that should ideally be paid will not be realized, governments must pay some of the necessary research and development costs. Such government support should come in many forms, including public procurement, jointly funded research programmes and prizes. In doing such things, governments should of course also provide a framework, for example, in the form of a cap-and-trade system or by introducing greenhouse gas taxes, such that the solutions are profitable alternatives in the market.

13.3.3 Technology development and implementation

It follows that the twin tasks of developing new technologies and getting them implemented requires *two* instruments: public research and development funding and optimal emissions pricing.

Thus, a cap-and-trade regime like that introduced by the Kyoto Protocol should be supplemented with a technology-based research and development agreement. Financing and other measures included in such a treaty should be verifiable, and each party to the treaty should be assured a share in research contracts, access to knowledge, and markets for technology. Such a research and development-based treaty may be self-enforcing and attractive to nations outside the major industrialized countries because research and development cooperation will attract participants interested in gains that yield energy security and climate benefits; in sharing in research contracts and technology cooperation, and in increased competitiveness and trade access.

Having acknowledged that emissions pricing may not alone suffice to stimulate the far-reaching end of technology development, several detailed questions must be answered. The private sector has much strength in terms of providing technological gains, particularly for applied as opposed to basic research. This implies that institutions, including international ones, must be creative and open to the many ways in which government and the private sector can work together. The range of instruments will likely include some used in the development of defence technologies, research and development subsidies (through tax benefits, for instance), specific research contracts and prizes for the first to market a product (Kremer 2000).

One important avenue is to set sector-specific targets and standards. Both the United States and Europe have mixed experiences with standards for cars, but it seems likely that long-term targets and standards for average emissions of greenhouse gases, say for urban buses as a class and for private vehicles as a class, can play a role in setting the stage for long-term development. Europe's efforts to establish 120 grams of carbon dioxide per vehicle kilometre as a standard, as well as the efforts of many countries that have standards for fuel efficiency, indicate that such instruments will be a part of the toolbox. Eskeland and Mideksa (2008) point out that such instruments play a role in providing credibility and commitment but should be accompanied by real price instruments (fuel taxes and/or emissions taxes). Their conclusion is that standards are helpful in circumventing political feasibility problems and aiding a transition.

Compared to the carbon pricing strategy, technology-oriented policies will involve calculated risks in picking winners, or in specifying targets too narrowly. On the one hand, we believe these risks are important and real, and that they highlight how the challenge of climate policy involves some political responsibilities and necessary strategic commitment. On the other hand, we believe that one central problem with neutral policy instruments, such as the textbook version of cap-and-trade, is that they do not chart very clearly how, in practice, society can live with lower emissions. A real-world political statement like 'buses will on average have lower emissions, and will play a greater role in our cities' conveys a strategic vision and may carry more responsibility than neutral, flexible instruments, since it communicates how one can live with lower emissions. Some research and development efforts will fail. However, we should not underestimate the value of learning from failures (Sanden and Azar 2005).

Another set of difficult issues relates to the trade-offs between rewarding investors (allowing a high price for emissions and technology) and achieving rapid dissemination (requiring a low product price for buyers). Marketing products most profitably requires including the large markets of industrialized countries; however, such focus should not leave developing countries with the impression that they will merely be paying high royalties to developed country patent holders. Quite probably, solutions to these problems will involve locating research and development centres and manufacturing to developing countries, combining efforts by universities and corporations from both industrialized and developing countries.

13.3.4 The complementarity between carbon pricing and research and development

Calling for a government role in facilitating greater emphasis on private sector research and development is not an alternative to carbon pricing. Rather, the two approaches are logically complementary and mutually supportive. Lund (1994), for example points out how emissions instruments like caps or taxes help guide research and development.

Consider the following perspective. First, let us think of countries in a cap-and-trade regime like the current EU emissions trading scheme. A question for investors is the price of emissions in future periods. An expectation that the price will be low depresses investments in climate-friendly technology, thus raising future emissions reductions costs. In the political process of setting goals for the future periods, estimates of emissions reduction costs will be key, and thus this belief will limit the stringency of future quotas.

However, successful research and development will reduce future costs of emissions reductions, thereby making tighter future caps politically feasible. Moreover, research and development efforts will be more effective if implementation instruments such as cap-and-trade create a market for new technologies. Similarly, research and development may strengthen broader participation in mitigation efforts globally by reducing future mitigation costs. More countries can confidently join talks about future emissions reductions when a forceful research and development treaty is in place.

The reasoning above means that research and development and cap-and-trade should not be seen as competing or alternative solutions. Rather, they are complementary tools in addressing the challenge of climate change.

13.3.5 Technology research and development, trade and commitment

Unilateral efforts to reduce emissions will be insufficient due to free-riding. Thus, we need international cooperation for greenhouse gas mitigation. In fact, the carbon-leakage literature (Bohm 1993; Golombek *et al.* 1995; Hoel 2001), building on the pollution-haven literature, points out that efforts by a limited set of countries may directly boost emissions in non-participating countries through stimulating their exports of emissions-intensive goods (like metals). Importantly, there is a debate about whether this phenomenon is empirically important (for example Eskeland and Harrison 2003; Taylor 2004; Lockwood and Whalley 2008). Nevertheless, many analysts have pointed out that the Kyoto Protocol fails to give rich countries the responsibility for emissions they cause when they consume products that cause emissions in the countries that produced them (examples are Pan *et al.* 2008; Peters and Hertwich 2008). Barrett (2008) points out that the choice between a consumption-based and production-based emissions accountability framework would be unimportant if there were broad participation in climate treaties, whereas it is important at the present stage of differentiated responsibilities.

The question of unilateral versus internationally coordinated efforts for greenhouse gas mitigation should be revisited when introducing technological change and research and development in the analysis. Golombek and Hoel (2006) examine the effects of unilateral emissions reductions under endogenous technological change and under research and development investments. They show that a unilateral rise in abatement causes change in abatement technologies, and through technological spill over this may raise abatement in other countries. This will not eliminate the free-rider problem but reduces the tendency that unilateral efforts raise emissions in other countries.

Treaties to induce research and development cooperation may of course directly address the positive externalities in knowledge creation. They may also address important commitment problems. The argument goes as follows: say a country A considers a strategic programme to develop carbon-lean cars. Such a programme could invest more, making deeper cuts in emissions from future cars, if a market in another country B were also available for the resulting cars (or car technology). For B, however, it may be tempting once those emissions-free cars are available, to implement car-sector policies that allow the new cars market access and success but not at low prices that do not fully reward A's research and development investment. This temptation to keep market prices low in country B would be reduced if A and B were investing in the car technology jointly.

Now imagine, however, that cooperation were not that deep, for instance because A was in the outset more apt to develop emissions-free cars than B. B may be better positioned and motivated to invest in low-emissions electric-power plants. In this case, B feels uncomfortable about investing heavily in power plant technology because A's policies in the power sector may only insufficiently reward B's low-emissions power-plant technology.

These two problems can be addressed through cooperation, and in two ways. First, if each country invests in different technologies or sectors, one country's incentive to renege and exploit the other's investment at a low price may be directly deterred by its concern for retaliatory actions against its own export products. In other words, because both countries sell environmentally friendly goods in the other's market, they must treat the other technology fairly, thus raising the feasibility of research and development investment. Second, commitment to equal treatment may be achieved by agreeing on a policy instrument that is sector neutral, such as a joint emissions trading system. In other words, if protectionism and other forms of opportunism are facilitated by sectorally differentiated environmental policy, then commitment to sector-neutral policy instruments can aid commitment and thereby efficient research and development investments. Part of the literature on border tax adjustments (Eskeland *et al.* 2008a, b; Lockwood and Whalley 2008) argues that trade in emissions-intensive goods poses a threat to the development of new technologies. For instance, since aluminium production may move to countries with low production costs, such movements curtail a rise in aluminium prices and emission prices that would be necessary to motivate research and development investments. Thus, restrained imports of emission intensive goods from non-mitigating countries can serve the dual purposes of reducing carbon leakage in the short term and stimulating technological change in the longer term. Nevertheless, such policies, often called border tax adjustments, would involve administrative, political and economic costs, leaving it open whether such policies will be recommendable (Eskeland *et al.* 2008b; Skodvin *et al.* 2008).

13.4 Conclusions and policy recommendations

Carbon pricing policies (taxes or quotas) are first best optimal in curbing carbon emissions. Hence, the behaviour of the private sector is straightforward. Faced with this single piece of price information they will adjust their behaviour, increase their consumption of low-carbon products, and increase their investments in low-carbon technologies.

However, carbon pricing is not implemented optimally in the real world, and a general weakness is the insufficient incentives given to the private sector in terms of research and development for far-reaching technological change. The reasons for this insufficiency are: that promises of socially optimal future carbon price increases would not be credible; that knowledge spill-over suppresses research and development investment in new technologies; that measures accompanying cap-and-trade, such as support for renewables and support for energy efficiency, result in lower prices for emissions and emissions-intensive goods (such as electricity), depressing further the expected rewards for technological development; and finally that research and development will be subject to scale economies, so that low participation keeps rewards for inventive activities at suboptimal levels. Consequently, governments, through cap-and-trade and related instruments, fail to facilitate the roles of both the universities and the private sector in terms of creating new technologies.

We think a strategy for facilitating the private sector's role in creating climate friendly technologies for the future should take into consideration the following three elements: international agreements to boost innovation, a broader palette of climate policy instruments and a substantial increase in public research and development support.

International agreements

International agreements are suited to boost research and development on climate friendly technologies. The climate convention, the Kyoto Protocol and the Bali statements mention the need to accelerate research and development for technological change. However, research and development on far-reaching technological change is presently at low levels and should be expanded unilaterally by the European countries, the United States or Japan. One may easily argue (Alfsen and Eskeland 2007) that such expansion is affordable and attractive. Climate change mitigation may be the main motivator for expanded research and development support in some (European) countries while energy security may be the main motivator to others. The efforts will still share a lot in terms of objectives and content. Even if lightly coordinated, there will be a sentiment of shared benefits allowing them greater political flexibility in the expansion.

In addition, an important concern will be committing to research in a way beneficial to developing countries, to ensure developing-country participation. This requirement, too, we believe will require innovative strategies. One perspective is that if an industrialized country and company (like the United States or Japan, and General Motors or Toyota, respectively) invest heavily in new technologies – electric vehicles, say – then a developing country and company like India and Tata may offer a partnership in terms of scientists, manufacturing capacity and markets. Their establishing a partnership may offer both effective research and development and an effective signal that rewards will be shared.

A broader palette of instruments

We argue that research and development agreements and cap-and-trade agreements are mutually supportive because research and development reduces future abatement costs and thus makes it feasible for politicians to agree on tighter future caps. Cap-and-trade strengthens a research and development programme because research and development becomes more efficient when a price on emissions can help stimulate inventive activities. Research and development and cap-and-trade should not be seen as alternatives or substitutes, but as mutually supportive elements in an effort to tackle climate change.

In terms of instruments – *how to stimulate far-reaching technological change* – we believe the literature is still young, and that each region or country will utilize many modes. An important concern will be about using taxpayers' money efficiently, and – we believe – this concern will have several important implications (Alfsen *et al.* 2008): first, to use instruments which are not funded by the public budget, such as standards and labelling; second, to use instruments that reward not efforts but results, such as prizes for a given solution (emissions-free transatlantic flights, for instance); third, to use public funds but in purchases that are anyway necessary and have professional procurement, such as in office buildings or in urban bus concessions; and finally, to use traditional research contracting, directly with institutes, with industrial corporations and in joint ventures.

More reliance on research and development support

Another important issue concerns the balance between incentivizing new knowledge and the rapid and cheap dissemination of existing knowledge. Protecting intellectual property rights will largely be necessary to stimulate innovation. In addition, sufficiently rapid implementation may require using funds to buy out patents, to subsidize 'marketization' (as with green certificates and feed-in tariffs) and other instrument combinations. Indeed, arguments for a balance between research and development subsidies and intellectual property rights are very similar to the idea that there is mutual support between research and development support and cap-and-trade (Alfsen and Eskeland 2007; Alfsen *et al.* 2008). Scotchmer points out that globalization shifts such a system in an undesirable direction towards less reliance on research and development support, pointing out that there are no international organizations for public spending on R&D that are analogous to the TRIPS negotiation for intellectual property rights (Scotchmer 2004: 347).

To sum up, in order to facilitate the private sector's role in reducing emissions of greenhouse gases the governments should provide not only one but several instruments. This climate policy mix should be designed so that the instruments mutually support each other, and with a greater emphasis on how to stimulate innovation.

Acknowledgements

The authors gratefully acknowledge research funding from the ICEPS Project of the RENERGI programme of the Norwegian Research Council as well as from the ADAM Project.

References

Alfsen, K. H. and G. S. Eskeland 2007. *A Broader Palette: The Role of Technology in Climate Policy*, Report to the Expert Group for Environmental Studies 2007:1. Stockholm: Ministry of Finance, Sweden.

Alfsen, K. H., H. Dovland and G. S. Eskeland 2008. *Elements for an Agreement on Climate and Energy Technology Development (ACT)*, Policy Note 2008: 1. Oslo: Cicero.

Barrett, S. 2008. 'Climate treaties and the imperative of enforcement', *Oxford Review of Economic Policy* **24**: 239–258.

Bohm, P. 1993. 'Incomplete international cooperation to reduce CO_2 emissions: alternative policies', *Journal of Environmental Economics and Management* **24**: 258–271.

Eskeland, G. S. and A. Harrison 2003. 'Moving to greener pastures? Multinationals and the pollution-haven hypothesis', *Journal of Development Economics* **70**: 1–23.

Eskeland, G. S. and T. Mideksa 2008. *Transportation Fuel Use, Technology and Standards: The Role of Expectations and Credibility*, World Bank Policy Research Working Paper. Washington, DC: World Bank.

Eskeland, G. S., T. Mideksa and N. Rive 2008a. 'European climate goals for 2020 and the role of the electricity sector', mimeographed, www.Cicero.uio.no.

Eskeland, G. S., T. Mideksa and N. Rive 2008b. 'European emissions, carbon leakage, a case for border tax adjustments?', mimeographed, www.Cicero.uio.no.

Golombek, R. and M. Hoel 2006. 'Unilateral emission reductions and cross-country technology spillovers', *The B.E. Journals in Economic Analysis and Policy* **4**(2). Available at http://works.bepress.com.

Golombek, R., C. Hagem and M. Hoel 1995. 'Efficient incomplete international climate agreements', *Resource and Energy Economics* **17**: 25–46.

Helm, D. 1998. 'Climate change policy: why has so little been achieved?', *Oxford Review of Economic Policy* **24**: 211–238.

Hepburn, C. and N. Stern 2008. 'A new global deal on climate change', *Oxford Review of Economic Policy* **24**: 259–279.

Hoel, M. 2001. 'International trade and the environment: how to handle carbon leakage', in H. Folmer, H. L. Gabel, S. Gerking and A. Rose (eds.), *Frontiers of Environmental Economics*. Cheltenham, UK: Edward Elgar, pp. 176–191.

Kremer, M. 2000. *Creating Markets for New Vaccines: Parts I and II*, NBER Working Paper 7716 and 7717. Cambridge, MA: National Bureau of Economic Research.

Lockwood, B. and J. Whalley 2008. *Carbon Motivated Border Tax Adjustments: Old Wine in New Bottles*?', NBER Working Paper 14025. Cambridge, MA: National Bureau of Economic Research.

Lund, D. 1994. 'Can a small country gain from a domestic carbon tax? The case with R&D externalities', *Scandinavian Journal of Economics* **96**: 365–379.

Montgomery, W. D. and A. E. Smith 2005. 'Price, quantity, and technology strategies for climate change policy', in M. Schlesinger, H. Kheshgi, J. Smith, F. de la Chesnaye, J. M. Reilly, T. Wilson and C. Kolstad (eds.), *Human-Induced Climate Change: An Interdisciplinary Assessment*. Cambridge, UK: Cambridge University Press, pp. 328–342.

Pan, J., J. Phillips and Y. Chen 2008: 'China's balance of emissions embodied in trade: approaches to measurement and allocating international responsibility', *Oxford Review of Economic Policy* **24**: 354–376.

Pigou, A. C. 1920. *Economics of Welfare*. London: Macmillan.

Peters, G. and E. Hertwich 2008. 'Post-Kyoto greenhouse gas inventories: production versus consumption', *Climatic Change* **86**: 51–66.

Prins, G. and S. Rayner 2007. 'Time to ditch Kyoto', *Nature* **449**: 973–975, doi:10.1038/449973a.

Sanden, B. and C. Azar 2005. 'Near term technology policies for long-term climate targets: economy wide versus technology specific approaches', *Energy Policy* **33**: 1557–1576.

Scotchmer, S. 2004. *Innovation and Incentives*. Cambridge, MA: MIT Press.

Skodvin, T., A. Gullberg and S. Aakre 2008. 'Target group influence as a determinant of political feasibility: the case of climate policy design in Europe', mimeographed, www.cicero.uio.no.

Stern, N. 2007. *The Economics of Climate Change: The Stern Review*. Cambridge, UK: Cambridge University Press.

Taylor, M. S. 2004. 'Unbundling the pollution haven hypothesis', *Advances in Economic Analysis and Policy* **4**: article 8.

Part III

Adaptation

14

Global adaptation governance: setting the stage

FRANK BIERMANN AND INGRID BOAS

14.1 Introduction

The Fourth Assessment Report of the Intergovernmental Panel on Climate Change has shown that past mitigation efforts have been too little and too late (IPCC 2007a). Climate change becomes a reality of world politics in the twenty-first century. This requires a new, additional focus in both academic research and policy planning. How can we build systems of global governance that will cope with the global impacts of climate change? What institutions are in need of redesign and strengthening? And to what extent, and in what areas, do we need to create new institutions and governance mechanisms?

These questions are at the centre of the third part of this book. While substantial research exists on local and national adaptation policies and institutions, the focus of this part is on *global* adaptation governance, an area that is still least explored. Global adaptation governance will affect most areas of world politics, including many core institutions and organizations. The need to adapt to climate change will influence, for example, the structure of global food regimes and the work of the UN Food and Agriculture Organization; global water regimes and the relevant organizations and programmes; global health governance and the agenda of the World Health Organization; global trade in goods whose production will be harmed or helped by climate change; the world economic system and the ability of the International Monetary Fund to address climate-related shocks to national and regional economies; and many other sectors from tourism to transportation, energy or even international security. Almost all areas of global governance will be challenged to cope with global warming and its impacts on human and natural systems. Yet not much research is available. In light of the most recent scientific findings, which indicate possibly accelerating climatic change, there is an urgent need for a new academic research programme on what we propose to call 'global adaptation governance'.

This chapter provides the conceptual frame for the subsequent more specific chapters on global adaptation governance. We discuss, first, the appropriate social science methodologies. We then sketch possible future climate impacts in a few core

Global Climate Governance Beyond 2012: Architecture, Agency and Adaptation, eds. F. Biermann, P. Pattberg and F. Zelli. Published by Cambridge University Press.

domains of world politics. For each domain, we discuss governance challenges and possible solutions, all of which require further analysis in more detailed studies in this still emerging field of international relations and environmental policy.

14.2 Global adaptation governance: policy planning for the unknowable

Research on global adaptation governance poses particular challenges for the analyst. To begin with, the scope and severity of the future impacts of climate change are unknown and, in fact, unknowable. Nonetheless, risks are high. Climate change is non-linear, full of surprises, feedbacks, tipping points and possibly extraordinary degrees of harm to socio-economic systems. Policy-making has always been marked by uncertainty and risk. Concerning climate change, however, uncertainties and risks are extreme. And as Jerneck and Olsson (this volume, Chapter 18) convincingly show, the impacts of global warming will be felt most severely by the poorest of the poor, who are least able to cope with, to adapt to, or to flee, climate change impacts. For the 'bottom billion' of humankind, climate change might evolve as one of the most existential threats of this century.

The methodological problem especially for students of global governance is that the potential future situations are unprecedented in scale and partially in type. For example, climate-induced migration might displace 20 times more people than all refugees recognized by the UN system today. Sea-level rise might erase small island nations and change the landscapes and means of production of many countries. Climate change might also cause the large-scale breakdown of ecosystems, with grave consequences for food production especially in Asia and Africa. Of course, flight, famine and floods have been part of life for humankind for as long as human memory can reach. Yet in modern times, in a highly interdependent global system that unites 6 billion people, the threats of large-scale migration, coastal devastation, drought and resource depletion pose challenges that the current systems of global governance, from the United Nations to the central regimes that govern commerce and cooperation, have never experienced before. In short, scholars of global adaptation governance cannot rely solely on past experience. They have to cope with testing possible future solutions to possible future problems that are unprecedented and merely predicted. Methodologically, this is no easy task.

As a consequence, students of global adaptation governance will need to develop new types of methodology and validation. Similar to the method of counterfactual analysis, one could refer to this methodological challenge as *futurefactuals* – methods to understand the effectiveness of existing or reformed global governance systems on future political events the effects of which remain largely uncertain.

Many issues vital for global adaptation governance are related to this question of future institutional effectiveness: how will current institutions perform if faced with new types and degrees of crises that may result from a substantially warmer world? And to the extent that we may conclude that current institutions will not live up to the needs of a warmer world – what other institutional designs and institutional arrangements could we envisage that better cope with the worldwide impacts of climate change? And then, how could we evaluate in rigorous research designs potentially conflicting claims about the expected performance of different institutional designs and institutional arrangements that are developed and proposed? We can learn here from the analysis of performance of mitigation institutions, in particular the burgeoning literature on environmental regime effectiveness (overviews in Young 2001; Mitchell 2003; Biermann and Pattberg 2008). Yet the transferability of these lines of research to global adaptation governance is to be questioned in each case, and carefully put to the test.

The study of global adaptation governance is inherently linked to the two other central themes of this book, architectures of global climate governance and agency in global climate governance.

Much research on global adaptation governance will need to address its architecture. In particular the question of fragmentation versus integration of governance architectures (Biermann *et al.*, this volume, Chapter 2) is crucial also in adaptation governance. One question is the institutional relation between the policy goals of adaptation and mitigation. For example, what are the benefits of early action on global adaptation in contrast to directing all efforts solely to mitigiation (Hof *et al.*, this volume, Chapter 15)? As shown by Ayers, Alam and Huq (this volume, Chapter 17), adaptation is often seen inferior to mitigation. Yet both adaptation and mitigation are linked in many respects, from the scientific analysis that needs to relate mitigation and adaptation to the moral and legal connection between failure in mitigation and responsibility for compensation. In concrete policies, however, adaptation and mitigation are often distinct policy areas especially at the local level. Saving energy by promoting renewable energies is politically quite distinct from building dikes and adapting agricultural production systems. At the global level, the question becomes even more complex (see Ayers, Alam and Huq, this volume, Chapter 17; Jerneck and Olsson, this volume, Chapter 18): should adaptation continue to be seen as (full) part of the processes of the climate convention and its institutional and organizational complex? Or should adaptation rather be mainstreamed in development cooperation or even be integrated into existing programmes for example under the UN convention on desertification? In short, global adaptation governance raises the question of finding the most effective governance architecture(s).

Agency in global adaptation governance is key, too. As for the traditional main agent in any governance activity – the state – the challenge is to adjust and adapt all core state functions to allow for swift and effective adaptation to possibly large-scale changes in the natural environment. Past debates and research programmes on the green state or the environmental state have to be augmented – if not, as in the case of some highly affected developing countries, replaced – by new discourses and research programmes on the 'adaptive state' (Biermann and Dingwerth 2004; Biermann 2007). In addition to the need for state actors to adapt their core functions to possibly new realities, non-state agency will also become important in global adaptation governance. Firms, environmentalists, human-rights activities, local representatives, scientists, all have assumed an increasing role in defining a new political agenda on global adaptation governance. Without the involvement of private actors at all levels, global adaptation governance, as many other areas of modern complex governance problems, is doomed to fail. Therefore, the question of agency is also a core problem of the new challenge of global adaptation governance.

In sum, to ensure effective systems of future global adaptation governance, we cannot merely rely on current and past experience. Scholarly analysis must explore in systematic study future institutional performance in merely predicted situations, and it must address questions of architecture and the role of agency at the same time.

14.3 Mapping world politics on a warmer planet

Over the next decades global warming might result in severe challenges in most core domains of world politics. There are seven areas where the impacts of global warming could be most severe and require most attention by students of global governance: we posit that these are global food governance, global water governance, global health governance, global energy governance, global refugee governance, global economic governance and global security governance. In this section, we map the challenges for these core areas and sketch what questions need to be addressed in order to adjust their governance systems to a substantially warmer world. For each domain we first sketch the magnitude of the problem, followed by an overview of research questions that are important from the perspective of global adaptation governance.

Our focus is climate change impacts in Africa, Asia and Latin America, where societies have the least means to adapt to this new environment and the strongest needs for international assistance and cooperation. Here are the developing nations where climate change impacts will be felt most severely and which are already highly vulnerable because of low adaptive capacities, poverty, weak performance of political institutions and many more factors (Ayers, Alam and Huq, this volume, Chapter 17; Jerneck and Olsson, this volume, Chapter 18). Even though industrialized countries will not remain unaffected, one can expect them to have a higher capacity to adapt and subsequently to be in less need of international assistance.

14.3.1 Global food governance

One area of world politics that will be severely affected by climate change is global food governance. Global warming could further exacerbate existing hunger for millions of people (Parry *et al.* 2004; Gregory *et al.* 2005; IPCC 2007b: ch. 5). By 2050, 34–212 million more people are projected to experience hunger due to climate change (Warren *et al.* 2006: 41–42). In a worst-case climate change scenario, an additional 551 million people could be at risk of hunger by 2080, mostly in Asia and Africa (23–200 million in Africa and 7–266 million in Asia: Warren *et al.* 2006: 41–42). The numbers for Latin America could be lower compared to the data for Africa and Asia; yet the percentage increase for Latin America might well be one of the highest (Warren *et al.* 2006: 41–42).

These estimates are alarming. Will global institutions be able to deal with these challenges? Even though the architecture of global food governance is rather decentralized (Gupta 2004a), the UN Food and Agriculture Organization serves in many respects as organizational core. Moreover, there are a number of institutional agreements on food security, such as the World Food Programme, as well as activities by the World Trade Organization, the World Bank and the UN Development Programme. Yet it is still uncertain whether these institutional arrangements require strengthening in their focus on food issues or whether it is more effective to create one overarching institution with global food governance impacts at the centre. In both cases, the question arises of how adequate financial support can be ensured (see also Klein and Persson 2008). If a distinct institution on food security would be deemed the most appropriate and effective solution, what would be its characteristics and tasks in a warmer world? How would this new institution function with respect to other global governance domains that are related to global food governance, such as global water governance or global refugee governance? Such questions need to be addressed in order to establish an effective system for global food governance (for more detail see Massey 2008).

14.3.2 Global water governance

Global water governance will also face additional challenges because of climate change. By 2085, 731–1459 million people could suffer from water stress due to climate change if one assumes a temperature rise of 1–2 °C and low population growth (Warren *et al.* 2006: 20). Millions of Asians depend for their water supply on glacier melt and could face increased water stress due to glacier retreat in the Himalayas (Barnett *et al.* 2005: 306). Glacier retreat will also affect Latin America, where millions of people – including those in large cities such as Lima and La Paz – depend on glacier melt from the Andes. This number could be

37 million in 2010 and 50 million by 2050 (Nagy *et al.* 2006: 20). Africa could also face increased water scarcity. Currently, 14 African countries experience problems related to water stress. In 20 years, ten countries could be added to this group (Tearfund 2006: 12).

How will international institutions on water governance cope with these challenges? Is the fragmented architecture of water governance (Gupta 2004b; Conca 2006; Dellapenna and Gupta 2008; Pahl-Wostl *et al.* 2008) up to this task? At present, global water governance falls under many different institutional arrangements that range from institutions specifically tailored to water governance to institutions that focus on development more generally. Institutions in this field include the World Water Council, the UN Development Programme (for example the UNDP Water Governance Facility), the UN Environment Programme, the UN Educational, Scientific and Cultural Organization and the World Bank, as well as many regional organizations. Does the fragmented institutional structure suffice or is there a need for an overarching institutional framework once water stress becomes aggravated? What is the role for private actors? How does this institutional architecture affect the people living in the poorest developing nations (see Jerneck and Olsson, this volume, Chapter 18)? And again, what are the interlinkages with other domains and how should these links be dealt with? All these questions require urgent examination (see in more detail also the Science and Implementation Plan of the Earth System Governance Project: Biermann *et al.* 2009: 69–73).

14.3.3 Global health governance

Millions of people are projected to face health problems as a result of climate change. Health problems are likely because of increased exposure to tropical diseases, along with increased cold and heat stress and rising malnutrition (Haines *et al.* 2006; Warren *et al.* 2006: 71–80). For example, the number of people exposed to malaria could rise by 18.9 per cent with a 3–4 °C temperature increase and by 8.8–14 per cent with a 1–2 °C increase (Warren *et al.* 2006: 74; see also Parry *et al.* 2001: 182 on projections of increased malaria exposure). How would global health governance institutions and mechanisms deal with this situation? The institutional architecture is here rather centralized around the World Health Organization, with some additional activities by the World Bank and others, including private foundations. What would be the institutional performance of the World Health Organization when millions of additional people suffer from tropical diseases? Would this require reforms in the organization's mandate, structure or funding? What would be its main tasks and what would be the main responsibility of local authorities? For instance, who would have to deal with the impacts of heat stress – global or local institutions? Or could this most effectively be organized through public–private partnerships? These issues need to be addressed in order to adjust the current global health governance to the new challenges of global warming (see also Global Environmental Change and Human Health 2007).

14.3.4 Global energy governance

Access to energy is essential to combat poverty and is a precondition for industrial development, the creation of employment and effective agricultural production (UN Energy 2005). Access to energy also improves education facilities, social, communication and health services and reduces the time spent on household activities (UN Energy 2005: 5–8). Many people in developing countries rely on non-commercial energy such as wood or animal residues as primary energy source (Seck 2007: 2–3; UNDP 2007: 303–307). Increased drought induced by global warming could force people to spend more time and energy on ensuring subsistence. Another challenge is the socio-economic and environmental impact of mitigation policies on developing nations. For instance, biofuel plantations meant to replace fossil fuels could have major negative impacts on local communities in many developing countries (Bastos Lima and Gupta 2009). Moreover, extreme weather events or gradual sea-level rise might destroy crucial elements of local energy infrastructures (German Advisory Council on Global Change 2007: 63–72, 103–105 and 103–116). Higher temperatures that increase evaporation levels together with less rainfall could also reduce water levels in reservoirs or rivers and hence hydro-energy production (Gaye 2007: 10).

As energy security is largely defined in national terms, there is no coherent institutional framework on global energy governance (Gupta and Ivanova 2009). The question arises whether current institutions are able to cope with the many additional problems in energy supply that may result from climate change. Would it increase effectiveness to establish a new overarching international institution with a core responsibility on energy? If yes, should this institution solely focus on energy and adaptation or should mitigation questions be included as well? What would be the costs and benefits from focussing on both adaptation and mitigation (see Hof *et al.*, this volume, Chapter 15 for more detail)? How is it ensured that mitigation concerns do not overrule energy and adaptation concerns (see Ayers, Alam and Huq, this volume, Chapter 17, on the empowerment of the adaptation agenda)? What will be the role of private actors?

14.3.5 Global refugee governance

Another area of world politics challenged by global warming is global refugee governance. Climate change impacts in the areas of food, water, health and energy, but also floods and extreme weather events could all trigger large-scale migration.

Millions of people might face situations in which the only option will be relocation and flight. By 2050 there could be around 200 million climate refugees (Biermann and Boas, this volume, Chapter 16). The main international institution currently protecting refugees is the 1951 Geneva Convention Relating to the Status of Refugees and the UN High Commissioner for Refugees. At present climate refugees do not fall under its core mandate, even though the UN High Commissioner for Refugees has some programmes to assist internally displaced persons that could include people who face environmental degradation. Given that the total number of refugees and migrants might increase 20 times due to climate change, the question about possible reform needs in this governance area is particularly urgent. For example, should the mandate of the Geneva Convention and the UN High Commissioner for Refugees be extended in order to include climate refugees? Or is this institution not capable of effectively dealing with millions of more refugees, which might require a different type of protection? If so, what kind of international institution is required? What would be its decision-making structure and core principles? These issues require further analysis in order to protect the many likely victims of climate change effectively (see Biermann and Boas, this volume, Chapter 16, for more detail).

14.3.6 Global economic governance

Global economic governance – including trade, investment and development cooperation – will also face new challenges. Climate change is certain to diminish economic growth in many developing countries. Many developing nations have an agriculture economy, with for instance about 60 per cent of the people in sub-Saharan Africa and South Asia being employed in this sector (Stern 2007: ch. 4.2). Many of these economies are likely to be hit hard because crop production relies on climatic conditions such as rainfall and temperature (Stern 2007: ch. 3.3). Agricultural production could decline in many developing nations. On the other hand, with a 1–3 °C temperature increase, agricultural production in many industrialized could even rise (Parry *et al.* 2005: 2129; IPCC 2007b: 285 and 296–298). Extreme weather events, drought or sea-level rise will threaten economic development also by making affected countries less attractive for investors. This reduces options for developing countries to improve their adaptive capacities (see also Stern 2007: ch. 4). Lower income levels might then also negatively affect other governance areas such as global refugee governance and global food governance (German Advisory Council on Global Change 2007: 97–99; Stern 2007: chs. 4.3, 4.5, 4.6 and 4.7).

How could systems of global governance respond? For example, should increases in poverty due to climate change fall under the responsibility of the climate convention or be mainstreamed into development cooperation? These alternatives have recently been examined by Gupta and van der Grijp (2010). They show that lack of funding in climate policy has forced policy-makers to explore whether climate change can be mainstreamed into existing development cooperation, but also that this is rather difficult (also Jerneck and Olsson, this volume, Chapter 18). More broadly, should distributive questions stronger underlie global economic governance in order to narrow the divide between North and South and increase adaptive capacities (see Jerneck and Olsson, this volume, Chapter 18)? For example, should the impacts of climate change be better reflected in the world trade regime (see also Zelli and van Asselt, this volume, Chapter 6)? Are the World Bank and the International Monetary Fund sufficiently equipped to deal with this new situation of possibly large-scale environmental degradation? How should they respond to climate change shocks that may have destructive effects on national or regional economies? All these matters require urgent examination.

14.3.7 Global security governance

Most of these problems could trigger interstate or intrastate conflict (German Advisory Council on Global Change 2007: 83–84 and 97–100). For example, large-scale climate change-induced migration could cause conflicts in the areas to which climate refugees flee (German Advisory Council on Global Change 2007: 119–122; Gleditsch *et al.* 2007). Food insecurity or water scarcity could trigger conflicts over diminishing resources (Homer-Dixon 1994, 1999; German Advisory Council on Global Change 2007: 83–84 and 97–100). In a worst-case scenario, sudden climate change could even lead to 'societal collapse, mega migration and intensifying competition for diminishing resources and widespread conflict' (Development, Concepts and Doctrine Centre 2007: 78–79).

At present, international conflicts are dealt with within the United Nations Security Council and by regional security institutions. But is the Security Council the most appropriate international forum to deal with conflicts induced by climate change (see also the statements of UN Member States in United Nations Security Council 2007)? If so, should it be reformed to give developing nations that experience most climate-change-related problems a larger role? Or should a separate climate change security institution be created with the main responsibility over addressing conflicts that may arise from climate change? Or should climate change not be framed and dealt with in terms of international security at all? Should it, instead, be dealt with in terms of international development or human rights, and subsequently fall under institutions in these domains? These questions require urgent consideration (see also discussion by Biermann and Boas, this volume, Chapter 16 on climate change migration as a security risk).

14.4 Conclusions

In sum, climate change threatens to cause major crises in many domains of world politics, especially in the areas of food, water, health, energy, migration, economics and security. Millions of people are expected to face hunger or severe problems due to water scarcity or tropical diseases. For many people in developing countries it will become increasingly difficult to ensure subsistence, which further increases poverty, health problems and gender inequality. Climate change and all the problems it might cause could compel hundreds of millions of people to flee their homes in the hope to escape the grim circumstances caused or aggravated by climate change. Moreover, climate change could influence the global economy and harm especially many developing countries. All these problems could lead to intra- or interstate conflicts due to increased political instability and disputes over scare resources.

Systems of global governance must adapt to this new future environment. Policy research needs to focus on how international institutions, agreements and mechanisms need to adjust to maintain and increase performance. Fresh research efforts in global adaptation governance are required to ensure a timely and effective response to alterations in the earth system.

The following chapters take on this challenge by systematically studying the state of the art in modelling impacts of climate change (Hof *et al.*, this volume, Chapter 15) and by outlining new ideas on protecting climate refugees (Biermann and Boas, this volume, Chapter 16), on the role of the developing countries (Ayers, Alam and Huq, this volume, Chapter 17) and of the dangers for, and special interests of, the poorest of the poor (Jerneck and Olsson, this volume, Chapter 18).

References

Barnett, T. P., J. C. Adam and D. P. Lettenmaier 2005. 'Potential impacts of a warming climate on water availability in snow-dominated regions', *Nature* **438**: 303–309.

Bastos Lima, M. G. and J. Gupta 2009. Biofuel and global change: the need for a multilateral governance framework. Unpublished manuscript, on file with authors.

Biermann, F. 2007. '"Earth system governance" as a crosscutting theme of global change research', *Global Environmental Change* **17**: 326–337.

Biermann, F. and K. Dingwerth 2004. 'Global environmental change and the nation state', *Global Environmental Politics* **4**: 1–22.

Biermann, F. and P. Pattberg 2008. 'Global environmental governance: taking stock, moving forward', *Annual Review of Environment and Resources* **33**: 227–294.

Biermann, F., M. M. Betsill, J. Gupta, N. Kanie, L. Lebel, D. Liverman, H. Schroeder, and B. Siebenhüner, with contributions from K. Conca, L. da Costa Ferreira, B. Desai, S. Tay, and R. Zondervan 2009. *Earth System Governance: People, Places, and the Planet: Science and Implementation Plan of the Earth System Governance Project*, ESG Report No. 1. Bonn: Earth System Governance Project of the International Human Dimensions Programme. www.earthsystemgovernance.org/publications/scienceplan.html

Conca, K. 2006. *Governing Water: Contentious Transnational Politics and Global Institution Building*. Cambridge, MA: MIT Press.

Dellapenna, J. and J. Gupta 2008. 'Toward global law on water', *Global Governance* **14**: 437–453.

Development, Concepts and Doctrine Centre (DCDC) 2007. *The DCDC Global Strategic Trends Programme 2007–2036*, 3rd edn, Crown Copyright/MOD 2007. London: The Stationery Office. www.mod.uk/NR/rdonlyres/94A1F45E-A830-49DB-B319-DF68C28D561D/0/strat_trends_17mar07.pdf.

Gaye, A. 2007. *Access to Energy and Human Development*, report prepared for the Human Development Report 2007/2008. New York: United Nations Development Programme.

German Advisory Council on Global Change 2007. *Climate Change as a Security Risk*. London: Earthscan.

Gleditsch, N. P., R. Nordås and I. Salehyan 2007. *Climate Change and Conflict: The Migration Link*, Coping with Crisis working paper series May 2007. New York: International Peace Academy.

Global Environmental Change and Human Health 2007. *Science Plan and Implementation Strategy*, Earth System Science Partnership (DIVERSITAS, IGBP, IHDP, and WCRP). Report No. 4. Global Environmental Change and Human Health Report No. 1. www.essp.org/index.php?id=13.

Gregory, P. J., J. S. I. Ingram and M. Brklacich 2005. 'Climate change and food security', *Philosophical Transactions of the Royal Society B* **360**: 2139–2148.

Gupta, J. 2004a. 'Global sustainable food governance and hunger: traps and tragedies', *British Food Journal* **5**: 406–416.

Gupta, J. 2004b. '(Inter)national water law and governance: paradigm lost or gained?', inaugural address as professor of policy and law on water resources and the environment. Delft, The Netherlands: UNESCO-IHE Institute for Water Education.

Gupta, J. and A. Ivanova 2009. 'The challenge of global energy governance', *Energy Efficiency* **2**(4): 339–352.

Gupta, J. and N. van der Grijp (eds.) (2010). *Mainstreaming Climate Change in Development Cooperation: Theory, Practice and Implications for the European Union*. Cambridge, UK: Cambridge University Press.

Haines, A., R. Kovats, D. Campbell-Lendrum and C. Corvalan 2006. 'Climate change and human health: impacts, vulnerability, and mitigation', *Lancet* **367**: 2101–2109.

Homer-Dixon, T. 1994. 'Environmental scarcities and violent conflict: evidence from cases', *International Security* **19**: 5–40.

Homer-Dixon, T. 1999. *Environment, Scarcity and Violence*. Princeton, NJ: Princeton University Press.

IPCC (Intergovernmental Panel on Climate Change) 2007a. *Climate Change 2007: The Fourth Assessment Report of the Intergovernmental Panel on Climate Change*. Cambridge, UK: Cambridge University Press.

IPCC 2007b. 'Food, fibre and forest products', in M. L. Parry, O. F. Canziani, J. P. Palutikof, P. J. van der Linden and C. E. Hanson (eds.), *Climate Change 2007: Climate Change Impacts, Adaptation and Vulnerability. Contribution of Working Group II to the Fourth Assessment Report of the Intergovernmental Panel on Climate Change*. Cambridge, UK: Cambridge University Press, pp. 273–313.

Klein, R. J. T. and A. Persson 2008. *Financing Adaptation to Climate Change: Issues and Priorities*, ECP Report No 8. Brussels: Centre for European Policy Studies.

Massey, E. 2008. 'Global governance and adaptation to climate change for food security', in F. Zelli (ed.), *Integrated Analysis of Different Possible Portfolios of Policy Options for a Post-2012 Architecture*, (ADAM Project report No. D-P3a.2b). Norwich, UK: Tyndall Centre for Climate Change Research, pp. 143–153.

Mitchell, R. 2003. 'International environmental agreements: a survey of their features, formation and effects', *Annual Review of Environment and Resources* **28**: 429–461.

Nagy, G. J., R. M. Caffera, M. Aparicio, P. Barrenechea, M. Bidegain, J. C. Giménez, E. Lentini, G. Magrin and coauthors 2006. *Understanding the Potential Impact of Climate Change and Variability in Latin America and the Caribbean: Executive Summary*. report prepared for the *Stern Review on the Economics of Climate Change*. London: HM Treasury and Cabinet Office. www.hm-treasury.gov.uk/stern_review_supporting_documents.htm.

Pahl-Wostl, C., J. Gupta and D. Petry 2008. 'Governance and the global water system: a theoretical exploration', *Global Governance* **14**: 419–435.

Parry, M., N. Arnell, T. McMicheal, R. Nicholls, P. Martens, S. Kovats, M. Livermore, C. Rosenzweig, A. Iglesias and G. Fischer 2001. 'Millions at risks: defining critical climate change threats and targets', *Global Environmental Change* **11**: 181–183.

Parry, M., C. Rosenzweig, A. Iglesias, M. Livermore and G. Fischer 2004. 'Effects of climate change on global food production under SRES emissions and socio-economic scenarios', *Global Environmental Change* **14**: 53–67.

Parry, M., C. Rosenzweig and M. Livermore 2005. 'Climate change, global food supply and risk of hunger', *Philosophical Transactions of the Royal Society B* **360**: 2125–2138.

Seck, P. 2007. *The Rural Energy Challenge in Senegal: A mission Report*, Report prepared for the Human Development Report 2007/2008, Occasional Paper No. 60. New York: United Nations Development Programme.

Stern, N. 2007. *The Stern Review on the Economics of Climate Change*. Cambridge, UK: Cambridge University Press. www.hm-treasury.gov.uk/stern_review_report.htm

Tearfund 2006. *Fleeing the Heat*. Teddington, UK: Tearfund.

UNDP (United Nations Development Programme) 2007. *Fighting Climate Change: human solidarity in a divided world Human Development Report 2007/2008*, New York: United Nations Development Programme.

UN Energy (United Nations Energy) 2005. *The Energy Challenge for Achieving the Millennium Development Goals*. New York: United Nations.

United Nations Security Council 2007. Security Council holds first-ever debate on impact of climate change on peace, security, hearing over 50 speakers. 5663rd Meeting, 17 April, 2007. New York: United Nations. www.un.org/News/Press/docs/2007/sc9000.doc.htm.

Warren, R., N. Arnell, R. Nicholls, P. Levy and J. Price 2006. *Understanding the Regional Impacts of Climate Change*, Research report prepared for the *Stern Review on the Economics of Climate Change*, Tyndall Centre Working Paper No. 90. Norwich, UK: Tyndall Centre for Climate Change Research.

Young, O. 2001. 'Inferences and indices: evaluating international environmental regimes', *Global Environmental Politics* **1**: 99–121.

15

Costs, benefits and interlinkages between adaptation and mitigation

ANDRIES HOF, KELLY DE BRUIN, ROB DELLINK, MICHEL DEN ELZEN AND DETLEF VAN VUUREN

15.1 Introduction

The thirteenth Conference of the Parties to the United Nations Framework Convention on Climate Change in 2007 decided that developing countries should be compensated for adaptation costs to climate change through the Adaptation Fund (first draft decision of the third session of the conference of the parties serving as the meeting of the parties to the Kyoto Protocol). This shows that adaptation to climate change has become important in international climate negotiations. Today, adaptation is widely recognized as an equally important and complementary response to climate change mitigation (for example, Commission of the European Communities 2007; IPCC 2007a; Agrawala and Fankhauser 2008).

Still, relatively little information is available to support more integrated climate policies that focus on both mitigation and adaptation (Klein *et al.* 2005). In particular, in integrated assessment models that aim at supporting climate policy by analysing their economic and environmental consequences and formulating efficient responses, explicit consideration of adaptation is still in its infancy (Tol 2005; Wilbanks 2005; Agrawala *et al.* 2008).

This chapter tries to fill the gap in integrated assessment models by integrating adaptation and residual damage functions from AD-RICE (de Bruin *et al.* 2009) with the FAIR model (den Elzen and van Vuuren 2007; Hof *et al.* 2008). This version of the FAIR model (from now on called AD-FAIR) enables an analysis of the interactions between mitigation, emissions trading, adaptation and residual damages (that is, damages not avoided by adaptation measures) on a global as well as regional scale. Furthermore, adaptation is modelled explicitly as a policy variable, providing insights in the economic consequences of adaptation. This information is vital for effective adaptation governance.

Because this study aims at introducing a novel approach of modelling adaptation in the FAIR model and at showing what results can be obtained from this approach,

Global Climate Governance Beyond 2012: Architecture, Agency and Adaptation, eds. F. Biermann, P. Pattberg and F. Zelli. Published by Cambridge University Press.

little attention has been given to analyse uncertainty (see Hof *et al.* 2008, for an extensive assessment of the uncertainties in cost–benefit analyses of climate change policies).

15.2 Methodology

15.2.1 Background on modelling adaptation in integrated assessment models

Adaptation aims to reduce the vulnerability of natural and human systems against actual or expected climate change effects (IPCC 2001). This implies that in order to model adaptation, first the impacts (or damages) of climate change have to be taken into consideration. Damage estimates of climate change involve scientific uncertainties (for example, impact of climate change on the number of storms or change in mortality) as well as value judgements (for example, how to monetize non-market damages and how to deal with uncertainty; see Azar 1998; Dietz *et al.* 2007; Weitzman 2007). Despite these large uncertainties and value judgements, several studies have estimated the damages related to climate change, and incorporated these estimates in integrated assessment models. The most notable examples are FUND (Tol 2002a, b) and DICE/RICE (Nordhaus 1994, 2008; Nordhaus and Boyer 2000). Both estimated the impacts and associated damages of climate change by identifying the most important sectors vulnerable to climate change. DICE/RICE includes the sectors agriculture, sea-level rise, other market sectors (forestry, energy systems, water systems, construction, fisheries and outdoor recreation), health, non-market amenity impacts, human settlements and ecosystems, and catastrophic events. In the DICE/RICE assessment, potential catastrophic events are by far the most important factor in total damages: for a 2.5 °C temperature increase compared to 1900, catastrophic events make up more than half of total estimated damages at the global scale (Nordhaus and Boyer 2000: 91). FUND identifies similar, but fewer, sectors than Nordhaus, omitting most of the other market sectors identified by Nordhaus and, more importantly, catastrophic events. Damages from increasing occurrence of extreme weather events are included neither in the FUND nor DICE/RICE model. Furthermore, both authors warn that their damage estimates are highly speculative. For example, Tol (2002a: 65) argues that ‘a lot needs to be done before one can place any confidence in the estimates’.

In addition to uncertainty, another limitation in the FUND and DICE/RICE damage curves is that adaptation is not explicitly taken into account as a policy variable. Instead, optimal adaptation is assumed in the construction of these curves – and the curves consist of the aggregated costs of remaining damages and adaptation costs. There have been few attempts to model adaptation as a decision variable in integrated assessment models, with Hope *et al.* (1993) among the first. However, they seem to be over-optimistic about the amount of damages that can be avoided by adaptation according to current existing empirical literature on this issue (de Bruin *et al.* 2009). In the recent AD-RICE model, de Bruin *et al.* (2009) used a more transparent method to model adaptation as a decision variable in the RICE integrated assessment model. As this is the most recent and best-documented attempt to explicitly model adaptation in an integrated assessment model, we have integrated the adaptation and residual damage curves from AD-RICE with the FAIR model.

15.2.2 Modelling climate–economy interactions

As the backbone of our study, we use the FAIR 2.1 model which includes 17 regions, described in detail in previous publications (den Elzen and Lucas 2005; den Elzen and van Vuuren 2007). Here, we will only provide a short description of the FAIR model.[1] FAIR uses a flexible set of assumptions and integrates information from detailed energy, climate and socio-economic models. It describes the interactions between multi-gas emissions, greenhouse gas concentrations and the climate system, as well as the interaction between the climate system and the economy through climate change damages and mitigation costs (including emissions trading), for different international burden sharing regimes. These elements are integrated in order to perform a cost–benefit analysis of climate policies (Hof *et al.* 2008).

Figure 15.1 gives a schematic overview of the AD-FAIR model. Regional climate mitigation targets, emissions trading and mitigation costs depend on the global climate mitigation target and the burden-sharing regime. Adaptation costs and residual damages depend on climate change impacts and the amount of adaptation measures taken to reduce these impacts. Climate impacts depend on global temperature increase, associated with a global emission pathway and parameters in the climate model, such as climate sensitivity. Mitigation, adaptation and residual damages are characterized by different dynamics. Mitigation reduces global temperature increase, and hence potential damages, in the long run. In the short run (that is, the coming 20–30 years) this effect is negligible, as the temperature increase for reduction pathway and baseline are very similar due to inertia in the climate system. Adaptation can reduce residual damages in the short run, but does not reduce climate change and therefore future potential damages. To estimate both the direct and indirect consumption losses of mitigation, adaptation and residual damages, we use a simple economic growth model based on a Cobb–Douglas production function for each region (Hof *et al.* 2008).

[1] See www.mnp.nl/en/themasites/fair/index.html.

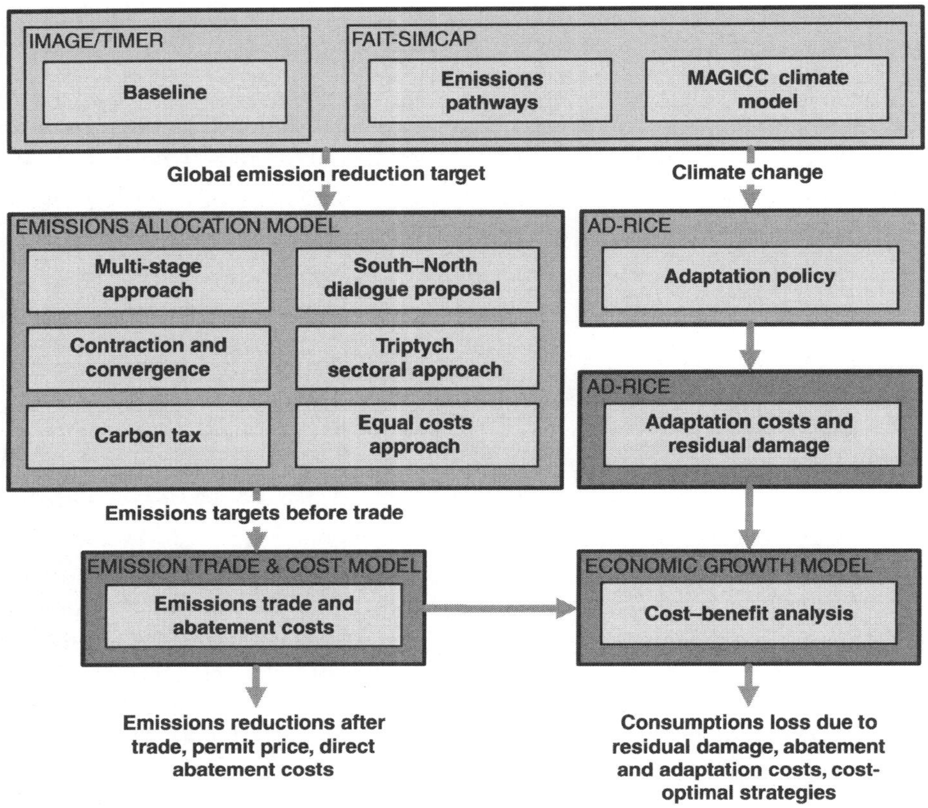

Figure 15.1 Schematic overview of the AD-FAIR model.

15.2.3 Modelling adaptation and residual damages: AD-RICE

The AD-RICE cost and residual damage functions in this chapter are based on the damage functions of the RICE model. These damage functions include both adaptation costs and residual damages. Because the parameterization of these regional damage functions plays a major role in our results, we here give some insight into these damage functions. The parameterization of the RICE damage functions is explained in detail in Nordhaus and Boyer (2000). RICE projects that damages will be small in East Asia, the United States and Japan. This is mainly caused by the assumed low willingness to pay to avoid catastrophic risks in these regions. Damages in the Middle East and South America are projected to be similar to the global average. The other regions – OECD Europe, South East Asia and especially West and East Africa and South Asia – have much higher damages of climate change according to RICE. In Africa, this is mainly due to negative health impacts.

In South Asia, the high risk of catastrophic impacts and the negative effects of climate change on agriculture cause high damages. Health impacts and the risk of catastrophic events are the main reasons for the high damages in South America.

Here we give a short summary of the AD-RICE methodology (see for a more detailed description de Bruin *et al.* 2009). First of all, the regional damage functions of the online available GAMS version of RICE99[2] are separated into a residual damage (*RD*) and adaptation cost (*PC*) component:

$$\frac{D_{r,t}}{Y_{r,t}} = \frac{RD_{r,t}(GD_{r,t}, P_{r,t})}{Y_{r,t}} + \frac{PC_{r,t}(P_{r,t})}{Y_{r,t}}. \tag{15.1}$$

The sum of residual damages and adaptation costs equals net damages *D*. Both residual damages and adaptation costs in region *r* at time *t* depend on the level of adaptation (*P*). Note that residual damages can be negative as well; in this case, adaptation can increase the benefits of climate change. Gross or potential damages *GD* (damages that would occur without any adaptation activities) take the following form:

$$\frac{GD_{r,t}}{Y_{r,t}} = \alpha_1 \Delta T_t + \alpha_2 \Delta T_t^{\alpha_3} \tag{15.2}$$

where $\alpha_2 > 0$ and $\alpha_3 > 1$ and ΔT stands for global temperature increase since 1900. Adaptation activities can reduce residual damages or increase climate change benefits:

$$RD_{r,t} = GD_{r,t}(1 - P_{r,t}),\ 0 \leq P_t \leq 1 \text{ if } GD_t > 0; \tag{15.3}$$

$$RD_{r,t} = GD_{r,t}(1 + P_{r,t}),\ 0 \leq P_t \leq 1 \text{ if } GD_t < 0. \tag{15.4}$$

The level of adaptation is chosen for every time period. No adaptation ($P=0$) means that gross damages are not reduced; in this case, residual damages equals gross damages. At the other extreme, with $P=1$ all gross damages are avoided by adaptation and residual damages equal zero. It is assumed that adaptation costs increase at a growing rate, since cheaper adaptation will be applied first, and more expensive and less effective options later:

$$\frac{PC_{r,t}}{Y_{r,t}} = \gamma_1 P_{r,t}^{\gamma_2},\ \text{where } \gamma_1 > 0 \text{ and } \gamma_2 > 1. \tag{15.5}$$

In our scenarios, adaptation efforts are determined by minimizing the regional sum of residual damages plus adaptation costs. In other words, the level of

[2] Available at www.econ.yale.edu/~nordhaus/homepage/dicemodels.htm.

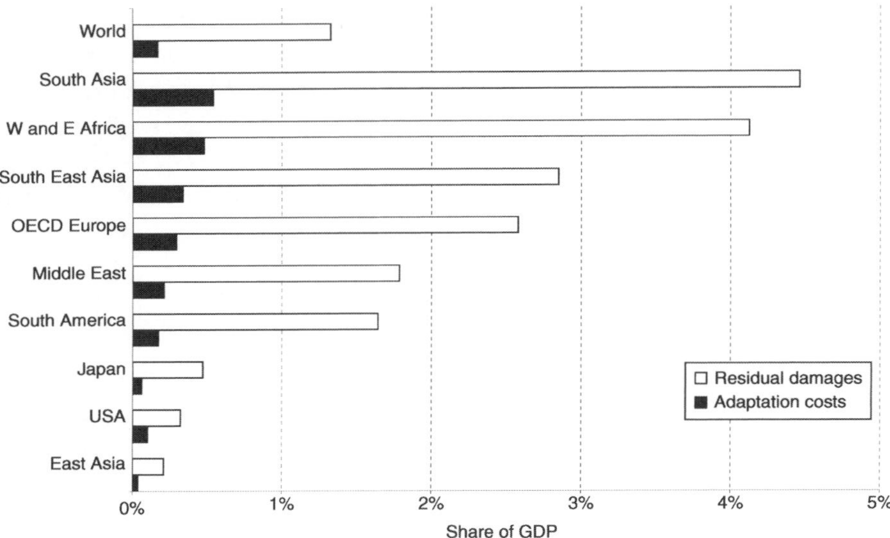

Figure 15.2 Regional distribution of residual damages and adaptation costs as percentage of GDP with a 2.5 °C increase in global temperature according to AD-RICE.

adaptation is chosen so that the marginal costs of adaptation equal the marginal benefits of reducing residual damages.

Figure 15.2 shows the result of the above methodology on separating the damages of RICE into adaptation costs and residual damages for a 2.5 °C global warming. As in de Bruin *et al.* (2009), adaptation costs are much smaller than residual damages in the case of optimal adaptation. This is due to the shape of the adaptation cost curves: up to a certain point adaptation is relatively easy and therefore cheap, but after this point adaptation costs rise sharply. So it would in principle be possible to adapt more, but at such high costs that the benefits of more adaptation (that is, the avoided residual damages) would be lower than the cost of more adaptation.

Because the results strongly depend on the regional damage estimates of RICE, Figure 15.3 shows how the regional damage projections of RICE compare to those of FUND 2.8[3] for a global warming of 2.5 °C. It is clear that there are some significant differences in damage estimates. The most apparent difference is that FUND projects lower climate change damages than RICE for all regions. As mentioned above, an explanation could be that FUND, unlike RICE, does not include catastrophic events and many of the other market sectors that are included

[3] The FUND code can be downloaded at www.fnu.zmaw.de/FUND.5679.0.html.

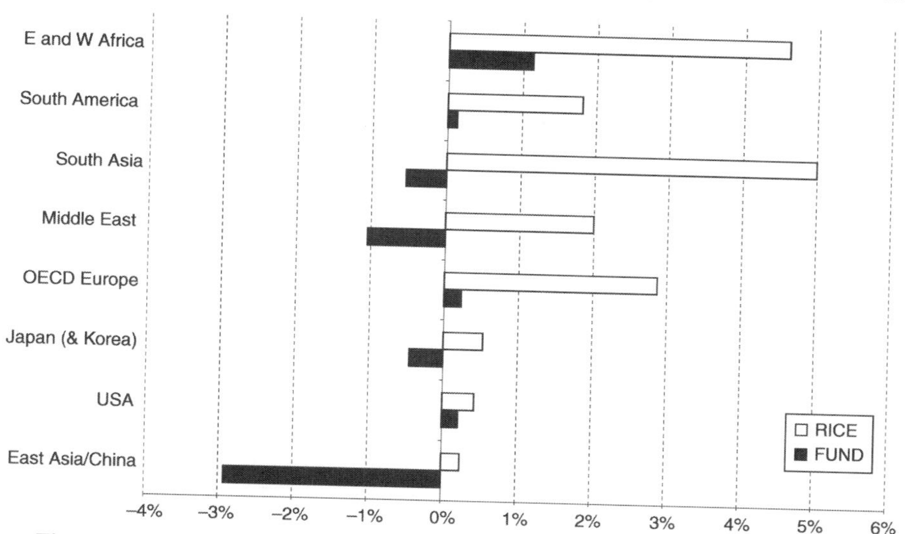

Figure 15.3 Regional damage estimates of a 2.5°C increase in global temperature according to FUND 2.8 compared to RICE.

in the damage estimates of RICE. Not only does FUND generally project lower damages, there are some regional differences in the estimates as well. For example, RICE projects negative impacts for the Middle East at a global warming of 2.5 °C, whereas FUND projects slightly positive impacts. The main reason for this is more optimistic projections for agriculture in the Middle East by FUND (Tol 2002b: 138). One similarity of FUND and RICE is that both project the lowest damages for East Asia and high damages for East and West Africa. In sum, however, this comparison shows how uncertain the damage estimates are and hence how careful our results have to be interpreted.

15.2.4 Calibration of the AD-FAIR model

As baseline for this study, we use the medium IPCC SRES baseline emissions scenario IMAGE&/TIMER SRES B2 (Nakicenovic *et al.* 2000; van Vuuren *et al.* 2007), extrapolated to the year 2250 as described in Hof *et al.* (2008). The B2 baseline describes a world in which the emphasis is on local solutions to economic, social and environmental sustainability. It is a world with moderate population growth and intermediate levels of economic development. Overall, these scenarios are a reasonable description of possible future developments (van Vuuren and O'Neill 2006). Population estimates are taken from the medium long-term UN population projections. Estimates of mitigation costs are based on a wide set of marginal abatement cost curves, differentiated over time as described in detail in den Elzen *et al.* (2007).

Table 15.1 *Calibration of the AD-RICE adaptation cost function and gross damage function applied to the 16 FAIR regions*

	Gross damage function parameters (15.2)			Adaptation cost function parameters (15.5)	
	α_1	α_2	α_3	γ_1	γ_2
Canada	0.0	0.0	1.01	0.0	2.094 287
USA	−0.001 037	0.000 383	4.057 580	0.015 816	10.205 560
Central America	0.0	0.005 709	1.852 274	0.331 922	5.058 278
South America	0.0	0.005 874	1.489 249	0.216 088	3.965 002
Northern Africa	0.0	0.005 709	1.852 274	0.331 922	5.058 278
Western Africa	0.0	0.021 521	1.205 976	0.750 540	5.121 911
Eastern Africa	0.0	0.021 521	1.205 976	0.750 540	5.121 911
Southern Africa	0.0	0.005 709	1.852 274	0.331 922	5.058 278
OECD Europe	−0.000 193	0.004 630	2.294 257	0.340 520	4.243 043
Eastern Europe	NA	NA	NA	NA	NA
Former USSR	NA	NA	NA	NA	NA
Middle East	0.0	0.006 966	1.532 830	0.315 328	5.190 274
South Asia	0.0	0.015 117	1.701 511	0.782 582	5.282 538
East Asia	−0.002 222	0.000 638	2.973 172	0.022 482	6.294 820
South East Asia	0.0	0.010 905	1.554 867	0.502 242	5.178 300
Oceania	0.0	0.0	1.01	0.0	2.094 287
Japan	−0.002 791	0.001 187	2.650 637	0.030 792	3.259 174

The MAGICC 4.1 model (Wigley and Raper 2001, 2002; Wigley 2003) is used to calculate the temperature implications of the emission pathways. The climate sensitivity is set at the best estimate according to the IPCC Fourth Assessment Report (IPCC 2007b) of 3 °C, meaning that a doubling of pre-industrial carbon dioxide concentrations will lead to a global average surface warming of 3 °C.

Calibration of the gross damage function (15.2) and adaptation cost curve (15.5) is done using the optimal control scenario of RICE, in such a way that it best replicates the results of the original RICE model. Table 15.1 provides the results of the calibration on the parameter values (see de Bruin *et al.* 2009 for the derivation of these parameters).

In the economic growth model, the savings rates per region in 2005 are taken from the World Development Indicators database (World Bank 2008) and are assumed to converge linearly to 21 per cent in 2100 in every region and stay constant afterwards. The initial capital stock in every region is based on the growth study datasets of the International Institute for Applied Systems Analysis (Miketa 2004); depreciation is set at 5 per cent annually.

Finally, we adopted the United Kingdom Green Book discounting method (United Kingdom Treasury 2003) for computing the discounted income losses due to mitigation costs, adaptation costs and residual damages over the time period

2005–2250 (Hof *et al.* 2008 discusses the implications of different discounting methods.).

15.2.5 Mitigation strategies

Several climate mitigation targets are analysed. These are taken from den Elzen and van Vuuren (2007) and cover a range of multi-gas emission reduction pathways corresponding with carbon dioxide-equivalent concentrations peaking between 500 and 800 ppm. With a climate sensitivity of 3°C, this implies that the global temperature increase of these emission reduction pathways ranges from 2 °C for the 500 ppm carbon dioxide-equivalent concentration peak to 4 °C for 800 ppm. Global emissions in the most stringent climate mitigation target that we analysed (concentrations peaking at 500 ppm carbon dioxide-equivalent) need to peak in 2015 at 28 per cent above 1990 level, after which emissions are reduced strongly to 53 per cent of 1990 level in 2050. In the 800 ppm pathway, emissions peak in 2040 at 66 per cent above 1990 level, after which emissions decline more gradually to 21 per cent above the 1990 level in 2100.

Emission burdens are allocated across regions using either the 'contraction and convergence' regime with convergence year 2050 (Contraction and Convergence 2050: Meyer 2000) or the 'multi-stage' regime (den Elzen *et al.* 2006; Gupta 1998). The 'contraction and convergence' regime is most often used in quantitative analysis because of its simplicity and straightforwardness (Hof *et al.*, this volume, Chapter 4). In this regime, emission burdens are allocated so that per capita emissions converge from their current values to a global average by 2050. In the 'multi-stage' regime, an increasing number of countries accept commitments over time based on per capita income and per capita carbon dioxide-equivalent emissions. First, countries accept emission intensity targets and as they become more developed, absolute reduction targets are set. The participation threshold levels are differentiated according to the stabilization level as described in den Elzen *et al.* (2008).

15.3 Analysis

This section looks at how adaptation costs relate to mitigation costs and residual damages. First of all, this serves as a check whether adaptation costs as projected by AD-FAIR are in line with the most recent estimates in literature. Furthermore, this section will provide insight in the size and development of the climate change cost components (that is, adaptation costs, mitigation costs and residual damages) over time and the effect of these costs on regional incomes for different mitigation strategies.

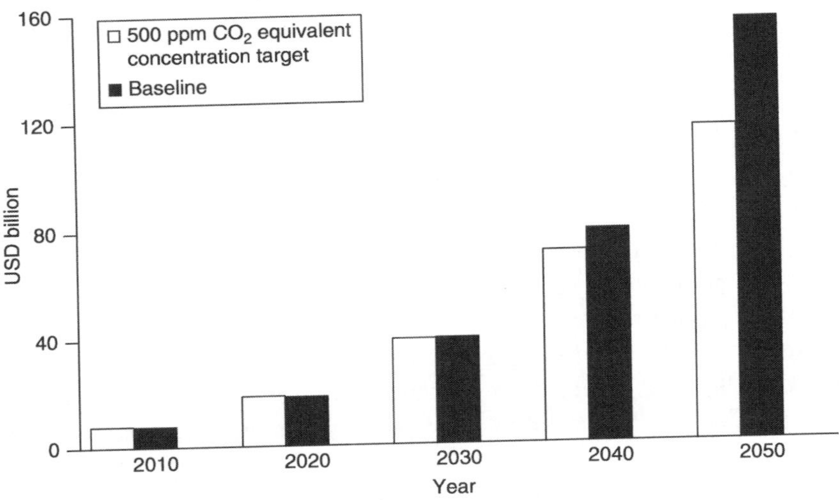

Figure 15.4 Projected adaptation costs from 2010 until 2050 with a concentration peak target of 500 ppm carbon dioxide-equivalent and a baseline scenario without any mitigation efforts.

15.3.1 Global climate change costs

Figure 15.4 shows the adaptation costs in the next decades in a scenario without any mitigation efforts compared to a scenario with a strong climate policy, based on the AD-RICE adaptation cost curves and assuming optimal adaptation. The underlying assumptions are listed in Section 15.2. In 2020, adaptation costs are estimated to amount to USD 18 billion in both scenarios (we use constant 2005 US dollar prices throughout this chapter). Adaptation costs will steadily increase over time, due to increased global warming and hence higher potential damages. Adaptation costs without any mitigation efforts are up to USD 40 billion in 2030, rising to USD 155 billion in 2050. With a strong climate policy that keeps the concentration level below 500 ppm carbon dioxide-equivalent, adaptation costs in 2050 are projected at USD 115 billion. These estimates are of the same order of magnitude as estimates of adaptation costs by the World Bank (2006) and UNFCCC (2007). The World Bank arrives at an order of magnitude of USD 10 to 40 billion per year for developing countries only. The climate convention secretariat estimates the investment and financial flows needed for adaptation to be USD tens of billions per year for the coming decades and potentially more than USD 100 billion per year in the longer run.

Figure 15.4 also shows that adaptation costs for a strong mitigation scenario and a scenario without any mitigation efforts are similar in the short term, and only start to diverge in the longer run (from 2040 onwards). The reason for this is that climate projections for scenarios with and without climate policy only start to diverge after

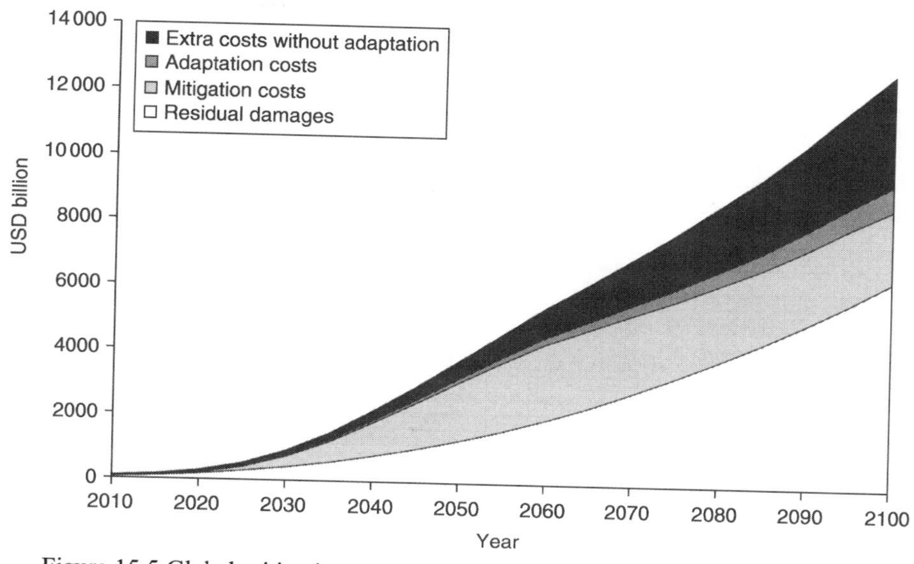

Figure 15.5 Global mitigation costs, adaptation costs, residual damages and extra costs if no adaptation is undertaken, with a concentration target of 550 ppm carbon dioxide-equivalent and a 'Contraction and Convergence 2050' regime.

2040 as a result of inertia in the climate system, combined with reduced sulphur emissions in the scenario with climate policy, which increases temperatures.

Figure 15.5 shows how the projected global adaptation costs compare with the other costs of climate change, that is, mitigation costs and residual damages, assuming an emission pathway leading to a concentration level peak of 550 ppm carbon dioxide-equivalent. Even with such a stringent climate mitigation target and optimal adaptation, residual damages are still the largest component of total climate change costs during most of the century. Especially in the second half of this century, residual damages are projected to increase sharply, reaching USD 6.5 trillion at the end of the century. However, residual damages will reach more than USD 11 trillion if no mitigation takes place, showing that mitigation does reduce damages substantially in the long run. Mitigation costs for a concentration peaking target at 550 ppm carbon dioxide-equivalent rise sharply from 2020 onwards, stabilizing at about USD 2.3 trillion in the second half of the century.

Figure 15.5 also shows the effect of not adapting optimally to climate change. There are many examples of suboptimal adaptation, due to many different reasons. Perhaps the most important of these is that there might not be enough information available about future climate change to adapt optimally to it. But even when this information is available, politicians might still underestimate the risks of climate change. Even though the share of adaptation costs in the total climate change costs is relatively small, adaptation plays a major role by reducing potential damages. The extra costs if no adaptation measures are taken (defined as the increase in residual damages minus the decrease of adaptation costs) are projected to amount to USD 30 billion globally in 2010 and increase sharply to USD 3.4 trillion in 2100. Investment in adaptation is therefore very effective: residual damages are on average reduced by about five dollars for every dollar invested in adaptation.

The above analysis could suggest that there is a trade-off between adaptation and mitigation, since both reduce residual damages. In order to analyse whether there is indeed a trade-off, we compare four different scenarios with each other. In the first scenario called 'reference' there are no mitigation nor adaptation measures; the second scenario consists of optimal adaptation, but no mitigation; the third consists of our most stringent mitigation path (leading to a concentration peak of 500 ppm carbon dioxide-equivalents); and in the final one both the most stringent mitigation measures and optimal adaptation are implemented.

Figure 15.6 shows the discounted climate change costs (with a fixed discount rate of 2.5 per cent) over the next two centuries for these four different scenarios. It shows that the mitigation-only and the adaptation-only scenarios both reduce the discounted costs substantially and to about the same degree compared to the reference case. The main difference seems to be that mitigation is more expensive,

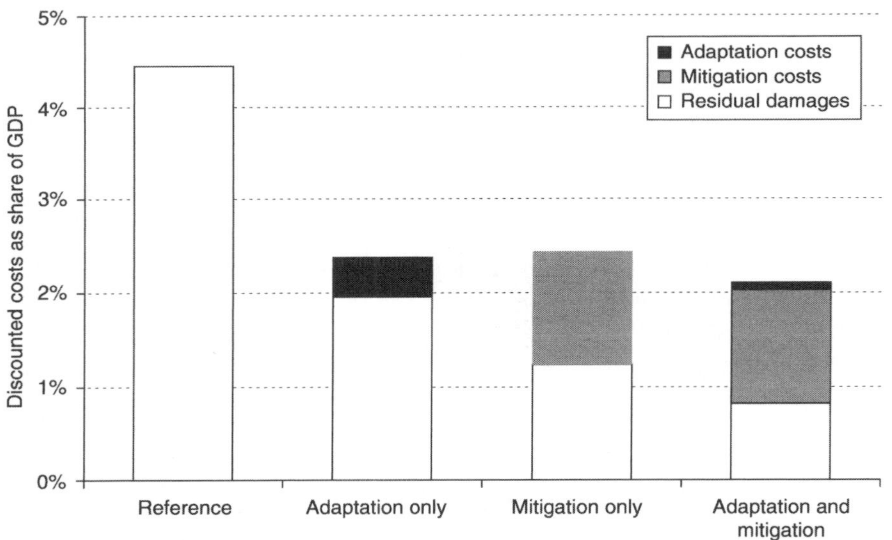

Figure 15.6 Global discounted climate change costs for four scenarios. Global discounted climate change costs for four scenarios: no mitigation and no adaptation (reference), no mitigation and optimal adaptation, no adaptation and stringent mitigation, and both optimal adaptation and stringent mitigation.

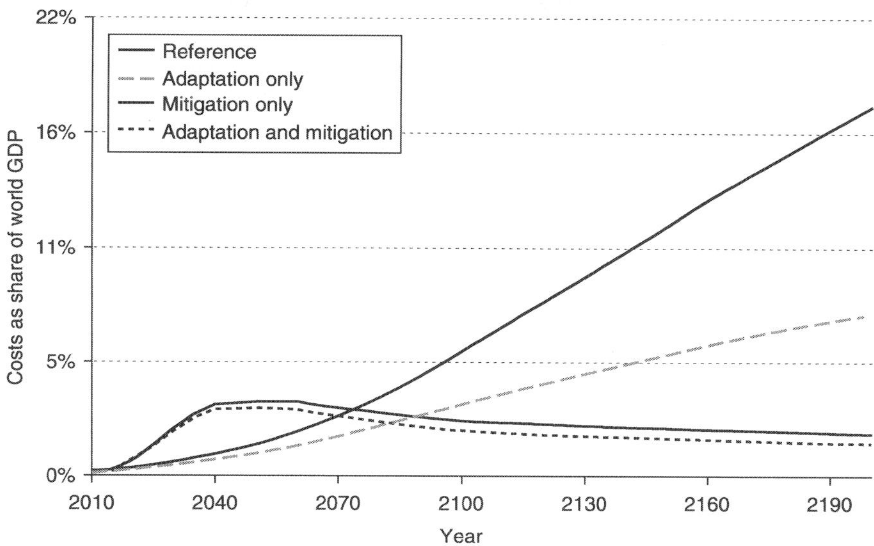

Figure 15.7 Global climate change costs over time for four scenarios. Global climate change costs over time for four scenarios: no mitigation and no adaptation (reference), no mitigation and optimal adaptation, no adaptation and stringent mitigation, and both optimal adaptation and stringent mitigation.

but reduces damages to a higher degree. Implementing both adaptation and mitigation, however, reduces costs even further. This indicates that adaptation and mitigation complement each other and cannot be regarded as substitutes. This becomes even clearer when looking at the total undiscounted costs over time (Figure 15.7): even though the adaptation-only and mitigation-only cases lead to similar discounted costs, the dynamics differ completely. For the adaptation-only case the costs are lower during most of this century, but steadily increase afterwards because climate change is not mitigated. In 2200, total climate change costs of the adaptation-only case are four times the mitigation-only case.

15.3.2 Regional climate change costs

Figure 15.8 shows the regional distribution of all climate change cost components in 2030 for a mitigation path leading to a peak concentration level of 550 ppm carbon dioxide-equivalent. Obviously, higher concentration targets imply lower mitigation costs, but higher adaptation costs and damages.

Residual damages are especially large in all lower income regions except East Asia, as is to be expected from the RICE damage functions. Interestingly, East Asia is even projected to benefit from modest climate change: climate change damages

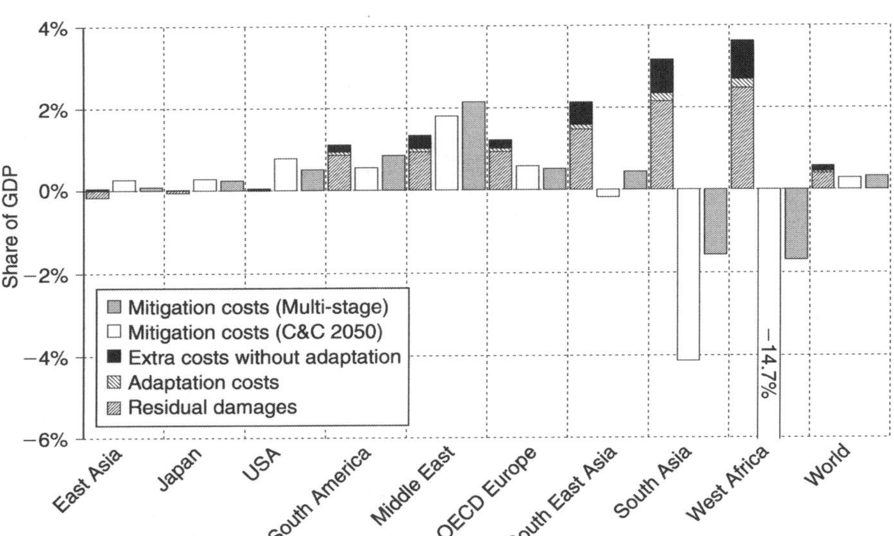

Figure 15.8 Climate costs components in 2030 as share of GDP with concentrations peaking at 550 ppm carbon dioxide-equivalent for selected regions assuming either a 'Contraction and Convergence 2050' (C&C 2050) or a 'multi-stage' regime.

are (slightly) negative in 2030 because of positive impacts on agriculture and non-market amenity value (such as climate-related time use) for small increases in temperature.

The regional differences in the distribution of mitigation costs are very large, both for a 'contraction and convergence 2050' burden-sharing regime and a 'multi-stage' regime. In regions less affected by climate change damages (United States, Japan and East Asia), the largest share of total climate change costs consists of mitigation costs in 2030 for both burden-sharing regimes. Mitigation costs in West Africa and South Asia are projected to be negative in the short to medium run, meaning that these regions will benefit from selling emission permits. The benefits from selling emission permits in these regions can even offset residual damages and adaptation costs with a 'Contraction and Convergence 2050' burden-sharing regime. In a 'multi-stage' regime, however, residual damages and adaptation costs far outweigh the revenues of selling emission permits in both South Asia and West Africa.

Globally, adaptation costs only amount to a small fraction of GDP. The regional differences are large, however. Adaptation costs as share of GDP are especially high in West Africa (about six times the world average in 2030), explained by the high potential damages in this region. Relative adaptation costs are also high in South Asia (about five times the world average in 2030) and South East Asia (about 3.5 times the world average). In absolute terms, about 60 per cent of total global adaptation costs are carried by low- and middle-income regions. Almost all of the remaining adaptation costs are carried by Europe, while the United States and Japan only account for a fraction of total adaptation costs. The low costs of adaptation in the United States, Japan and East Asia are largely explained by the projected low climate change damages of RICE.

The total climate change costs (the sum of adaptation, residual damage and mitigation costs) relative to GDP are the highest in the Middle East, followed by South East Asia in 2030 for both burden-sharing regimes. South America and West Africa will also face relatively high climate change costs in a 'multi-stage' regime, while OECD Europe will face high climate change costs in a 'Contraction and Convergence 2050' regime. In most low- and middle-income regions adaptation is very important - total climate change costs increase by almost 1 per cent of GDP for West Africa and South Asia in 2030 and by 0.5 per cent of GDP for South East Asia if no adaptation takes place. The reason is that in most low- and middle-income regions, potential damages are relatively high, increasing the need for adaptation.

15.3.3 Regional discounted income losses of climate change

The discounted climate change income losses as percentage of discounted income (from now on simply called discounted income loss) over a range of climate mitigation targets provide us with useful information. First, a climate mitigation target can be identified for which the discounted income loss is minimized, providing an indication of the optimal climate mitigation target. However, this strongly depends on the chosen assumptions as the discount rate and damage estimates, as shown earlier by Hof *et al.* (2008). With our assumptions and for a 'multi-stage' burden-sharing regime, the global and regional discounted income loss for different concentration peak levels is shown in Figure 15.9. Global discounted income loss is minimized at a concentration peak target of around 540 ppm carbon dioxide-equivalent, resulting in a discounted income loss of 2.3 per cent.

The regional differences in the climate mitigation target for which discounted income loss is minimized are large. The United States minimizes discounted income loss at a concentration peak higher than the evaluated range of 500 to 800 ppm carbon dioxide-equivalent, due to their low damage estimates. The minimum for East Asia is at around 600 ppm carbon dioxide-equivalent, OECD Europe at about the same level as the global average, and South Asia and East Africa minimize their discounted income loss at a concentration peak level below our analysed range of 500 to 800 ppm carbon dioxide-equivalent, as residual damages in these regions increase rapidly for less stringent climate mitigation targets. These results indicate that it could be difficult for the world to agree to one single climate mitigation target

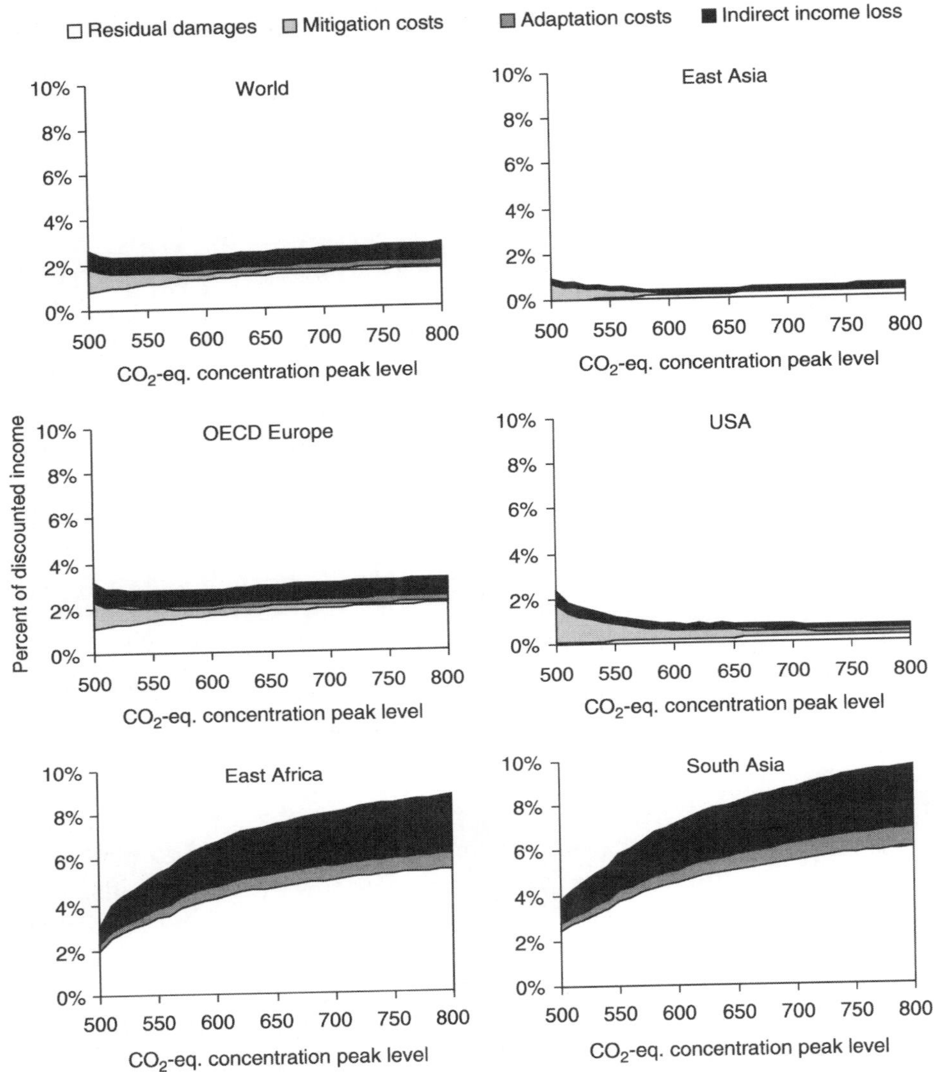

Figure 15.9 Discounted income loss in selected regions due to different climate change costs components for a range of concentration peak levels (in ppm) and a 'multi-stage' burden sharing regime, as percentage of discounted income. Mitigation costs in East Africa and South Asia are slightly negative and included in the residual damage estimates.

(a result which is well known from the game-theoretic literature on international agreements: see Barrett 1994; Finus *et al.* 2006).

Second, the variation in discounted income loss between different climate mitigation targets tells us something about the costs of not reaching the climate mitigation target with lowest costs. For the global economy, these costs are relatively small: for all concentration peak levels between 500 and 800 ppm carbon dioxide-equivalent, discounted income losses lie in the relatively narrow range of 2.3 per cent to 2.8 per cent. For Europe and East Asia, differences in discounted income loss are relatively small as well. For the United States, discounted income loss for concentration peak levels below 600 ppm carbon dioxide-equivalent are projected to increase quite strongly due to rapidly increasing mitigation costs, although the relatively low damages (as estimated by the RICE model) imply that total costs in the United States remain below the world average even for a 500 ppm carbon dioxide-equivalent concentration target. In East Africa and South Asia, the discounted income loss is much higher than the world average, due to higher projected residual damages. Moreover, income losses also increase strongly for higher concentration targets as the impacts of climate change become more severe. This indicates that adaptation is especially important in high-impact regions, which are mostly developing regions.

Finally, it is worth noting that the slope of the residual damage curve in Figure 15.9 for all regions decreases with increasing concentration levels. This may sound counterintuitive, since damages increase exponentially with temperature increase, and higher concentrations imply higher temperatures. The explanation lies in the timing of mitigation. If we want to achieve a very stringent mitigation target, very early emission reductions are required. This means that residual damages will be lowered relatively early as well. On the other hand, if we want to achieve a less stringent concentration target, emissions in the short run can stay the same and should be reduced only in the longer run. Therefore, residual damages are only lowered in the long run as well. Because of discounting, the further residual damages occur in the future, the smaller the impact on the present value of these damages. This explains why the residual damage curve in Figure 15.9 is concave.

15.4 Conclusions

In this study, we used an integrated assessment model to analyse the interactions between adaptation costs, mitigation costs and emissions trading, and residual damages. Our analysis is subject to a number of qualifications and caveats, the most important being the large uncertainties of the damage estimates by the RICE model of Nordhaus and Boyer (2000), on which we base our method for modelling adaptation. Therefore, the results should be interpreted with sufficient care. However, we can still draw a number of conclusions.

First, we have shown that adaptation and mitigation are not substitutes of each other. Adaptation and mitigation have completely different dynamics. Adaptation can effectively reduce climate change damages in the shorter run, but is much less effective in the long run since it does not reduce climate change itself. Mitigation is very effective in reducing climate change damages in the long run. Implementing both adaptation and mitigation gives the best results according to our model.

Second, even though the costs of adaptation are small compared to residual damages and mitigation costs, adaptation is important in the context of reducing potential damages, especially in lower-income regions. Keeping in mind that our results depend on very uncertain estimates of damage and adaptation costs, we project that with optimal adaptation efforts more than a quarter of the potential damages are avoided by adaptation in the long run. Relatively small investments in adaptation could avoid substantial amounts of damages, especially in lower income regions where potential damages are projected to be higher.

Third, the total amount of adaptation in a region depends only in the longer term on the climate mitigation target. Without any mitigation, climate change will be stronger, increasing the need for adaptation. Regardless of the mitigation strategy, adaptation costs will increase strongly over time, as climate will change even if emissions are cut back drastically. Our model projections show that global adaptation costs will increase from USD 6 billion in 2010 to USD 125 billion in 2050.

Fourth, climate change costs differ substantially between regions. For regions such as East Africa and South Asia, income losses are much higher than the global average and rise steeply for higher concentration targets as well. This indicates that both adaptation and mitigation are important especially for these developing regions.

To sum up: adaptation will increase sharply over time even if strong mitigation measures are taken, and adaptation is especially important in developing regions. This indicates that the chances that developing countries join a climate mitigation regime could be higher if adaptation, and especially adaptation funding, is incorporated in such a regime.

References

Agrawala, S. and S. Fankhauser 2008. 'Putting climate change adaptation in an economic context', in S. Agrawala and S. Fankhauser (eds.), *Economic Aspects of Adaptation to Climate Change: Costs, Benefits and Policy Instruments*. Paris: OECD, pp. 19–28.

Agrawala, S., F. Crick, S. Jetté-Nantel and A. Tepes 2008. 'Empirical estimates of adaptation costs and benefits: a critical assessment', in S. Agrawala and S. Fankhauser (eds.), *Economic Aspects of Adaptation to Climate Change: Costs, Benefits and Policy Instruments*. Paris: OECD, pp. 29–84.

Azar, C. 1998. 'Are optimal CO_2 emissions really optimal? Four critical issues for economists in the greenhouse', *Environmental and Resource Economics* **11**: 301–315.

Barrett, S. 1994. 'Self-enforcing international environmental agreements', *Oxford Economic Papers* **46**: 878–894.

Commission of the European Communities 2007. *Green Paper: Adapting to Climate Change in Europe – Options for EU Action*. Brussels: Commission of the European.

de Bruin, K., R. Dellink and S. Agrawala 2009. *Economic Aspects of Adaptation to Climate Change: Integrated Assessment Modelling of Adaptation Costs and Benefits*. Paris: OECD.

den Elzen, M. G. J. and P. L. Lucas 2005. 'The FAIR model: a tool to analyse environmental and costs implications of climate regimes', *Environmental Modeling and Assessment* **10**: 115–134.

den Elzen, M. G. J. and D. P. van Vuuren 2007. 'Peaking profiles: achieving long-term temperature targets with more likelihood at lower costs', *Proceedings of the National Academy of Sciences of the USA* **104**: 17 931–17 936.

den Elzen, M. G. J., M. M. Berk, P. Lucas, P. Criqui and A. Kitous 2006. 'Multi-stage: a rule-based evolution of future commitments under the climate change convention', *International Environmental Agreements: Politics, Law and Economics* **6**: 1–28.

den Elzen, M. G. J., P. L. Lucas and D. P. van Vuuren 2008. 'Regional abatement action and costs under allocation schemes for emission allowances for achieving low CO_2-equivalent concentrations', *Climatic Change* **90**: 243–268.

den Elzen, M. G. J., M. Meinshausen and D. P. van Vuuren 2007. 'Multi-gas emission envelopes to meet greenhouse gas concentration targets: costs versus certainty of limiting temperature increase', *Global Environmental Change* **17**: 260–280.

Dietz, S., C. Hope and N. Patmore 2007. 'Some economics of "dangerous" climate change: reflections on the Stern Review', *Global Environmental Change* **17**: 311–325.

Finus, M., E. van Ierland and R. Dellink 2006. 'Stability of climate coalitions in a cartel formation game', *Economics of Governance* **7**: 271–291.

Gupta, J. 1998. *Encouraging Developing Country Participation in the Climate Change Regime*. Amsterdam: Institute for Environmental Studies (IVM), Vrije Universiteit.

Hof, A. F., M. G. J. den Elzen and D. P. van Vuuren 2008. 'Analysing the costs and benefits of climate policy: value judgements and scientific uncertainties', *Global Environmental Change* **18**: 412–424

Hope, C., J. Anderson and P. Wenman 1993. 'Policy analysis of the greenhouse effect: an application of the PAGE model', *Energy Policy* **21**: 327–338.

IPCC 2001. *Climate Change 2001: Impacts, Adaptation and Vulnerability. Contribution of Working Group II to the Third Assessment Report of the Intergovernmental Panel on Climate Change*. Cambridge, UK: Cambridge University Press.

IPCC 2007a. *Climate Change 2007: Impacts, Adaptation and Vulnerability. Contribution of Working Group II to the Fourth Assessment Report of the Intergovernmental Panel on Climate Change*. Cambridge, UK: Cambridge University Press.

IPCC 2007b. *Climate Change 2007: The Physical Science Basis. Contribution of Working Group I to the Fourth Assessment Report of the Intergovernmental Panel on Climate Change*. Cambridge, UK: Cambridge University Press.

Klein, R. J. T., E. Lisa, F. Schipper and S. Dessai 2005. 'Integrating mitigation and adaptation into climate and development policy: three research questions', *Environmental Science and Policy* **8**: 579–588.

Meyer, A. 2000. *Contraction and Convergence: The Global Solution to Climate Change*, Schumacher Briefings No. 5. Dartington, UK: Green Books.

Miketa, A. 2004. *Technical Description on the Growth Study Datasets*. Environmentally Compatible Energy Strategies Program, Laxenburg, Austria: International Institute for Applied Systems Analysis. www.iiasa.ac.at/Research/ECS/data_am/index.html.

Nakicenovic, N., J. Alcamo, G. Davis, B. de Vries, J. Fenhann, S. Gaffin, K. Gregory, A. Grübler, T. Y. Jung, T. Kram, E. Emilio la Rovere, L. Michaelis, S. Mori, T. Morita, W. Pepper, H. Pitcher, L. Price, K. Riahi, A. Roehrl, H. Rogner, A. Sankovski, M. Schlesinger, P. Shukla, S. Smith, R. Swart, S. van Rooyen, N. Victor and Z. Dadi 2000. *IPCC Special Reports: Special Report on Emissions Scenarios*. Cambridge, UK: Cambridge University Press.

Nordhaus, W. D. 1994. *Managing the Global Commons: The Economics of Climate Change*. Cambridge, MA: MIT Press.

Nordhaus, W. D. 2008. *A Question of Balance: Weighing the Options on Global Warming Policies*. New Haven, CT: Yale University Press.

Nordhaus, W. D. and J. Boyer 2000. *Warming the World: Economic Models of Global Warming*. Cambridge, MA: MIT Press.

Tol, R. S. J. 2002a. 'Estimates of the damage costs of climate change: part I – benchmark estimates', *Environmental and Resource Economics* **21**: 47–73.

Tol, R. S. J. 2002b. 'Estimates of the damage costs of climate change: part II – dynamic estimates', *Environmental and Resource Economics* **21**: 135–160.

Tol, R. S. J. 2005. 'Adaptation and mitigation: trade-offs in substance and methods', *Environmental Science and Policy* **8**: 572–578.

United Kingdom Treasury 2003. *The Green Book: Appraisal and Evaluation in Central Government*. London: The Stationery Office.

UNFCCC 2007. *Climate Change: Impacts, Vulnerabilities and Adaptation in Developing Countries*. Bonn: UNFCCC.

van Vuuren, D. P. and B. O'Neill 2006. 'The consistency of IPCC's SRES scenarios to 1990–2000 trends and recent projection', *Climatic Change* **27**: 9–46.

van Vuuren, D. P., M. G. J. den Elzen, B. Eickhout, P. L. Lucas, B. J. Strengers and B. van Ruijven 2007. 'Stabilising greenhouse gas concentrations. assessment of different strategies and costs using an integrated assessment framework', *Climatic Change* **81**: 119–159.

Weitzman, M. L. 2007. 'A review of the *Stern Review on the Economics of Climate Change*', *Journal of Economic Literature* **45**: 703–724.

Wigley, T. M. L. 2003. *MAGICC&/SCENGEN 4.1: Technical Manual*. Boulder, CO: UCAR, Climate and Global Dynamics Division.

Wigley, T. M. L. and S. C. B. Raper 2001. 'Interpretation of high projections for global-mean warming', *Science* **293**: 451–454.

Wigley, T. M. L. and S. C. B. Raper 2002. 'Reasons for larger warming projections in the IPCC Third Assessment Report', *Journal of Climate* **15**: 2945–2952.

Wilbanks, T. J. 2005. 'Issues in developing a capacity for integrated analysis of mitigation and adaptation', *Environmental Science and Policy* **8**: 541–547.

World Bank 2006. *Clean Energy and Development: Towards an Investment Framework*. Washington, DC: The World Bank.

World Bank 2008. *World Development Indicators*. http://go.worldbank.org/U0FSM7AQ40.

16

Global adaptation governance: the case of protecting climate refugees

FRANK BIERMANN AND INGRID BOAS

16.1 Introduction

Climate change is likely to cause or increase regional migration, for example when coastal areas are given up because of sea-level rise and more frequent and severe storm surges. The future problem of 'climate refugees' is often associated with smaller island states, such as the Maldives or Tuvalu. Indeed, in 2006 the Republic of the Maldives (2006) organized a first meeting of representatives of governments, humanitarian organizations and United Nations agencies on this issue. Yet climate-related migration could also evolve into a much larger, global crisis far beyond threats to a few island nations. According to some estimates, by 2050 more than 200 million people might flee their homes due to climate change (Myers and Kent 1995; Myers 2002; see also the discussion in Stern 2007: ch. 3.5). Such estimates have a large margin of error and depend on underlying assumptions about population growth, economic development, temperature increase or the degree and timing of climate change impacts such as sea-level rise (critically on such estimates see Suhrke 1994: 478; Black 2001: 2–8; Castles 2002: 2–3). Yet most scenarios agree on a general trend: that global warming might force millions of people to migrate to other places.

Most of these climate refugees are likely to come from developing nations (Barnett *et al.* 2005: 306; Nagy *et al.* 2006; Warren *et al.* 2006: 18, 20, 41–42). Most victims of sea-level rise are expected in Asia: 26 million people in Bangladesh and 73 million people from China could be forced to flee their homes because of sea-level rise (Myers 2002: 211). In Africa and to a lesser degree in Latin America many people may also experience flooding (Myers and Kent 1995: 148; Nicholls *et al.* 1999: 80; Myers 2002: 611; Nicholls 2003: 16; Brooks *et al.* 2006: 6). For many smaller island states, their very existence may be threatened by sea-level rise (German Advisory Council on Global Change 2006: 46, 50).

Global Climate Governance Beyond 2012: Architecture, Agency and Adaptation, eds. F. Biermann, P. Pattberg and F. Zelli. Published by Cambridge University Press.

How to deal with this emerging crisis of large-scale internal and transnational climate refugees? At present, the political responses to this problem are uncertain and hardly studied in policy research.

Contestation and uncertainty begins with the appropriate terminology: should these people be called 'refugees' in the first place? The term 'environmental refugee' was popularized by the United Nations Environment Programme in the 1980s (El-Hinnawi 1985). In 1992, Agenda 21 – the influential intergovernmental programme of action agreed upon by almost all governments at the 1992 United Nations Conference on Environment and Development – also used the term 'environmental refugees' repeatedly (United Nations 1992a: ch. 12, especially 12.4, 12.46 and 12.47). The notion of 'climate refugees' appears to find acceptance in some national political debates too. For example, Australia's Labor Party had proposed an international coalition to accept climate refugees from the Pacific (Albanese and Sercombe 2006) – in response to the (then) Australian government's position that rejected the notion of climate refugees (Renaud *et al.* 2007: 20–21) – and Australia's Greens party tabled in 2007 even a Migration Amendment (Climate Refugees) Bill (NSW Greens Party 2007). However, some intergovernmental agencies – such as the International Organization for Migration and the UN High Commissioner for Refugees – prefer the term 'environmentally displaced persons' (International Organization for Migration 1996; UNHCR 2002: 13; Keane 2004: 215). They reject the term 'environmental refugee' or 'climate refugee' because of the legal rights that the intergovernmental system currently bestows upon 'refugees'. In their view, the term 'refugee' should remain limited to transboundary flight, mainly because the 1951 Geneva Convention Relating to the Status of Refugees is restricted to persons who cannot avail themselves of the protection of their home state for fear of persecution.

We support the use of the term climate 'refugee' for two main reasons. First, the distinction between transboundary and internal flight that is a core element of the traditional 'refugee' concept does not help much since climate change will cause both transnational and internal flight. Some island nations will effectively cease to exist, and some countries, especially those affected by drought, will be overburdened by the degree of the national predicament. These people will have to find refuge outside their home country. Some climate refugees might thus cross borders while most will stay within their country – it seems difficult to argue that a global governance mechanism for their protection should bestow a different status, and a different term, depending on whether they have crossed a border. Second, we see no convincing reason to reserve the stronger term 'refugee' for a category of people that stood at the centre of attention after 1945, and to invent less appropriate terms – such as 'climate-related environmentally displaced persons' – for new categories of people who are forced to leave their homes now, with similar grim consequences. Why should inhabitants of some atolls in the Maldives who require resettlement for reasons of a well-founded fear of being inundated by 2050 receive less protection than others who fear political persecution? The term 'refugee' has strong moral connotations of societal protection in most world cultures and religions. By using this term, the protection of climate refugees will receive the legitimacy and urgency it deserves.

What are the most appropriate governance responses to the looming crisis of climate refugees? What is the potential performance of current governance mechanisms in this field? And if they are deemed ineffective, what other governance mechanisms could be developed? These questions are the focus of this chapter.[1] We first review in Section 16.2 our methodology, and then analyse in Section 16.3 three existing governance areas: refugee governance, human rights governance and security governance. In all three areas, current governance mechanisms are not sufficient to deal with the estimated millions of climate refugees. In Section 16.4, we thus develop a proposal for a new legal regime on the recognition, protection and resettlement of climate refugees.

16.2 Methodology

This study is based on policy analysis, combined with a meta-analysis of existing studies that seek to estimate likely numbers and origins of climate refugees. As we have laid out in Chapter 14 in more detail, such policy analysis is no easy task, since it needs to assess the effectiveness of possible future governance mechanisms against a possible future crisis, given that both large streams of climate refugees and governance mechanisms to respond to this crisis are not yet existing. We thus needed to assess first the potential effectiveness of current governance systems and then options for reform within its current institutional set-up or the development of new and additional governance mechanisms specifically tailored to the needs of climate refugees. We employ three criteria for the performance of governance mechanisms: political effectiveness in terms of ability to protect climate refugees; political feasibility of reform options; and equity, considering both procedural equity in decision-making and the distribution of costs and benefits. In the following sections, we assess on the basis of these criteria the current and the likely future performance of governance mechanisms to the looming climate refugee crisis in order to determine the most effective political response.

16.3 Analysis

16.3.1 Introduction

The study of the effectiveness of any system of governance to protect climate refugees needs to begin with a careful identification of type of problems that these

[1] Parts of this chapter draw on Biermann and Boas (2008).

people are faced with. Four characteristics of climate refugees make them a group of migrants clearly distinct from other groups:

First, climate refugees are predictable within limits. Even though extreme weather events and droughts are not foreseeable as such, the need of migration for many affected regions is evident, especially in some island nations, poorer low-lying areas and arid regions affected by increasing water scarcity. Unlike the movement of political refugees, climate-related migration could therefore be a planned, voluntary, organized process of resettlement, assuming appropriate national and global governance structures. Second, unlike political or war refugees, climate refugees are unlikely to return to their homes. Even though the option of return has been unrealistic for many victims of political persecution or civil war, the assumption of eventual return has been one foundation of refugee governance. Climate refugees, on their part, do not require temporary asylum, but a new home. Third, climate refugees are likely to migrate in large numbers and collectively. While current refugee governance builds on the individual person persecuted by a public authority in his or her country, climate migration is in principle a collective phenomenon that entails entire villages, cities, provinces and at times entire nations, and that hence calls for collective responses.

Fourth, climate refugees differ from political and war refugees through the moral and legal embedding of their situation. Political and war refugees are victims of their home state or of a regionalized conflict, with no direct responsibility for their plight with the countries that eventually offer refuge. The moral responsibility for climate change is different. The historic causation of climate change is in particular the responsibility of a few rich countries in North America and Europe, especially when emissions per capita and past emissions are considered. Conversely, almost all climate refugees are likely to come from countries that are least responsible for climate change and least able to finance and implement adaptation programmes. This situation creates a moral link between impoverished climate refugees and the richer countries that have the means to provide funds and refuge for the victims of climate change. Climate refugees thus have a status different from political refugees and a special moral claim vis-à-vis industrialized countries. As argued by Barnett, refusing climate refugees to enter would be 'particularly for industrialized countries, morally difficult to sustain since it is their emissions that will have caused the problem' (Barnett 2003: 12).

The legal significance of this moral link is open to debate. It is a general principle of law, also of the law of nations, that a wrongful act creates duties of liability including the compensation of victims. Applied to climate change impacts, climate refugees may be entitled to demand compensation payments or other assistance from industrialized countries (see in general, though not related to climate refugees per se, the discussion in Biermann 1995; Müller 2002: 71, 73; German Advisory

Council on Global Change 2007: 174). At present, however, governments of industrialized countries reject any claims of legal liability towards possible victims of climate change. A common legal argument is that it is difficult to relate past emissions of industrialized countries to current or future impacts of climate change (discussed in Grubb 1995: 474; Paavola and Adger 2002: 13). However, global climate change has barely begun, and increasing pressure from climate victims, and their governments, might result in innovative compromises that grant special status to populations that need to be relocated due to sea-level rise or increased severity of extreme weather events.

In sum, any global governance architecture to recognize, protect and support the resettlement of climate refugees must account for these four characteristics of climate refugees: the broad predictability of their plight, the impossibility of their return, the collectivity of their flight, and the special moral and possibly legal responsibility of the rich countries in the North.

We now review three governance areas that could play a role in protecting climate refugees: refugee governance, human rights governance and security governance. In all three domains, climate refugees are today part of the broader debate, although not yet in the more narrow sense of policy-making. We discuss in these three domains the current governance structure, options for reform within the current governance structure and the creation of new structures and additional institutions specifically to deal with climate refugees.

16.3.2 Refugee governance

One potential political response to the looming climate refugee crisis would be to mandate the current refugee institutions to also protect climate refugees. At present, the 1951 Geneva Convention Relating to the Status of Refugees, its 1967 Protocol (UNHCR 2007a) and the UN High Commissioner for Refugees protect about 10 million recognized refugees. Yet climate refugees are excluded from this protection, since the current legal definition of refugees covers only persons who flee from individual state-led persecution (UNHCR 2007b; discussed in McGregor 1994, 126; Williams 2008: 507–509). The UN High Commissioner for Refugees has some programmes for people who are not recognized as refugees, such as internally displaced persons, which could include climate refugees (UNHCR 2006: 5 and 12, 2007c). Nonetheless, this term does not impose strong obligations on states (Keane 2004: 217).

One option for reform would thus be to extend the mandate of the 1951 Geneva Convention and of the UN High Commissioner for Refugees to cover climate refugees. At the Maldives meeting in 2006, delegates proposed therefore an amendment to the 1951 Geneva Convention Relating to the Status of Refugees that would extend the mandate of the UN refugee regime to cover also climate refugees (Maldives 2006). Yet such an amendment leads into the wrong direction. It does not promise to resolve effectively the emerging climate refugee crisis. To start with, the political feasibility of this proposal is highly uncertain. Already today, the UN refugee regime is under constant pressure from industrialized countries that seek restrictive interpretations of its provisions. It is unrealistic that governments will extend the same level of protection to 20 times more climate refugees, which is equal in numbers to half the population of the European Union. Second, an extension of the current UN refugee regime to include climate refugees will raise difficult moral issues. It will create an unnecessary tension and trade-offs between the persons protected today under the Geneva Convention and the new additional streams of climate refugees. It is also highly doubtful whether the current institutional apparatus of the UN High Commissioner for Refugees, and the personalized refugee regime under the Geneva Convention, would be able effectively to protect and support a much larger stream of climate refugees.

More importantly, the proposal of an extension of the UN refugee regime misses the core characteristics of the climate refugee crisis. Climate refugees do not have to leave their countries because of a totalitarian government. In principle, they still enjoy the protection of their government and their country. The protection of climate refugees is therefore essentially a development issue. It requires large-scale, long-term planned resettlement programmes for groups of affected people, mostly within their country. Often this will be in concert with adaptation programmes for other people who are not evacuated but can still be protected, for instance through strengthened coastal defences. It is therefore not the UN High Commissioner for Refugees but other international agencies such as the UN Development Programme or the World Bank that are called upon to deal with the emerging problem of climate refugees. Climate refugees must surely be protected and require international assistance. Yet this help will differ from the protection of political refugees who flee a totalitarian government. The legal and political responses to both types of refugees must thus be different.

16.3.3 Human rights governance

Second, climate-related migration is increasingly framed as an issue of human rights. Several development and human rights organizations now argue that climate change has severe impacts on the livelihoods of the poor and that human rights concerns should guide climate policies (for example International Council on Human Rights Policy 2008; Leckie 2008: 18–19; Oxfam International 2008). In March 2008, the UN Human Rights Council adopted a resolution on human rights and climate change (UNHRC 2008) and decided to further develop this issue through a detailed study by the UN High Commissioner for Human Rights (OHCHR 2009). This study will be made available to the fifteenth conference of the parties to the climate convention in 2009.

Framing climate change as a potential infringement of human rights seems justified and can help to raise awareness at both local and international levels. The approach can also galvanize political action, for example in the human-rights-oriented proposal by Oxfam International (2008: 3–4) which promotes development of international and national adaptation programmes. Yet from a governance perspective, it is questionable to what extent the legal institutions and supporting bureaucracies in the field of human rights can deliver the rule-setting and rule-implementing mechanisms that are needed. To start with, many provisions in this area lack strong enforcement. For instance, the Office of the UN High Commissioner for Human Rights (1998: article 2) currently supports internally displaced persons, including persons displaced due to environmental degradation. Nevertheless, governments are not obliged to any particular activities, and the main responsibility to protect environmentally displaced persons remains with the home state rather than with the international community (see discussion in OHCHR 1998: principle 3; also Keane 2004: 217). Moreover, the core mandate of the Human Rights Council is monitoring and promoting human rights (UN General Assembly 2006), yet with no authority to command detailed international action. In essence, the same critique as with the refugee regime applies also here: links with the climate regime are limited, detailed standards of assistance difficult and supporting bureaucracies and agencies too weak to sufficiently deal with the threat of millions of additional refugees.

16.3.4 Security governance

Finally, the spectre of large-scale climate-related migration has led to an emerging 'securitization' of the problem. The German Advisory Council on Global Change, for example, has warned that '[i]f global temperatures continue to rise unabated, migration could become one of the major fields of conflict in international politics in future' (2007: 174). A recent report by the European Commission argued that climate-related migration should be seen as a security issue and addressed within the context of European defence planning (Council of the European Union 2008). Similar perceptions can be found in domestic discourses on long-term defence planning in a number of countries, including the United Kingdom (DCDC 2007) and United States (Military Advisory Board 2007). In 2007, the Nobel Peace Prize was awarded jointly to US Senator Albert Gore and the Intergovernmental Panel on Climate Change on the grounds of links between climate change and the preservation of peace. Last but not least, the effects of climate change on world peace and security have been formally addressed by the United Nations Security Council in April 2007 (United Nations Security Council 2007; see for a review of the debate, Sindico 2007). In short, climate change and its impacts, including the expected millions of climate refugees, are increasingly seen as part and parcel of security policy in industrialized countries. Climate change is becoming 'securitized'.

Yet the question arises, from a governance perspective, whether security institutions are also the best option to address the plight of climate refugees and related problems. In principle the Security Council is the most powerful international institution. A decision by the Security Council that a refugee situation threatens world peace and security would trigger a range of options under Chapter VII of the UN Charter, including military intervention. Moreover, the issue would gain salience in political debate. Some observers, such as the German Advisory Council on Global Change (2007: 195–196), have thus welcomed the increasing attention of the Security Council for climate change.

Yet an extended mandate of the Security Council does not promise to be feasible, effective or equitable. The Council is marked by strong influence of a few major countries, all of which belong to the group of the largest greenhouse gas emitters. Poorer and vulnerable developing countries are less represented. For this reason, many developing countries question the authority of the Security Council in this field. Actions by the Security Council could also impinge on the internal affairs of developing countries, since climate-related migration and related conflicts are more likely to occur in the South. In the 2007 Security Council debate, developing countries thus opposed any role of the Security Council in climate change (see United Nations Security Council 2007; Sindico 2007).

16.4 Conclusions and policy recommendations

For these reasons, we argue for a different approach.[2] The solution will neither be an extension of the legal definition of refugees under the Geneva Convention and the UN High Commissioner for Refugees, nor an active governance role for human rights institutions, nor an extended mandate of the UN Security Council. Instead, we argue for a separate legally binding agreement on the recognition, protection and resettlement of climate refugees under the United Nations Framework Convention on Climate Change. This could be either an independent 'climate refugee protocol' to the convention or a distinct part of a larger new agreement, such as an adaptation protocol. A climate refugee protocol could build on the political support from almost all countries as parties to the climate convention. It could draw on widely agreed principles, such as common but differentiated responsibilities and the

[2] This section draws on Biermann and Boas (2008).

reimbursement of full incremental costs. It could support the protection of climate refugees by linking their protection with the overall climate regime, including future advances in climate science in defining risks for people in certain regions.

We propose five principles for such an agreement on the recognition, protection and resettlement of climate refugees. The first principle is that at its core must not be programmes on emergency response and disaster relief, but on planned and voluntary resettlement and reintegration over periods of many years and decades. Spontaneous flights, as in the case of political turmoil or war, can then be avoided. The second principle is that climate refugees must be seen, and treated, as permanent immigrants to the regions or countries that accept them. Climate refugees cannot return to their homes, as political refugees can (at least in theory). The third principle is that the climate refugee regime must be tailored not to the needs of individually persecuted people (as in the current UN refugee regime), but of entire groups of people, such as populations of villages, cities, provinces or even entire nations, as in the case of small island states. Fourth, an international regime for climate refugees will be targeted less at the protection of persons outside their states but rather at the support of governments, local communities and national agencies to protect people within their territory. The governance challenge of protecting and resettling climate refugees is thus essentially about international assistance and funding for the domestic support and resettlement programmes of affected countries that have requested such support. Fifth and finally, the protection of climate refugees must be seen as a global problem and a global responsibility. In most cases, climate refugees will be poor, and their own responsibility for the past accumulation of greenhouse gases will be small. By a large measure, the rich industrialized countries have caused most emissions in past and present, and it is thus these countries that have most moral, if not legal, responsibility for the victims of global warming. This does not imply transnational migration of 200 million climate refugees into the North. Yet it does imply the responsibility of the industrialized countries to do their share in financing, supporting and facilitating the protection and the resettlement of climate refugees.

How could a protocol (or other type of legally binding agreement) on the recognition, protection and resettlement of climate refugees work in practice?

In our proposal, the most important governance mechanism would be a list of specified administrative areas (such as villages, islands or districts) under the jurisdiction of member states whose population is determined to be ‘in need of relocation due to climate change’ or ‘threatened by having to relocate due to climate change’. Any state party to the protocol – and in fact only state parties – would be entitled to propose areas under its jurisdiction for inclusion into the list of affected areas. The protocol would provide for an executive committee on the recognition, protection and resettlement of climate refugees that would function under the authority of the meeting of the parties (which could meet back-to-back with the conference of the parties to the climate convention). In line with the sovereignty principle of the United Nations, the executive committee would determine the inclusion of affected areas, as well as the type of support measures, only upon formal proposal from the government of the affected country. Regarding decision-making procedures, the executive committee could include an equal number of affected countries and donor countries, and its decisions could require the simple majority of donor countries and the simple majority of affected countries (so-called double-weighted majority rule). This rule would allow both the affected developing countries and the donor countries to hold a collective veto right over the future evolution and implementation of the regime.

In our proposal, if certain groups of people, for example from a number of coastal villages, were included in this 'list of populations in need of relocation due to climate change', they would gain certain rights and would benefit from the support mechanisms under the protocol. This could include financial support; inclusion in voluntary resettlement programmes over several years together with the purchase of new land; retraining and integration programmes; and, in the special case of small island states, organized international migration. Since richer countries will be able to support their own affected populations, the rights under the protocol that we propose should be restricted to inhabitants of developing countries (in technical terms: countries that are not listed in Annex I to the climate convention).

Resettlement of millions of people will require additional, and most likely substantial, funds. Institutionally, the best governance mechanism would be a separate fund, such as a Climate Refugee Protection and Resettlement Fund. The funding for climate refugees could be integrated into existing funding mechanisms of the climate convention, such as the Global Environment Facility, the Clean Development Mechanism or the Special Climate Change Fund (Paavola and Adger 2002: 12; Richards 2003: 6–7 for an overview and discussion of these funds). Yet the governance of the climate refugee fund should be independent and stand under the authority of the meeting of the parties to the climate refugee protocol. In order to generate the funds needed, the Climate Refugee Protection and Resettlement Fund could be coupled with novel income-raising mechanisms that are currently proposed, such as an international air travel levy (Müller and Hepburn 2006). A key question for this new facility will be the amount of funding required by the international community, and the funding principle underlying their protection.

For mitigation programmes under the climate convention, industrialized countries have committed to reimburse developing countries the agreed full incremental costs, a concept originally developed in the 1990 London amendments of the ozone regime (Biermann 1997: 179–218). Similar provisions apply to adaptation.[3] In addition, the climate convention obliges industrialized countries to assist the most vulnerable countries in meeting adaptation costs (article 4.4) and gives special rights to least developed countries (article 4.9). This suggests applying the principle of reimbursement of full incremental costs also to the protection and resettlement of climate refugees at least in situations where the causal link with climate change is undisputed, namely sea-level rise. For other situations in which climate change is only one factor to account for environmental degradation – for example in the case of water scarcity – a principle of additional funding instead of full reimbursement is probably more appropriate. In any case, the costs of the voluntary resettlement and reintegration of millions of people who have to leave their islands, coastal plains or arid areas will be substantial and probably in the order of billions of Euros over the coming decades. Even if novel mechanisms will be introduced, such as an international air travel levy, the final responsibility for funding will rest with the governments of the industrialized countries and possibly richer developing countries.

The climate refugee protocol that we propose will not create new international bureaucracies. Instead, the resettlement of millions of climate refugees over the course of the century will be the task of existing agencies. Given the complexity of climate-related flight, the best model will be to mandate not one single agency but rather a network of agencies as 'implementing agencies' of the protocol. A crucial role will lie with the UN Development Programme and the World Bank, both of which could serve as implementing agencies for the climate refugee protocol in the planned voluntary resettlement of affected populations. The UN Environment Programme, even though it lacks a strong operational mandate, may provide further assistance in terms of scientific research and synthesis, information dissemination, legal and political advice and other core functions of this programme. A small coordinating secretariat to the protocol on climate refugees would be needed, possibly as a subdivision of the secretariat of the Climate Secretariat in Bonn. In addition, the UN High Commissioner for Refugees will play a role, even though it is unlikely to be the main agency given the special characteristics of the climate refugee crisis. Yet the expertise of the High Commissioner in view of emergencies, as well as its legal and technical expertise in dealing with refugee crises, will be indispensable also for the protection of climate refugees.

[3] Article 4 paragraph 3 of the climate convention reads: 'The developed country Parties and other developed Parties included in Annex II shall provide new and additional financial resources ... including for the transfer of technology, needed by the developing country Parties to meet the agreed full incremental costs of implementing measures that are covered by paragraph 1 of [article 4] and that are agreed between a developing country Party and the international entity or entities referred to in Article 11, in accordance with that Article.' Paragraph 1 of article 4 includes in letter (e) the commitment of developing countries to 'cooperate in preparing for adaptation to the impacts of climate change and develop and elaborate appropriate and integrated plans for coastal zone management, water resources and agriculture, and for the protection and rehabilitation of areas, particularly in Africa, affected by drought and desertification, as well as floods' (United Nations 1992b).

Scientists predict serious impacts of climate change that could compel millions of people to leave their homes beginning sometime in the next decades. Yet the existing institutions and organizations are not sufficiently equipped to deal with this looming crisis. Reforms towards a system of 'global adaptation governance' are thus needed. As we have laid out in this chapter, some of the possible reform options are less promising. In particular, we argue against an extension of the definition of refugees under the 1951 Geneva Convention Relating to the Status of Refugees. We also do not see much use in a role for human rights institutions from a governance perspective, or in an extended mandate of the UN Security Council, which might even be counterproductive. Instead, we call in this chapter for a new legal instrument specifically tailored for the needs of climate refugees.

The broad predictability of climate change impacts requires, and allows for, preparation and planning. We have thus framed our proposal deliberatively not in terms of emergency response and disaster relief, but of planned and organized voluntary resettlement programmes. There is no need to wait for extreme weather events to strike and islands and coastal regions to be flooded. All areas that we cannot protect over the long-term through increased coastal defences, for practical or economic reasons, need to be included early in long-term resettlement and reintegration programmes that make the process acceptable and endurable for the affected people. This, however, calls for *early action* in terms of setting up effective and appropriate global governance mechanisms.

References

Albanese, A. and B. Sercombe 2006. Labor calls for international coalition to accept climate change refugees. Press release, 9 October 2006. www.anthonyalbanese.com.au/news/1398/index.html.

Barnett, J. 2003. 'Security and climate change', *Global Environmental Change* **13**: 7–17.

Barnett, T. P., J. C. Adam and D. P. Lettenmaier 2005. 'Potential impacts of a warming climate on water availability in snow-dominated regions', *Nature* **438**: 303–309.

Biermann, F. 1995. *Saving the Atmosphere. International Law, Developing Countries and Air Pollution*. Frankfurt am Main: Lang.

Biermann, F. 1997. 'Financing environmental policies in the South: experiences from the multilateral ozone fund', *International Environmental Affairs* **9**: 179–218.

Biermann, F. and I. Boas 2008. 'Protecting climate refugees: the case for a global protocol', *Environment* **50**(6): 8–16.

Black, R. 2001. *Environmental Refugees: Myth or Reality?*, New Issues in Refugee Research Working Paper No. 34. Geneva: United Nations High Commissioner for Refugees.

Brooks, N., R. J. Nicholls and J. Hall 2006. *Sea Level Rise: Coastal Impacts and Responses*, external expertise for the WBGU report *The Future Oceans: Warming Up, Rising High, Turning Sour*. Berlin: German Advisory Council on Global Change.

Castles, S. 2002. *Environmental Change and Forced Migration: Making Sense of the Debate*, New Issues in Refugee Research Working Paper No. 70. Geneva: United Nations High Commissioner for Refugees.

Council of the European Union 2008. *Climate Change and International Security*, Doc: 7249/08, 3 March 2003. Brussels: Council of the European Union. www.euractiv.com/29/images/SolanaCCsecurity%20reportpdf_tcm29-170886.pdf.

Development, Concepts and Doctrine Centre (DCDC) 2007. *The DCDC Global Strategic Trends Programme 2007–2036*, 3rd edn. London: The Stationery Office. www.mod.uk/NR/rdonlyres/94A1F45E-A830-49DB-B319-DF68C28D561D/0/strat_trends_17mar07.pdf.

El-Hinnawi, E. 1985. *Environmental Refugees*. Nairobi: United Nations Environment Programme.

German Advisory Council on Global Change 2006. *The Future Oceans: Warming Up, Rising High, Turning Sour*. Berlin: German Advisory Council on Global Change.

German Advisory Council on Global Change 2007. *World in Transition: Climate Change as a Security Risk*. Berlin: German Advisory Council on Global Change.

Grubb, M. 1995. 'Seeking fair weather: ethics and the international debate on climate change', *International Affairs* **71**: 463–469.

International Council on Human Rights Policy 2008. *Climate Change and Human Rights: A Rough Guide*. Geneva: International Council on Human Rights Policy. www.ichrp.org/files/reports/36/136_report.pdf.

International Organization for Migration 1996. *Environmentally Induced Population Displacements and Environmental Impacts Resulting from Mass Migration*, international symposium, Geneva, 21–24 April 1996. Geneva: International Organization for Migration.

Keane, D. 2004. 'The environmental causes and consequences of migration: a search for the meaning of "environmental refugees"', *Georgetown International Environmental Law Review* **16**: 209–223.

Leckie, S. 2008. 'Human rights implications', in M. Couldrey and M. Herson (eds.), *Climate Change and Displacement*, Forced Migration Review No. 31. Oxford, UK: Refugee Studies Centre. www.fmreview.org/FMRpdfs/FMR31/FMR31.pdf.

Maldives, Republic of (Ministry of Environment, Energy and Water) 2006. *Report on the First Meeting on Protocol on Environmental Refugees: Recognition of Environmental Refugees in the 1951 Convention and 1967 Protocol Relating to the Status of Refugees*, Male, Maldives, 14–15 August 2006. Male, Maldives: Republic of Maldives (Ministry of Environment, Energy and Water). On file with authors.

Military Advisory Board 2007. *National Security and the Threat of Climate Change*. Alexandria: The CNA Corporation.

McGregor, J. 1994. 'Climate change and involuntary migration: implications for food security', *Food Policy* **19**: 120–132.

Müller, B. 2002. *Equity in Climate Change: The Great Divide*. Oxford, UK: Oxford Institute for Energy Studies.

Müller, B. and C. Hepburn 2006. *IATAL: An Outline Proposal for an International Air Travel Adaptation Levy*. Oxford, UK: Oxford Institute for Energy Studies.

Myers, N. 2002. 'Environmental refugees: a growing phenomenon of the 21st century', *Philosophical Transactions of the Royal Society of London B* **357**: 609–613.

Myers, N. and J. Kent 1995. *Environmental Exodus: An Emergent Crisis in the Global Arena*. Washington, DC: Climate Institute.

Nagy, G. J., R. M. Caffera, M. Aparicio, P. Barrenechea, M. Bidegain, J. C. Giménez, E. Lentini, G. Magrin and coauthors 2006. *Understanding the Potential Impact of Climate Change and Variability in Latin America and the Caribbean: Executive Summary*, report prepared for the *Stern Review on the Economics of Climate Change*. London: The Stationery Office. www.hm-treasury.gov.uk/stern_review_supporting_documents.htm.

Nicholls, R. J. 2003. *Case Study on Sea-Level Rise Impacts*, OECD workshop on the benefits of climate policy: improving information for policy-makers. Paris: Organization for Economic Cooperation and Development.

Nicholls, R. J., F. M. J. Hoozemans and M. Marchand 1999. 'Increasing flood risk and wetland losses due to global sea-level rise: regional and global analyses', *Global Environmental Change* **9**: 69–87.

NSW Greens Party 2007. Climate refugees bill introduced into parliament. Press release, 21 June 2007. www.nsw.greens.org.au/media-centre/news-releases/climate-refugees-bill-introduced-into-parliament.

Office of the United Nations High Commissioner for Human Rights (OHCHR) 1998. *Guiding Principles on Internal Displacement*, Document E/CN.4/1998/53/Add.2, Geneva, 11 February 1998. Geneva: Office of the United Nations High Commissioner for Human Rights.

OHCHR 2009. *Report of the Office of the United Nations High Commissioner for Human Rights on the Relationship between Climate Change and Human Rights*, Document A/HRC/10/61, New York, 15 January 2009. New York: United Nations.

Oxfam International 2008. *Climate Wrongs and Human Rights: Putting People at the Heart of Climate Change Policy*, Oxfam briefing paper No. 117. Oxford, UK: Oxfam International.

Paavola, J. and W. N. Adger 2002. *Justice and Adaptation to Climate Change*, Tyndall Centre Working Paper No. 23. Norwich, UK: Tyndall Centre for Climate Change Research.

Renaud, F., J. J. Bogardi, O. Dun and K. Warner 2007. *Control, adapt or flee: how to face environmental migration? Interdisciplinary Security Connections 5*. Bonn: United Nations University Institute for Environment and Human Security (UNU-EHS).

Richards, M. 2003. *Poverty Reduction, Equity and Climate Change: Global Governance Synergies or Contradictions*? London: Overseas Development Institute. www.odi.org.uk/iedg/publications/climate_change_web.pdf.

Sindico, F. 2007. 'Climate change: a security (council) issue?', *Carbon and Climate Law Review* **1**: 29–34.

Stern, N. 2007. *The Stern Review on the Economics of Climate Change*. Cambridge, UK: Cambridge University Press. www.hm-treasury.gov.uk/stern_review_report.htm.

Suhrke, A. 1994. 'Environmental degradation and population flows', *Journal of International Affairs* **47**: 473–496.

United Nations 1992a. *Agenda 21: The United Nations Programme of Action from Rio*. Rio de Janerio: United Nations. www.un.org/esa/sustdev/documents/agenda21/english/agenda21toc.htm.

United Nations 1992b. *United Nations Framework Convention on Climate Change*. New York: United Nations.

United Nations General Assembly 2006. *Human Rights Council*, Resolution 60/251, 15 March 2006. New York: United Nations. www2.ohchr.org/english/bodies/hrcouncil/docs/A.RES.60.251_En.pdf.

United Nations High Commissioner for Refugees (UNHCR) 2002. 'Environmental migrants and refugees', *Refugees* **127**: 12–13.

UNHCR 2006. *Internally Displaced People: Questions and Answers*. Geneva: United Nations High Commissioner for Refugees. www.unhcr.org/basics/BASICS/405ef8c64.pdf.

UNHCR 2007a. *Convention and Protocol Relating to the Status of Refugees*. Geneva: United Nations High Commissioner for Refugees. www.unhcr.org/protect/PROTECTION/3b66c2aa10.pdf.

UNHCR 2007b. *Statute of the Office of the United Nations High Commissioner for Refugees*. Geneva: United Nations High Commissioner for Refugees. www.unhcr.org/protect/PROTECTION/3b66c39e1.pdf.

UNHCR 2007c. *Mission Statement*. Geneva: United Nations High Commissioner for Refugees. www.unhcr.org/publ/PUBL/4565a5742.pdf.

United Nations Human Rights Council (UNHRC) 2008. *Human Rights and Climate Change*, Resolution 7/23, 28 March 2008. Geneva: United Nations Human Rights Council. http://ap.ohchr.org/documents/E/HRC/resolutions/A_HRC_RES_7_23.pdf (last visit 9 April 2009).

United Nations Security Council 2007. *Security Council holds first-ever debate on Impact of Climate Change on Peace, Security, Hearing over 50 speakers*, 5663rd meeting, 17 April 2007. New York: United Nations. www.un.org/News/Press/docs/2007/sc9000.doc.htm.

Warren, R., N. Arnell, R. Nicholls, P. Levy and J. Price 2006. *Understanding the Regional Impacts of Climate Change*, research report prepared for the *Stern Review on the Economics of Climate Change*, Tyndall Centre Working Paper No. 90. Norwich, UK: Tyndall Centre for Climate Change Research. www.tyndall.ac.uk/publications/working_papers/twp90.pdf.

Williams, A. 2008. 'Turning the tide: recognizing climate change refugees in international law', *Law and Policy* **30**: 502–529.

17

Global adaptation governance beyond 2012: developing-country perspectives

JESSICA AYERS, MOZAHARUL ALAM AND SALEEMUL HUQ

17.1 Introduction

It became apparent in the most recent report of the Intergovernmental Panel on Climate Change (IPCC) that climate change impacts are already being felt (IPCC 2007) and that those most vulnerable to these impacts are the poorest communities within poor countries, notably in the small-island developing states, the least industrialized countries, and countries whose economies heavily rely on climate-sensitive activities, particularly in Africa (Huq and Ayers 2007; IPCC 2007: 9). Yet it is also apparent, as we lay out in this chapter, that climate governance under the United Nations Framework Convention on Climate Change (the 'climate convention') still fails to adequately address climate change adaptation needs in developing countries.

Adaptation to climate change describes the adjustment in natural or human systems in response to actual or expected climatic stimuli or their effects, which moderates harm or exploits beneficial opportunities (IPCC 2007). Adaptation is seen as one of two options for managing climate change; the other is mitigation, which involves the limiting of greenhouse gas emissions, particularly of carbon dioxide and methane. Although adaptation and mitigation are inherently linked in the climate system (the more effective mitigation is undertaken now, the less need for adaptation in the future), until very recently they have been viewed as separate or even competing policy options under the convention (Swart and Raes 2007). Mitigation has been treated as an issue for industrialized countries who hold the greatest responsibility for climate change, while adaptation is seen as a priority for developing countries where mitigation capacity is lower and vulnerability is high (Dodman *et al.* 2009). Adaptation has historically been seen as a marginal policy option, mitigation's 'poor cousin' in the climate policy arena (Pielke *et al.* 2007).

Such perspectives have recently changed, and adaptation is now a prominent aspect of the climate policy agenda, with negotiations at the thirteenth conference of

Global Climate Governance Beyond 2012: Architecture, Agency and Adaptation, eds. F. Biermann, P. Pattberg and F. Zelli. Published by Cambridge University Press.

the parties to the climate convention in Bali in 2007 bringing adaptation on to equal footing with mitigation, by highlighting it as one of four 'building blocks' that are required in response to climate change (along with mitigation, technology cooperation and finance). Adaptation will inevitably form a key part of future climate governance. Yet there is a risk that the adaptation policy rhetoric will not be translated into equally substantive action, and that relative to mitigation, adaptation in future climate governance will continue to be insufficiently supported by appropriate fiscal and institutional mechanisms.

It seems important that this inattention to adaptation is rectified under a future adaptation architecture. Closer analysis is needed as to why adaptation is so poorly managed at the international level, and why the adaptation agenda is so persistently undermined. This chapter explores these questions and suggests that we cannot look for solutions in conventional international relations arguments, which explain inattention to adaptation simply as a reflection of North–South power relations within the climate convention. While there are inequalities within the international arena between Northern and Southern representation, these do not fully explain the particularly poor governance of adaptation as a developing country priority.

We suggest here that the current climate governance architecture is not conducive for fair and effective action on adaptation for developing countries. Successful adaptation requires input from a range of stakeholders, most importantly from developing-country actors and vulnerable communities on the ground, to inform strategies that are effective across scales. Yet having initially been developed to negotiate and manage reductions in global emissions of greenhouse gases, climate governance structures have traditionally focused strongly on technological and economic issues, relying on top–down aggregate modelling to inform mitigation strategies that take a global systemic approach to limiting greenhouse gases over the long term. Such an approach requires external expertise and does not fit with the reality of adaptation, which requires the integration of local and place-based knowledge.

This chapter begins by describing the evolution of adaptation under the climate convention and shows that the emergence of adaptation from a mitigation-orientated policy process has resulted in a discourse for adaptation that is scientific, technical and environmental, and a 'technology-based' view of adaptation that is large-scale and top–down and not necessarily appropriate to responding to actual vulnerability on the ground. Next, we show that the contextual nature of vulnerability requires a flexible understanding of adaptation that can be applied to an array of adaptive responses. This necessitates a range of inputs and approaches from different stakeholders, from the knowledge of vulnerable local communities and non-governmental organizations regarding traditional and appropriate systems of adaptation in environmentally sensitive environments, through to the technical expertise of scientists and engineers.

We then demonstrate the barriers to incorporating such diverse expertise into international adaptation policy and the consequences for democratic, and ultimately for effective, adaptation policy-making. We propose that future global adaptation governance must present a way of 'democratizing the discourse' on adaptation, opening up policy deliberation to a range of diverse inputs and collaborative dialogue, particularly from the South. Finally, we consider whether this is possible under the climate convention, or whether new and alternative means of governing adaptation are necessary. We propose policy options that might enable such an approach, including the possibilities for managing adaptation outside the climate convention.

17.2 Methodology

This chapter is based on inputs from a variety of sources. We undertook a historical analysis of climate change policy documents since the inception of the climate convention to track the emergence of the adaptation agenda and its framing as a 'developing country' issue. This included looking for shifts in reports of the Intergovernmental Panel on Climate Change (for example, shifts in the ways in which vulnerability and adaptation are referred in successive reports) as well as policy changes reflected in the documentation of the climate convention. We also reviewed developing-country submissions to the climate convention on issues related to adaptation, to track developing-country demands over the same time period, considering the extent to which these have been taken into account by the climate convention and its associated bodies.

We also attended numerous meetings and conferences under the climate convention and have been actively engaged in debates on adaptation governance. These experiences, in addition to numerous discussions with developing-country delegates and non-governmental partners, gave us important insights into developing-country perspectives on the negotiations surrounding adaptation. Views and opinions gleaned from interviews, informal discussions and participant observation at meetings have been supported where possible by both published and grey literature, from developed and developing country authors, on adaptation in a future climate governance architecture.

17.3 Analysis

17.3.1 The emergence of the adaptation agenda under the climate convention

When climate change was formally taken up by the UN General Assembly in 1988, it was cast as a 'pollution' problem (Burton *et al.* 2008: 25), following quickly on the heels of the issues of acid rain and stratospheric ozone depletion, both transboundary systemic issues to be managed by international cooperation to mitigate the causes of pollution 'upstream' (Schipper 2006; Ayers and Dodman 2010). Discussions were dominated by how to abate greenhouse gas emissions (mitigation), and adaptation to the impacts of climate change was given relatively little attention. Burton *et al.* (2008) suggest that adaptation was initially 'regarded as at best a distraction but more seriously as an impediment to the mitigation agenda' (Burton *et al.* 2008: 26).

Over the last two decades adaptation has gained gradual prominence in both climate change science and policy, alongside mitigation. Huq and Toulmin (2006) suggest tracking this progress through three 'eras' of the climate and development discourse, which run from the 1990s to 2000, from 2001 to 2007 and from 2007 onwards. The first era is marked by the establishment of the Intergovernmental Panel on Climate Change by the United Nations Environment Programme and the World Meteorological Organization, tasked to evaluate the risk of climate change. The IPCC published its first report in 1990, which established climate change as a global, long-term environmental problem which necessitates urgent action. This stimulated the creation of the 1992 climate convention. Although mitigation and adaptation were both set out in the Convention, the focus remained on mitigation, and there is no specific definition of adaptation in the Convention. In this era, adaptation was considered a marginal policy option by both climate scientists and decision-makers. Attention was given to adaptation in terms of assessing how much mitigation was needed, rather than being an important issue in its own right (Schipper 2006; Ayers and Huq 2008).

The second era began with the Third Assessment Report of the IPCC, which recognized climate change as a development problem. It was shown that the efforts to reduce greenhouse gases had not been able to 'solve' climate change, so impacts would occur, and the developing countries and particularly the least developed countries would be most vulnerable (Huq and Ayers 2009). Adaptation therefore began to be associated with developing country interests. Following the publication of the Third Assessment Report, an agenda item taking up adaptation was introduced in the climate convention's Subsidiary Body for Scientific and Technological Advice, and a work programme on adaptation was adopted.

This was translated into policy at the seventh conference of the parties in Marrakech in 2001, where the Marrakech Accords were concluded. These included three new funds, the so-called Marrakech Funds: the Least Developed Countries Fund, established under the convention to support the 49 least developed countries to adapt to climate change and initially used to support the design of National Adaptation Programmes of Action; the Special Climate Change Fund to support climate change activities including mitigation and technology transfer, but intended to prioritize adaptation; and the Kyoto Protocol Adaptation Fund to support concrete adaptation projects in developing countries that are party to the protocol. This fund sits under the Kyoto Protocol and is financed from a levy on the Clean Development Mechanism. The Marrakech Accords also included a decision that called for financing adaptation under the Global Environment Facility (GEF)[1] Trust Fund, resulting in the development of a Strategic Priority on Adaptation under the GEF. However, while this Strategic Priority on Adaptation was consistent with the global environmental approach adopted by the GEF – requiring the generation of global benefits by supporting the adaptation in broader areas such as vulnerable ecosystems – the three new Marrakech funds are development-focused and enable investments in agriculture and food security, water resources, disaster preparedness and health. The second era, then, took steps to strengthen action on adaptation, although Kates suggests that these steps were nominal: 'Of the 728 pages of substantive text, about two thirds are devoted to impacts, one third to mitigation and only 32 pages to adaptation' (Kates 2000: 1). The third era is shaped by the IPCC's Fourth Assessment Report of 2007. This report has shown that climate change impacts are already happening, because for the first time the IPCC has used observations over the last ten years rather than working only on predictions.

As adaptation gained prominence in negotiations and policy, its context shifted from being tied to discussions over impacts thresholds – the more that adaptation can be used to reduce impacts that might be considered dangerous, the higher impacts threshold of greenhouse gas concentrations can be accepted (Burton 2004) – to being increasingly branded as a developing country issue and reflective of the profound global inequality of climate change, that is, that those who will suffer the most from the impacts of climate change are least responsible.

The emergence of adaptation from a mitigation-dominated climate change agenda has created policy frameworks for adaptation based on a 'pollutionist understanding' of adaptation (Burton 2004). Adaptation within the policy context of the climate convention is understood and described in terms that are scientific, technical and environmental, which has penetrated climate policy frameworks. Klein (2008) refers to this as the 'technology-based' view of adaptation, which has resulted in adaptation measures such as dams, early-warning systems, seeds and irrigation schemes that are based on specific knowledge of future climate conditions. Correspondingly, international climate governance structures under the

[1] The Global Environment Facility was established as the financial mechanism to the climate convention at the first conference of the parties in 1995. Three out of the four funds for adaptation under the climate convention (the Least Developed Countries Fund, Special Climate Change Fund, and Strategic Priority on Adaptation) are all managed by the Global Environment Facility. The Adaptation Fund is the only fund that is managed by an independent board.

climate convention treat adaptation in the narrowest sense, as an issue of specifically climate change (rather than vulnerability more broadly), with adaptation actions limited to changes that are proven to be anthropogenic and distinct from climatic variability (Ayers and Dodman 2010).

17.3.2 Implications for adaptation governance under the climate convention

On the one hand, this concept of adaptation framed specifically in relation to climate change impacts has helped support arguments on behalf of developing countries for fair and equitable international funding arrangements for adaptation. Such arguments state that the responsibility for assisting the vulnerable developing countries to adapt to the impacts of climate change must be based on the 'polluter pays principle', pointing towards responsibility-based rather than burden-based criteria. Funding for adaptation is therefore the responsibility of high-income, high-emitting countries, to be paid to countries most vulnerable to the impacts of these emissions. This funding should be additional to existing aid commitments, because climate change is an additional responsibility of industrialized countries and an additional burden for vulnerable developing countries.

This principle is specifically recognized by the climate convention in article 4, which states that industrialized countries have committed to helping 'particularly vulnerable' developing countries meet the costs of adaptation and that this funding is additional to existing aid commitments (UNFCCC 1992). In practice, this requires distinguishing between funding for building resilience to climate change (which is additional to development assistance) versus funding for building resilience to climate variability more generally (which could be included in development assistance contributions) (Ayers 2009). This distinction is supported by developing countries at the level of the climate change negotiations, in order to prevent industrialized countries from incorporating adaptation funding into development assistance and thereby avoiding providing new and additional funding for adaptation under the climate convention. For example, at the June 2008 meeting of the subsidiary bodies to the climate convention, developing countries called for the measurable, reportable and verifiable use of new and additional funding for climate-change-specific activities (as opposed to more general resilience building) (Klein 2008).

However, taking such a 'pollutionist' view of adaptation that distinguishes between the impacts of climate change specifically, versus vulnerability to climatic variability more generally, has not been productive in terms of progressing action on adaptation in developing countries on the ground. First, a climate-impacts view of adaptation discourages investment in adaptation because of the inevitable uncertainty tied into measuring and predicting climate change patterns. This has resulted in a historic reluctance to commit to action on adaptation on the basis that it is difficult to adapt specifically to climate change if we do not know exactly what we are adapting to. So the argument follows, pre-emptive action against an uncertain threat could be maladaptive.[2]

Although adaptation is now prioritized under the convention, the lack of commitment is reflected by the insufficient funds pledged to meet the costs of adaptation in developing countries. While the costs of adaptation have been estimated at USD 50 million per year (Oxfam 2007), in 2008 the total resources pledged to the Least Developed Countries Fund, the Special Climate Change Fund and the Strategic Priority on Adaptation Trust Fund totalled USD 298 million – 172.8 million to the Least Developed Countries Fund, 75.6 million to the Special Climate Change Fund and 50 million to the Strategic Priority on Adaptation Trust Fund (Global Environment Facility 2008). Further, donors delaying meeting their pledged commitments, so the actual funds in the Least Developed Countries Fund were, at the time of writing, merely USD 91.8 million, 59.9 million in the Special Climate Change Fund and 50 million in the Strategic Priority on Adaptation Trust Fund (Global Environment Facility 2008). This leaves almost USD 100 million pledged to the climate convention that is still outstanding (Ayers 2009). Although the Adaptation Fund does promise to generate much more significant sums (the revenue generated from the CDM levy alone is projected to be between USD 160 and 190 million, potentially much more depending on the volumes traded and prices set: Müller 2007), this does not reflect commitment to prioritizing adaptation, given that the fund is not based on bilateral donations by industrialized countries.

Second, technology-based measures that specifically react to the impacts of climate change can only be partially effective if they do not also address non-climatic factors that are the underlying 'drivers' of vulnerability. Defining adaptation in specifically climate change terms ignores the now widely accepted role for development in contributing to building resilience (Ayers and Huq 2009). Vulnerability to climate change is determined not only by the impacts of climate change on people and the resources on which they depend, but also by the entitlement of individuals over these and alternative resources (Adger 1999; Sen 1999). This has been repeatedly demonstrated through debates in development studies in relation to disaster risk reduction (see for example Wisner *et al.* 2004; Janssen *et al.* 2006; Smit and Wandel 2006), which connect the risks people face with the specific and contextual reasons behind their vulnerability in the first place (Wisner *et al.* 2004). Accordingly, technology-based measures can only be partially effective if

[2] Maladaptations are actions or investments that enhance rather than reduce vulnerability to impacts of climate change. This can include the shifting of vulnerability from one social group or place to another. It also includes shifting risk to future generations and/or to ecosystems and ecosystem services (Ayers 2009).

they do not also address non-climatic factors that are the underlying 'drivers' of vulnerability (Ayers and Dodman 2010).

The narrow approach to adaptation adopted in the climate convention is therefore limited to the extent that it can actually contribute to building resilience to climate change on the ground in developing countries. Ayers and Huq (2009) use the example of funding for adaptation projects under National Adaptation Programmes of Action through the Global Environment Facility to illustrate this point. One of the projects identified by the National Adaptation Programme of Action of Tuvalu is coastal infrastructure to protect the shoreline from erosion, a problem regardless of climate change and so an existing vulnerability, but one exacerbated by climate change, so also an additional cost. The project team for the National Adaptation Programme of Action had extreme difficulties calculating the 'climate change' component of the infrastructure needs. Furthermore, the government lacks the funds to finance infrastructure that would reduce vulnerability of communities to coastal erosion regardless of climate change. Thus, 'the offer to fund, as it were, the "top section" of the infrastructure required to respond to "additional" impacts of climate change, is absurd in light of the fact that co-financing to pay for the lower section cannot be found' (Ayers and Huq 2009: 5).

Third, such a technical discourse on adaptation excludes non-technical expertise that is essential to achieving effective adaptation on the ground. While there are incidences where large investments in infrastructure are necessary for adaptation, all too often such investment is inappropriate, and even counterproductive, when it comes to reducing the vulnerability of poor people on the ground. A critique of government adaptation efforts in Uganda from the local NGO Development Network of Indigenous Voluntary Association notes that 'Local coping strategies provide the foundation for people's own ideas on how to survive during harsh times' (Twinomugisha 2005: 6). Yet, despite this seemingly common-sense statement, 'government approaches to adaptation are based on top-down … models, which have little relevance for local Ugandans. … Climate change adaptation in the context of government policies seems distant from local realities' (Twinomugisha 2005: 7).

The existing climate governance is therefore not conducive to incorporating bottom–up approaches rooted in community-based patterns of resource management that contribute to building resilience to existing climatic variations. Attempts have been made to redress this; for example, in recognition of the need for locally appropriate national adaptation policies, the climate convention states that the process of the National Adaptation Programmes of Action, funded by the Least Developed Country Fund under the climate convention,

> Takes into account existing coping strategies at the grassroots level, and builds upon that to identify priority activities, rather than focusing on scenario-based modelling to assess future

vulnerability and long-term policy at state level. In the NAPA [National Adaptation Programmes of Action] process, prominence is given to community-level input as an important source of information, recognizing that grassroots communities are the main stakeholders (http://unfccc.int/adaptation/napas/items/2679.php).

This is formally implemented, with each national adaptation programme of action being exposed to public review and comment, endorsed by the relevant national government and only then published (Mace 2006). However, despite the fact that national adaptation programmes of action have now been completed for 39 least-developed countries (as of January 2009), no projects under these programmes have yet been completed. This is in part because funds were not made available from the Least Developed Countries Fund to fully develop project proposals from national adaptation programmes of action, which creates a disjoint when the process shifts from country level to international cooperation. A review of national adaptation programmes of action in Africa by Elasha-Osman and Downing (2007) noted that one major problem was the lack of institutional support at lower levels for using information gleaned by the national adaptation programmes of action, delaying execution of activities. There is no mechanism under the climate convention for implementing national adaptation projects once identified, and no guarantee that undertaking national adaptation programmes of action will result in any activities being funded.

Thus, the current politics of climate change adaptation are dominated by an inappropriate definition of climate change risk (to climate change, rather than vulnerability more generally), which does not incorporate the locally and contextually specified nature of climate change vulnerability. The hegemonic status of a global and technical climate change discourse means that policies are failing to represent diverse voices and to inform adaptation strategies that are most useful to beneficiaries.

Finally, the fact that the adaptation agenda gained legitimacy slowly within international negotiations, and then only by being framed as a developing country priority (rather than an offshoot to failed mitigation), means it has also been negotiated in relation to other 'developing country issues' such as capacity-building and technology transfer. The labelling of adaptation as a developing-country issue ignores complexities that arise from the fact that 'developing countries' as a group have hugely divergent interests when it comes to adaptation. Schipper (2006) suggests that for high-emitting developing countries, adaptation priorities detract attention from looming mitigation targets. There are also many differences between the needs of small-island developing states and inland least-developed countries, with sea-level rise meaning inevitable retreat for populations of the former, while for the latter building resilience may rely more heavily on strengthening progress against existing development indicators less easily associated with climate change. The management of adaptation under the climate convention alongside other developing country issues does not have room to negotiate such nuances.

Following on, linking adaptation to other developing country issues reflects the fact that the text of the climate convention has no article specifically on adaptation (adaptation is mentioned only five times in the convention text). If we compare mitigation to adaptation, mitigation has a clear definition, baselines and targets. There is no adaptation baseline and little attention to how progress in adaptation should be measured. There are no targets for adaptation. Whereas mitigation has clear funding regimes, adaptation is funded through many different funds, all of which are voluntary (Burton 2004). This has resulted in a piecemeal approach to policy-making on adaptation under the convention, including a lack of understanding over what adaptation is (and is not), and a failure to produce any firm definition or guidance on adaptation. As noted by Schipper (2006), 'the lack of specific definition of adaptation, even more confused by its association with other aspects of the climate convention, posed a significant constraint to furthering policy on adaptation'. This continues to penetrate the post-2012 discussion on the adaptation agenda. While adaptation continues to be discussed under a number of different convention items under both the climate convention and the Kyoto Protocol, a comprehensive formal proposal consisting of all issues on adaptation under the future regime is missing. Policy-makers are continuing to struggle with how best to address adaptation in any comprehensive way.

17.4 Conclusions and policy recommendations

17.4.1 The potential for a revised global adaptation governance architecture

Significant progress has been made on empowering the adaptation agenda within the climate convention adaptation architecture. Yet, the result has been a framing of adaptation that is confused and inappropriate for addressing the myriad of developing-country concerns. What is needed, then, is a way of reframing the adaptation agenda to ensure that developing-country priorities can be met in a comprehensive and consistent manner. Is this possible within existing climate convention frameworks? We suggest that it is, but several changes in how adaptation is framed and negotiated must be made.

First, adaptation must be taken as seriously as mitigation. While this is now the case in the policy rhetoric, it is different in implementation, where mitigation is marked by concrete actions, targets and mandated funds to support those actions, whereas adaptation is less institutionalized. A more comprehensive and operational approach to adaptation must be taken. Commitments must be made on adaptation that reflect the urgency with which it is discussed. This should be manifest through financial commitments that are substantial and mandatory (as opposed to the existing situation where they are insignificant relative to the problem and voluntary). Further, a legal framework for adaptation is needed, as exists for mitigation, to make commitments on adaptation binding. This could either take the form of an independent legal instrument, or be incorporated into an expanded Kyoto Protocol.

Second, international negotiations where developing-country interests are bundled together, although successful in achieving attention to the adaptation agenda as a whole, have limited potential to address diverse national adaptation needs. More effective systems are needed to account for local and regional circumstances where the actual impacts are encountered, adaptive capacities are built, and responses are implemented. Linkages must be strengthened between local, regional and global scales, and spaces must be created within the negotiations for such insights.

Third, and a related issue, the technical discourse of adaptation must become more flexible and better incorporate a development-based approach to building resilience to current climate variability as well as climate change. It is essential that policy dialogue on adaptation is opened up to a new range of expertise that originate from insights into vulnerable communities, generated by local stakeholders and development and disaster risk reduction practitioners, rather than being restricted to impacts-based scientific inputs alone.

This is particularly difficult, especially since the climate convention has adopted a policy framework based on 'neutral' scientific assessments (hence the establishment of the IPCC); the communication of these results to policy-makers (the climate convention); and then the initiation of a state-led environmental regime based on international targets that were regularly updated and strengthened (the Kyoto Protocol).

The powerful notions that science is neutral, expert networks are benign and representative, and governments act rationally according to expert advice, penetrate the climate convention discourse (Ayers and Huq 2008). Adaptation research currently offers very little in the way of such formal, scientific and 'neutrally expert' outputs. Rather, expertise in vulnerability reduction comes from local community-based case studies, and indigenous knowledge of locally appropriate solutions to climatic variability and extremes. While a number of studies have dealt with local adaptation in case studies (see for example Moss *et al.* 2001; Morduch and Sharma 2002), Smit and Wandel suggest that the motivation behind such research has been to 'identify what can be done in a practical sense, in what way and by whom, in order to moderate the vulnerability to the conditions that are problematic to the community' (2006: 285). There has so far been little attention to scaling up these examples, or consideration of where they lie with regards to larger scale and more technical approaches (Ayers and Dodman 2010). Given that the IPCC limits its influential reports to published and peer-reviewed research, there is no space for the integration of informal indigenous knowledge or unpublished yet very relevant evidence from community-based organizations and researchers.

It is therefore important that if adaptation is to be effectively managed under the climate convention in a future regime, spaces are created for the integration of local and indigenous 'expertise' into the adaptation discourse. Burton *et al.* (2008) suggest several ways in which this might be achieved, two of which we highlight here. The first is through adaptation modelling, which would generate adaptation knowledge that is seen as 'legitimate' under the current norms of the climate convention processes. However, this does not overcome the problem of exclusion of non-technical expertise in adaptation policy-making. The second approach is to allow the accumulating evidence from local case studies of adaptation to 'permeate the negotiations' and gradually alter the mindset of negotiators. Although such locally specific research can be difficult to categorize, at the same time the variety of circumstances would encourage a reframing of adaptation into a more flexible definition.

Yet, it may be that the climate convention and an extended Kyoto Protocol can never give adequate attention to adaptation, given the original intention of the treaty, which is to focus on reducing the source of climate change rather than adapting to the changes. As argued by Schipper (2006), 'mitigation has been given more attention since the climate convention was drafted, not only out of political choice, but because mitigation was considered more important even from the beginning. This is the primary reason why the climate convention does not reflect a great emphasis on adaptation' (Schipper 2006: 91).

17.4.2 Alternative options for a future architecture for adaptation

Perhaps, then, it is inevitable that adaptation will never been seen as a policy priority under the climate convention and its Kyoto Protocol, even a revised one. The focus of future climate policy under the climate convention will continue to be extending the life of the Kyoto Protocol and enforcing emissions reductions for the world biggest emitters. In this case, it may be unlikely that attention and financial assistance for adaptation are prioritized under the climate convention in the future regime until significant progress is made against the mitigation agenda. Thus, it may be that achieving the above for adaptation may be easier under a different type of international architecture.

One option for such a different type of architecture could be an 'adaptation protocol'. Such a protocol could be based on a more flexible definition of adaptation and contain operationizable targets and guidance on adaptation funding and action. Also the proposal of Biermann and Boas for a climate refugee agreement goes into the same direction, even though more focused on one particular aspect of adaptation (Biermann and Boas, this volume, Chapter 16). Another (not necessarily independent) option would be to integrate climate change adaptation into existing frameworks that are more closely aligned with the goals of adaptation than the climate convention, such as the Hyogo Framework for Disaster Risk Reduction (Huq and Ayers 2009), or other international development frameworks.

Alternatively, some argue that we do not need to look towards a global policy to support adaptation, given that action on adaptation is at some point always locally specific and that generalizable guidelines for adaptation seem difficult to achieve. Many argue that adaptation therefore necessitates a local approach because it is a local issue. As Mace notes, 'The need for adaptation arises from a global cause, but the remedy must yield local benefits' (Mace 2006: 64). However, while perhaps appealing for simplicity's sake given the context-specific nature of adaptation, we suggest that a global framework for adaptation *is* important nonetheless. This is not the least because of the fundamental equity principle that rich countries are responsible for the costs of adaptation in poor countries.

In addition, it is important that adaptation and mitigation are not managed totally independently, given that they are inherently linked in the climate system. While it is dangerous to suggest that the two should be linked to the extent that an 'optimal mix' of adaptation and mitigation is possible (potentially discouraging investment in mitigation until the limits of adaptation are breached), recent attention to exploring the synergies between mitigation and adaptation suggests that an integrated approach could go some way in bridging the gap between the development and adaptation priorities of the South and the need to achieve global engagement in mitigation (Swart and Raes 2007; Willbanks and Sathaye 2007; Ayers and Huq 2008). Further, as the impacts of climate change become evident, the 'adaptation is local' discourse is proving decreasingly valid. Adaptation impacts are starting to manifest themselves regionally, and have implications for migration and other transboundary issues (Biermann and Boas, this volume, Chapters 14 and 16) that warrant more strategic approaches that cross political boundaries (Burton 2008). Problematic as it is, it is essential that a comprehensive, inclusive and overarching international framework for cooperation on adaptation be in place in the future climate governance architecture.

In conclusion, significant progress has been made on empowering the adaptation agenda within the climate governance architecture; yet, the result has been a framing of adaptation that is confused and inappropriate for addressing the myriad of developing country concerns related to climate change. First, under existing frameworks, adaptation remains an undervalued policy option relative to mitigation.

Second, the type of adaptation favoured by the Climate Convention is not conducive to building the broader resilience necessary to reduce the vulnerability of developing countries. Third, the adaptation discourse under the Climate Convention, being as it is technical and 'pollutionist', does not open itself up to alternative types of expertise that are locally generated and non-technical.

Such conclusions may suggest that it is impossible effectively to manage adaptation under the Climate Convention and a future architecture that is likely to give rise to an extended Kyoto Protocol. However, instead of looking to new institutional arrangements, we suggest that refining the adaptation agenda under the Climate Convention is necessary and possible. Rather than view the currently authoritative adaptation discourse as inherent and therefore impenetrable, we suggest that by exploring how and why it became authoritative, we can find opportunities to redefine adaptation and create frameworks that are more appropriate. The review of the emergence of the adaptation agenda in this chapter does show significant progress from conceptualizing adaptation as an alternative and possibly even destructive agenda item relative to mitigation, to one of equal importance and tied closely to developing country interests. From here, it is important that this progress does not stall and the current framing of adaptation is not cemented too readily. Instead, it is essential that more deliberative policy-making processes are created for adaptation that are better able to engage with vulnerable communities and citizens to create bottom–up, locally meaningful adaptation strategies. This requires a reframing of the adaptation discourse that is more open to non-technical expertise generated from indigenous and locally based knowledge.

References

Adger, W. N. 1999. 'Social vulnerability to climate change and extremes in coastal Vietnam', *World Development* **27**, 249–269.

Ayers, J. 2009. 'Financing urban adaptation', in J. Bicknell, D. Dodman and D. Satterthwaite (eds.), *Adapting Cities to Climate Change*. London: Earthscan.

Ayers, J. and D. Dodman 2010 (in press). 'Climate change adaptation and development: the state of the debate', *Progress in Development Studies*.

Ayers, J. and S. Huq 2008. 'The value of linking mitigation and adaptation: a case study of Bangladesh', *Environmental Management*: doi:10.1007/s00267–008–9223–2.

Ayers, J. and S. Huq 2009 (in press). 'Supporting adaptation through development: what role for ODA?', *Development Policy Review*.

Burton, I. 2004. *Climate Change and the Adaptation Deficit*, Adaptation and Impacts Research Group Occasional Paper No. 1. Toronto: Meteorological Service of Canada, Environment Canada.

Burton, I. 2008. 'Beyond borders: the need for strategic global adaptation', in *Sustainable Development Opinion*, December 2008. London: International Institute for Environment and Development.

Burton, I., T. Dickinson and Y. Howard 2008. 'Upscaling adaptation studies to inform policy at the global level', *Integrated Assessment Journal* **8**: 25–37.

Dodman, D., J. M. Ayers and S. Huq 2009. 'Building resilience', in Worldwatch Institute (ed.), *State of the World 2009: Into a Warming World*. Washington, DC: Worldwatch Institute, pp. 75–77.

Elasha-Osman, B. and T. Downing 2007. *Lessons Learned in Preparing National Adaptation Programmes of Action in Eastern and Southern Africa*, European Capacity Building Initiative Policy Analysis Report. Oxford, UK: University of Oxford.

Global Environment Facility (GEF) 2008. *Status Report on the Climate Change Funds as of 4 March 2008*, Report from the Trustee, GEF/LDCF.SCCF.4/Inf.2. Washington, DC: Global Environment Facility.

Huq, S. and J. Ayers 2007. 'Critical list: the 100 nations most vulnerable to climate change', in *Sustainable Development Opinion*. London: International Institute for Environment and Development.

Huq, S. and J. Ayers 2009. 'Linking adaptation and disaster risk reduction', in L. Brainard, A. Jones and N. Purvis (eds.), *Climate Change and Global Poverty: A Billion Lives in the Balance*. Washington, DC: Brookings Institution Press, pp.

Huq, S. and C. Toulmin 2006. 'Three eras of climate change', in *Sustainable Development Opinion*. London: International Institute for Environment and Development.

IPCC 2007. 'Summary for policymakers', in M. L. Parry, O. F. Canziani, J. P. Palutikof, P. J. van der Linden and C. E. Hanson (eds.), *Climate Change 2007: Impacts, Adaptation and Vulnerability. Contribution of Working Group II to the Fourth Assessment Report of the Intergovernmental Panel on Climate Change*. Cambridge, UK: Cambridge University Press, pp. 7–22.

Janssen, M. A., M. L. Schoon, W. Ke and K. Borner 2006. 'Scholarly networks on resilience, vulnerability and adaptation within the human dimensions of global environmental change', *Global Environmental Change* **16**: 240–252.

Kates, R. W. 2000. 'Cautionary tales: adaptation and the global poor', *Climatic Change* **45**: 5–17.

Klein, R. T. J. 2008. 'Mainstreaming climate adaptation into development policies and programmes: a European perspective', in European Parliament (ed.), *Financing Climate Change Policies in Developing Countries*. PE 408.546 IP/A/CLIP/A/CLIM/ST/2008–13. Brussels: European Parliament.

Mace, M. 2006. 'Adaptation under the UN Framework Convention on Climate Change: the international legal framework', in W. N. Adger, J. Paavola, S. Huq and J. Mace (eds.), *Fairness in Adaptation to Climate Change*. Cambridge, MA: MIT Press, pp. 53–76.

Mordoch, J. and M. Sharma 2002. 'Strengthening public safety nets from the bottom up', *Development Policy Review* **20**: 569–588.

Moss, S., C. Pahl-Wostl and T. Downing 2001. 'Agent-based integrated assessment modelling: the example of climate change', *Integrated Assessment* **2**: 17–30.

Müller, B. 2007. *The Nairobi Climate Change Conference: A Breakthrough for Adaptation Funding*. Oxford, UK: Oxford Institute for Energy Studies.

Oxfam International 2007. *Adapting to Climate Change: What's Needed in Poor Countries and Who Should Pay*, Oxfam Briefing Paper No. 104. Washington, DC: Oxfam.

Pielke, R., G. Prins, S. Rayner and D. Sarewitz 2007. 'Lifting the taboo on adaptation', *Nature* **445**: 597–598.

Schipper, L. 2006. 'Conceptual history of adaptation in the UNFCCC process', *Review of European Community and International Environmental Law* **15**: 82–92.

Sen, A. K. 1999. *Development as Freedom*. Oxford, UK: Oxford University Press.

Smit, B. and J. Wandel 2006. 'Adaptation, adaptive capacity and vulnerability', *Global Environmental Change* **16**: 282–292.

Swart, R. and F. Raes 2007. 'Making integration of adaptation and mitigation work: mainstreaming into sustainable development policies?', *Climate Policy* **7**: 288–303.

Twinomugisha, B. 2005. 'Indigenous adaptation', *Tiempo* **57**: 6–9.

United Nations Framework Convention on Climate Change (UNFCCC) 1992. *United Nations Framework Convention on Climate Change*. Bonn: UNFCCC.

Willbanks, T. J. and J. Sathaye 2007. 'Integrating mitigation and adaptation as a response to climate change: a synthesis', *Mitigation and Adaptation Strategies for Global Change* **12**: 957–962.

Wisner, B., P. M. Blakie, T. Cannon and I. Davis 2004. *At Risk: Natural Hazards, People's Vulnerability and Disasters*, 2nd edn. London: Routledge.

18

Shaping future adaptation governance: perspectives from the poorest of the poor

ANNE JERNECK AND LENNART OLSSON

18.1 Introduction

Adaptation to climate change impacts has eventually become widely accepted in science and climate policy as an inevitable social process in need of extensive funding and explicit action (Stern 2006; Pielke *et al.* 2007). From a geopolitical perspective the promotion of adaptation is a step forward to overcome political tensions between a focus on mitigation in richer countries versus a focus on adaptation in poorer countries.

From a sustainability point of view it is of great concern to ensure that adaptation reaches and involves vulnerable people in places exposed to climate change impacts and responses. The acceptance and promotion of adaptation in science and policy circles alike may represent a promise for such vulnerable communities where adaptation is to be translated into short-term practices and longer-term livelihood strategies. For adaptation to be successful, the local level must be bridged in an effective way with the global level where funds and mechanisms are available. The bridging between the local and the global creates challenges for science and policy. A first challenge is to organize participatory learning processes where local experiences and priorities in vulnerable communities can be used as a knowledge basis for adaptation. A second challenge is to mobilize *sufficient* global funds for adaptation and make them *available* for local adaptation needs. Even though the global policy framework is adjusting to generate funds for adaptation, this is at much lower levels than for mitigation (Müller 2006, 2007; Möhner and Klein 2007). The low funding for adaptation is unfortunate for several reasons. First, adaptation is *a legal obligation* of the rich countries towards poorer countries to be paid as a compensation for historical inequalities in emissions. Second, adaptation is *a moral obligation* of the world's less vulnerable people towards the more vulnerable people across the world who are increasingly exposed to natural hazards, climate variability and climate change now and in the future (Paavola and Adger 2006).

Global Climate Governance Beyond 2012: Architecture, Agency and Adaptation, eds. F. Biermann, P. Pattberg and F. Zelli. Published by Cambridge University Press.

Among those who are vulnerable to climate change impacts and climate change responses certain people are even more vulnerable and also difficult to reach with policies and measures. This group we call the poorest of the poor.

In the context of the poorest of the poor, mitigation is not a priority because their contribution to the global emission of greenhouse gases is minuscule and their capacity to reduce emissions is low. This makes adaptation their main priority. But for the purpose of successful promotion of adaptation and alleviation of vulnerability it is important to look for creative synergies between mitigation and adaptation as well as between other desirable goals and matching policies.

Across the globe there are 923 million hungry people (FAO 2008), who are also extremely vulnerable to climate change impacts (UNEP 2007). In numbers this is equivalent to the combined populations of the United States, EU and Japan. The already large number of poor people is expected to increase further and remain large for a long time (IFAD 2001), while people exposed to climate change are expected to become even more vulnerable due to increasing incidence of extreme climate events (IPCC 2007). We have noticed that certain livelihood strategies, like risk aversion and risk minimization as often practiced by poor smallholders and subsistence farmers who are vulnerable to climate change, may no longer be sufficient for getting by in poverty, let alone getting out of poverty or escaping vulnerability (Jerneck and Olsson 2008). In relation to the poorest of the poor it can therefore be argued that we should understand adaptation to climate change as a process of *profound* social change away from livelihoods threatened at their roots by climate change. But climate change impacts are not the only problem. Climate change responses in richer parts of the world may harm people already vulnerable to climate change even further. As an example, the massive expansion in biofuel production for consumption in industrialized countries contributed largely to the rapidly rising food prices in 2008, which pushed another 100 million people into food insecurity (Mitchell 2008).

The policy lesson is obviously that we need comprehensive adaptation to both climate change impacts and climate change responses. The science implication is that we need an improved understanding of nature–society dynamics and global–local links relating to vulnerability. A problem-driven approach would ask questions like: who is vulnerable; in what ways and why; how can vulnerability be overcome through adaptation; and will we need both long-term social changes and short-term solutions? Such research resonates with the emerging fields of sustainability science and earth system governance (Biermann 2007).

This leads us to our main objective, which is to identify possible adaptation policies and mechanisms from the perspective of the poorest of the poor. For that purpose we must identify and locate the most vulnerable as well as define and discuss how they are impacted. The people we are identifying are diverse and distributed across the globe. Despite heterogeneity there are common denominators because the poorest of the poor are highly vulnerable to climate change for similar reasons such as spatial and social marginalization, weak links to the economy and exposure to combinations of multiple stressors. Due to inequalities and social stratification, occurring at all levels from the global to the local, it is misleading to equate the poorest of the poor with the total population of the poorest regions or, as some observers do (Collier 2007), with the poorest countries. We therefore seek to provide a much needed analytical definition of the poorest of the poor.

18.2 Methodology

Our argument is informed by literature and rooted in field work in sub-Saharan Africa and South East Asia where we have explored livelihoods in diverse groups of people living in, depending on or associated with vulnerable rural and urban communities. The policy content of our research is scrutinized for its relevance and effectiveness from the perspective of the poorest of the poor.

Importantly, the paradigms of development and sustainability are normative and involve social goal setting to be matched by strategies for social change. This calls for responsible politicians and reflexive researchers who state their values explicitly (Mikkelsen 2005). In the early development debate, Gunnar Myrdal insisted on the need to integrate the intellectual analysis with a moral task (Myrdal and Streeten 1958). More recently, Wallerstein, the founder of world systems theory, argued that research is a threefold task of intellectual, moral and political activities and contributions (Wallerstein 2007). We acknowledge the legal and moral obligations in regard to adaptation as a bottom line in the academic debate, and seek to envision new options for improved livelihoods for the poorest of the poor.

18.3 Analysis

Our analysis of adaptation in the context of the poorest of the poor is structured around social, natural and political aspects of vulnerability and framed by a poverty discussion. The analysis aims at policy formulations on adaptation for the poorest of the poor.

18.3.1 Poverty: politics, dynamics, escape routes

Societies have struggled with poverty for centuries, but despite much political and social effort poverty is widespread, persistent and increasing and is likely to remain so for decades (IFAD 2001). The Millennium Declaration is the latest global policy addressing poverty with a set of social goals to be met within a limited period. There is much evidence, however, that the goals will not be met in the expected time-frame (Sweetman 2005). A lack of integrated solutions addressing several goals simultaneously may be part of the explanation. The fact that the poorest people suffer from environmental degradation and high vulnerability to climate change suggests that sustainable management of nature is important for poverty reduction (UNEP 2007). Hence, if poverty goals were combined with vulnerability goals there is a chance to overcome both challenges. This suggests that a focus on vulnerability to climate change may help solve the persisting problem of poverty being the crux of development in theory, practice and aid. This would imply that in our time there is scope for more synergetic efforts and policies for poverty eradication than in the past. But this calls for a somewhat renewed theoretical and political understanding of poverty.

In nineteenth-century Europe and twentieth-century Asia, agrarian poverty was overcome by technological and institutional change relating to industrialization and a flow of landless people into rural and urban industrial wage work and services (Cameron and Neal 1996). But in the twenty-first century agricultural modernization and labour-intensive industrialization is no longer a universal development strategy. In the absence of agricultural transformation many societies historically experienced huge international migration into relatively richer territories as an alternative route out of poverty. Presently, in spite of restrictive legislation on labour migration, there are over 200 million international migrants whose remittances, an estimated USD 250–270 billion, by far exceeded the total official development assistance in 2005 (Global Commission on International Migration 2005). Despite this huge transnational redistribution of money it is estimated that poverty will persist because the poorest of the poor have few connections that could provide remittances from outside their nature–society system. Moreover, with climate change it is expected that forced migration from vulnerable areas will increase resulting in a rapidly rising number of environmental refugees (Biermann and Boas, this volume, Chapter 16).

In the early post-war development debate, poverty was understood in economic terms, counted in absolute and relative numbers, and to be alleviated through economic growth and trickle-down effects (Martinussen 1997). Lately poverty is redefined in the development debate into a human and actor-oriented view that sees the poor not only as victims but also as agents of their own change relying on legal frameworks provided by society (de Soto 2000). Such definitions relate to human-rights-based approaches that give equal importance to participation (process) and outcome (results). This newer focus orients our analysis towards theories on values, freedoms, entitlements, capabilities and opportunities (Sen 1999) and provides an empirical focus on what society can offer the poor in terms of legal systems and rights (Commission on Legal Empowerment of the Poor 2008).

Owing to extensive research into poverty, much is known about social causes and dynamics as well as about various escape routes based on income, loans, human rights, awareness-raising or empowerment (Razavi 1999; Lister 2004). Identity in terms of age and gender is crucial to poverty as seen when women and men live side by side in poor households and communities but experience poverty differently because it has gendered roots, characteristics and consequences (Razavi 1999). The equal opportunities propagated in the discussed theories all resonate well with the ethics of environmental and intergenerational justice in the sustainability debate. This is extremely important for marginalized populations and especially for the poorest of the poor.

In the pursuit of modernity, human activity has not only resulted in the achievement of widely spread wealth but also in the undermining of conditions of production that may ultimately put survival on this planet at risk (Angel and Rock 2005; Wallerstein 2007). It is a fact that the deteriorating natural conditions are extremely unequally distributed across the globe (Martinez-Alier 2002). Moreover, with climate change poverty is expected to increase disproportionately putting the livelihoods of the already poor in the South at greater risks (UNEP 2007). This is how poverty has become a sustainability problem in need of other solutions than the conventional, especially since many poor people live in areas that are already and, more importantly, will become even more vulnerable to natural hazards in the future owing to climate change reinforced by multiple stressors (RIVM 2008). This means that we need to pose other questions to the dynamics of poverty and think differently about solutions. Historically poverty was defined both as an individual ordeal and a societal problem (Lister 2004). Indeed, poverty is a lived experience and thus truly human and social. But we also stress that the physical context and natural conditions must be explicit in the poverty analysis because when poverty is reframed from a social problem into a problem rooted in the linked nature–society system then other dynamics and patterns of perpetuation become visible.

18.3.2 Poverty: numbers, places, people

One billion people across the world – often referred to as the bottom billion (Sachs 2005; Collier 2007; World Bank 2007) – live in *extreme poverty*. But definitions of poverty and deprivation vary between organizations, observers and scholars. The World Bank (World Bank 2000, 2007) defines the poorest in money terms as those living on one dollar a day (to be precise, USD 1.25 as of 2008 due mainly to a surge in food prices). The FAO (2008) defines the poorest in food terms and estimates that in 2008 there were 923 million food-insecure people unable to supply food throughout the year. Economists like Paul Collier (2007) refers to the bottom billion as the sum of people in the poorest countries – mainly sub-Saharan Africa – while Jeffrey Sachs (2008) refers to the poor as people in 'large regions' like sub-Saharan Africa rather than in countries. According to the International Fund for Agricultural Development, those who are poor today are likely to stay poor for decades. The Rural Poverty Report of 2001 (IFAD 2001) estimated the number of absolute poor as 1.2 billion people, of which 75 per cent live in rural areas. In absolute numbers Asia has more poor people, but in relative terms the share is bigger in sub-Saharan Africa where nearly every second person is classified as poor in monetary terms (World Bank 2007).

Yet we argue that definitions of poverty in quantitative terms (cash, calories, consumption) are insufficient from a sustainability and vulnerability perspective. It has recently been documented that every third person in India is socially vulnerable because one single event like an accident, death in the family, major hospitalization or even temporary loss of income can bring such a person into destitution (Sengupta *et al.* 2008). This explains why multidimensional approaches that estimate poverty and deprivation not only in terms of assets, income and consumption expenditures – which may vary rapidly with prices – but also in social status relating to health, work, education, formal employment and human dignity are compelling (Sengupta *et al.* 2008). The problem with purely quantitative definitions is also apparent in the context of climate change where we need to explicitly consider aspects of location, demography, socio-economic conditions (Morton 2007), access to a legal framework (Commission on Legal Empowerment of the Poor 2008) and social networks when we locate the poorest of the poor.

Sub-Saharan Africa is home to many poor and vulnerable people who are particularly dependent on primary resources like soils, vegetation and surface water the quantity and quality of which are threatened by climate impacts. Since the poor cannot compensate technologically through advanced agriculture, infrastructure and purification for the loss of ecosystem products and services like food, flood protection and clean water, they are extremely dependent on sound ecosystems which are therefore exceptionally valuable (RIVM 2008). This shows how a nature–society approach takes us closer to an understanding of the concept of the poorest of the poor.

18.3.3 Poverty: the food poor

Food security can be defined in terms of physical and economic access at all times to safe, sufficient and nutritious food (Pinstrup-Andersen and Herforth 2008). According to a similar definition, the number of food-insecure people in the world was estimated to 730 million in 1987 rising to 923 million in September 2008 (FAO 2008). Except for China the number of food-insecure people in poorer countries has increased since the 1970s and is expected to continue to rise (Pinstrup-Andersen and Herforth 2008).

Rural areas are home to 70 per cent of the world's poor and food-insecure people (World Bank 2007) making poverty and food insecurity disproportionately rural (Pinstrup-Andersen and Herforth 2008). Many rural areas suffer from weak agricultural conditions due to poor soil quality and low rainfall combined with deficient infrastructure and few alternative income opportunities (World Bank 2007). This means, ironically, that for those who depend for their livelihoods on ecosystem services the proximity to land and nature does not necessarily make you more food secure. The fact that the number of food insecure is expected to increase is especially serious given that the basis of future food production is undermined due to soil erosion, waterlogging, lowering of groundwater levels, biodiversity loss and habitat loss caused by unsustainable land use and reinforced by climate variability and climate change (Pinstrup-Andersen and Herforth 2008). Due to the strong interaction between poverty, food insecurity and unsustainable natural resource use – where poverty may reinforce land degradation that in turn may reinforce food insecurity and poverty – there is a call for joint action on these matters.

Food-insecure subsistence farmers are particularly vulnerable to climate variability and climate change impacts. Subsistence farmers rely on their own production and generally have weak links to the market economy on which they depend for food purchase in times of shortage. So in case of a food crisis emerging from climate impacts their already vulnerable livelihoods will be exposed to further pressure. This is serious especially since climate stressors are very likely to increase substantially in the near future (IPCC 2007; Lobell *et al.* 2008). Such food-insecure subsistence farmers are difficult to identify and locate geographically. They can be described as farmers who depend on 'farming and associated activities which together form a livelihood strategy where the main output is consumed directly, where there are few if any purchased inputs, and where only a minor proportion of output is marketed' (Morton 2007). Yet poor subsistence farmers are also subject to social differentiation, which creates segments of even poorer people (Jerneck and Olsson, unpubl. data).

18.3.4 Poverty: defining the poorest of the poor

It is agreed that the poorest countries and the poorest people are the most vulnerable to climate change (IPCC 2007). By the concept of the poorest of the poor, we refer to a fraction of the poor who are vulnerable to climate change and at the same time unable to benefit from a particular action aimed at reducing vulnerability and securing or improving their livelihoods. The reasons for their inability may vary from context to context. The fact that climate change is expected to severely harm the poorest of the poor, who are by definition notoriously difficult to reach, implies that climate change policy has a special responsibility to seek new ways of assisting the poorest of the poor to overcome destitution and deprivation.

The concept of the poorest of the poor is an analytical tool for understanding the difficulties of reaching certain groups of people in any society, not only in developing regions. Take the example of Hurricane Katrina in New Orleans in August 2005. It is well known that the poor were particularly hard hit by the flooding following the hurricane. Further analysis shows that people without their own means of transport were the most vulnerable because evacuation routes out of the flood risk areas relied on car ownership (Reed 2006; Laska 2008). A next step in the analysis may suggest that among people without cars those who were unemployed were the hardest hit because they were less likely to have a social network including somebody owning a car. This illustrates how the concept of the poorest of the poor can be used as an analytical and contextual tool to iteratively identify a group of people who is notoriously difficult to reach with policies and mechanisms for adaptation. We argue that the contextual dimension of the concept of the poorest of the poor includes not only material assets but also immaterial assets like problem awareness.

The common vocabulary in discussions on adaptation to climate change includes the two concepts *adaptive* and *system* (Smit and Skinner 2002; Grothmann and Patt 2005; Smit and Wandel 2006). In relation to the poorest of the poor there are two serious problems with this vocabulary. The first is related to the notion of *adaptive* or *adaptiveness* in the meaning of taking informed and pro-active actions to avoid damage from a climate change impact. The notion of *adaptive* gets problematic, in fact nonsensical, when we define the poorest of the poor as people who are *unable* to take decisions because of structural constraints such as food insecurity, deprivation of rights, lack of assets and options, or simply incapacity to act. The second problem is related to the use of *system*. The most fundamental description of a system is its boundaries within which it is supposed that adaptiveness should take place. Owing to the condition that the poorest of the poor are unable to grasp adaptive opportunities offered by the system they will in fact fall outside the system boundaries. Once outside, the notion of adaptiveness *within* the system is problematic. In any system there will of course be individuals who fail to support themselves. But the conclusion here is that it is unacceptable that *groups* of people, the poorest of the poor, are unable to overcome vulnerability because their situation and conditions are not even identified or addressed by the system. Consequently, it is crucial to approach the poorest of the poor explicitly.

To conclude this section it can be stated that in the context of climate change and developing countries we assume that the poorest of the poor coincide to a very large extent with people who are food insecure; who cannot afford to run an extra risk; or who are unable to grasp opportunities offered by the system. The reason why food-insecure people largely overlap with the poorest of the poor is that their constant superordinated priority of feeding the household cannot be postponed in favour of other goals. This non-negotiable priority prevents them from taking actions with long-term benefits. We call their condition the 'food imperative'. As an illustration it can be seen in western Kenya that poor subsistence farmers are reluctant to practise agro-forestry if the land and labour needed compete with immediate food production. This is so despite the high awareness among farmers that agro-forestry will improve soils and yields in the long term (Jerneck and Olsson, unpubl. data).

18.3.5 Social change in the wake of natural changes

There is an important link between the poorest of the poor and climate change impacts which goes through their location in and dependence on climate-sensitive natural systems. The most important condition, though, is their narrow margins to a range of climate impacts. If we understand adaptation to climate change as the lack of resilience of these livelihoods to the impacts of climate change (Tompkins and Adger 2004) then the likely response is to increase resilience. But if the long-term bases of such livelihoods are threatened, as in the examples below, then such responses run the risk of postponing more fundamental change. Take the example of fishing communities facing increasing risks of loss of vessels to hurricanes. An insurance scheme could compensate for the increasing risks due to climate change, but would be detrimental in the long term if the fish stock is declining due to other stressors. In the aftermath of the great tsunami in 2004 a renowned marine ecologist, Daniel Pauly (2005), asked the international relief community to invest in long-term alternative livelihoods with a more secure future instead of rebuilding the fishing fleets in the impoverished fishing communities. His underlying argument was that a restored fishing fleet with new and efficient vessels would result in a complete collapse of the fisheries (Pauly 2005).

It is a dominating view on adaptation, at least among leading international development agencies, that adaptation is a matter of technological solutions including funding (Heller and Mani 2002; Tschakert and Olsson 2005) or of managing risk through financial mechanisms such as different kinds of insurance schemes (Gurenko 2006; Linnerooth-Bayer and Mechler 2006). In the context of the poorest of the poor we argue that in some instances adaptation should be a process of profound social change as mentioned above, rather than technical improvements to existing livelihoods. Here we provide several examples of such livelihoods where the long-term basis for their existence might be threatened by climate change or a combination of multistressors. In such cases we can argue that a profound change away from existing livelihoods, rather than an increased resilience of those livelihoods, is to be preferred.

One example is dryland agriculture. Climate change may cause more frequent, longer and/or more intense droughts in semi-arid regions (Schmidhuber and Tubiello 2007). Livelihoods based on rain-fed agriculture may thus become unviable. Potential adaptation in the form of a social change could for example involve a gradual shift from a crop-based agriculture to livestock-based agriculture.

Another example is low-lying coastal regions. Low-lying coastal regions across the world are often fertile and highly productive in terms of agriculture and other land uses. In such regions the poorest of the poor may occupy the most marginal and probably also the most vulnerable areas. Low-lying coastal regions are particularly vulnerable to climate change for several and often reinforcing reasons: long-term sea-level rise; increasing frequency and intensity of storms; and socio-economic processes that increase vulnerability for the poorest of the poor because more people are shifted to marginal and risky areas resulting in congestion and the prevention of escape routes.

Third, river mouths share many features with low-lying coastal regions but there are additional reasons for high vulnerability such as changing natural hydrological regimes that may increase the risk of flooding from terrestrial systems. Moreover, changing land use and water consumption may reduce sediment loads and thereby alter the balance of coastal erosion and sedimentation.

A fourth example is hydrological regimes. Livelihoods based on hydrological regimes threatened by climate change may be very vulnerable. Livelihoods based on waters originating from high mountains where precipitation is accumulated as snow during winter and gradually released as meltwater during spring and summer are particularly threatened by warming. According to Barnett *et al.* (2005) over 1 billion people in the world are currently dependent on such hydrological regimes.

Also fishing livelihoods are particularly vulnerable to multiple stressors, several of which are gradually eroding the very basis of the livelihood (Jackson *et al.* 2001). Coastal small-scale fisheries are vulnerable to climate change impacts such as higher frequency and intensity of hurricanes plus other stressors such as ocean acidification, destruction of coastal ecosystems, pollution and over-fishing (Pauly 2005).

From this we conclude that in cases where there are reasons to believe that the long-term sustainability of the main resource base of a livelihood is threatened, by climate change or other processes, adaptation must imply a reorientation away from the dependency on a threatened resource.

18.4 Conclusions and policy recommendations

From our analysis we conclude that in rural areas the poorest of the poor live mainly in high-risk regions and cultivate lands that are marginal, infertile and located in degraded ecosystems. In urban areas they live mainly in or close to high-risk areas that are often polluted, unhealthy and degraded. These vulnerable groups must be reached through policies and mechanisms that are designed to identify such areas and the people within; not only the larger system of which they constitute a greater part. Adaptation for the poorest of the poor should therefore be contextual, targeted and focused. We argue that adaptation should at least involve special social schemes and mechanisms and in some instances even profound social change away from risky livelihoods. Hence, the best solutions are not always technological.

We now turn to the practical policy consequences of our findings. In the climate change debate global challenges must be addressed in international policy making and be translated into specific national and local circumstances (RIVM 2008). We see three different approaches to promote adaptation for the poorest of the poor: mainstreaming climate change into development assistance; identifying synergies with other mechanisms, such as climate change mitigation, biodiversity or desertification; or stand-alone adaptation policies. Below we discuss the pros and cons of these alternatives.

18.4.1 Adaptation through mainstreaming into development and official development assistance

Some argue that development policy, including donor strategies and poverty reduction strategy papers, needs more explicit attention to climate change impacts and natural hazards on vulnerable peoples and areas (Agrawala and van Aalst 2008). Such calls for mainstreaming of climate change and climate change impacts into development may at first seem reasonable. They are methodologically problematic, however, and from the point of view of the poorest of the poor they may not be optimal. Methodologically the development discourse is rooted in economic and social theories that disregard three crucial links within sustainability, namely those between nature and society, between poor and rich populations, and between past, present and future societies (Jerneck and Olsson 2008). As we see it, the broad paradigm of sustainability relating to climate change can therefore not be mainstreamed into the more narrow development paradigm focusing on socio-economics issues. It can even be argued that mainstreaming contributes to postponing transitions to sustainability because it nurtures and maintains the development paradigm rather than replaces it with a sustainability approach.

Moreover, empirically there is ample evidence that development policies aimed at alleviating the burden of the poor often fail to reach the poorest of the poor because economic opportunities are more often grasped by the not so poor. This can be illustrated by the green revolution in India (Pande 2007) and by micro-finance. Micro-credit schemes have received much attention and are becoming increasingly popular among governments and non-governmental organizations as a means to alleviate poverty and promote socio-economic development (Klasen 2004). Empirical evidence indicates that micro-finance and micro-credit schemes do reach poor people, but in rural areas in particular there are severe difficulties in reaching the poorest fraction of the population (Rogaly 1996; Dunford 2000; Eversole 2000; Hulme 2000; Navajas *et al.* 2000; Rutherford 2000). This means that the mentioned poverty alleviation policies may unintentionally even increase the social stratification between the poorest and the not so poor.

The trickle-down theory is popular but also controversial in the development debate. There is ample evidence that the trickle-down effect did not work in practice as a dynamic force from the very top to the lowest levels of society (Martinussen 1997). If, however, the trickle-down idea is applied to a much lower level we may see that the poor often serve the poorer with income opportunities. As an illustration, food-secure farmers who are able to produce a surplus for the local market may hire food-insecure local farmers as day labourers (Jerneck and Olsson unpubl. data), whereas poor people in the urban informal economy may provide income opportunities for others who are poorer for example in food provision and transport (Jerneck 2009). These options are to be seen mainly as short-term practices for *getting by* in poverty rather than strategic options for *getting out* of poverty (Lister 2004); they may result in perpetuated poverty.

18.4.2 Adaptation through synergies with other mechanisms

Mitigation and adaptation represent two fundamental but profoundly different societal response options for reducing the risks associated with anthropogenic climate change. Due to their respective distinct actors and timescales they are complementary. While mitigation reduces both the root causes of anthropogenic climate change and its impacts on all climate-sensitive systems, adaptation does not. It is therefore not surprising that mitigation has received more attention than adaptation (Füssel 2007).

It is very important to note, however, that when we turn our focus to poor people in poor and middle-income countries and regions, not to mention the poorest of the poor, then the *potentials of mitigation* are decisively lower while the *need for adaptation* is decisively higher compared to (post)industrial regions. An important exception, though, is where synergies of mitigation and adaptation are sought. Even if mitigation is not a main priority for the poorest of the poor it is still important to seek synergies because successful mitigation may reduce the need for adaptation and vice versa. This is an argument for closely linking mitigation to adaptation in global climate policy. Moreover, since the funding arrangements for climate change are not designed to facilitate adaptation (Klein *et al.* 2003), it can be argued that instead of solely focusing on adaptation we should search for synergies between mitigation and adaptation the main reason for this being the international funding available on mitigation (Tschakert and Olsson 2005). If policies and measures for mitigation and adaptation were combined into joint programmes this could create opportunities for financing the needs for adaptation which prevail mainly in the South while simultaneously mitigating further climate change the benefit of which would accrue across the globe as a public good. This calls for attention to combined solutions.

In the context of the poorest of the poor, however, the implementation of climate policy differs between mitigation and adaptation in substance and mechanisms. The implementation and efficacy of synergetic measures could be limited by institutional complexity. Besides, the appeal of joint solutions may be lowered by the difficulties in evaluating their actual benefits especially if it cannot be excluded that half the money invested in adaptation and the other half in mitigation would reduce more damage than a joint effort (Klein *et al.* 2003). Concerning costs it is important to recognize that adaptation requires *additional* financial support. A renaming of official development assistance into official adaptation assistance does not help the poor out of vulnerability since the costs of adaptation have to come in addition. Until such thinking takes root there is need for additionality and it should even be 'a key requirement for recipient countries in any international adaptation regime' (Füssel 2007). A main supportive argument for this is that 'many measures undertaken to adapt to climate change have important ancillary benefits' (Füssel 2007).

From this we can conclude that we prefer combined mitigation and adaptation solutions but in case adaptation efforts are jeopardized by the costs of mitigation efforts we would underline the importance of adaptation. And importantly, if we think that there is a moral obligation for industrial and (post)industrial countries to assist people who are impacted the most by climate change then adaptation should be prioritized in case we cannot act constructively in combining mitigation and adaptation.

When food-insecure people are threatened not only by typical climate change impacts like drought and flooding but also by multiple stressors like land degradation, land fragmentation, market failure, social unrest and political conflict or disintegration then adaptation is not sufficient and therefore not successful. In such situations there is need for profound social change that is more encompassing than adaptation.

In the current climate policy only two mechanisms provide options for synergies of mitigation and adaptation and address the needs of developing countries explicitly, namely the Clean Development Mechanism and the forthcoming mechanism of Reducing Emissions from Deforestation and Forest Degradation in Developing Countries (REDD), often referred to as avoided deforestation. Both are discussed below.

18.4.3 The Clean Development Mechanism

In general the Clean Development Mechanism (CDM) has not reached the least developed countries (Silayan 2005), let alone the poorest of the poor (Olsen 2007; Sutter and Parreno 2007). The most important reason for this shortcoming is that the CDM is entirely market-based, and the poorest people are seldom integrated into a market. We propose two areas where the CDM could potentially play a much more active role in promoting sustainable development for the poorest of the poor. Both examples are based on the idea of targeting the type of land or areas that are often occupied by the poorest of the poor (on the CDM, see also Stripple and Lövbrand, this volume, Chapter 11).

The Kyoto Protocol mentions agricultural soils as a potential carbon sink, but so far it is not included as an eligible component in the mechanisms. Yet one mechanism in particular has the potential to specifically reach the poorest of the poor, namely soil carbon sequestration in agricultural soils. Soil carbon sequestration recognizes that land degradation, particularly in the tropics, is as urgent an environmental issue as climate change. The sequestration of carbon in degraded soils, if properly managed, has the potential to counter degradation and therefore increase productivity, resilience and sustainability of these agro-ecosystems, and in the longer term increase food security and vulnerability (Olsson and Ardö 2002). The poorest of the poor often occupy the most degraded soils and practice so-called low-input agriculture with no or very little external inputs in the form of energy, machinery and agrochemicals. These agro-ecosystems may have a higher potential for net carbon accumulation than do intensive forms of agriculture where the inputs already have a high carbon cost (Schlesinger 1999). Another good reason is that low-input agriculture is less damaging to biodiversity than intensive forms of agriculture, and when integrated with trees in agro-forestry systems such farming systems can even improve biodiversity (Sanchez 2000; Delali *et al.* 2004). This is how we see that the poorest of the poor, in terms of being farmers on degraded lands, have a comparative advantage for being targeted.

Municipal waste and food waste in particular represent one of the largest sources of anthropogenic methane emissions and are predicted to increase by over 40 per cent from 2005 to 2025 (Adhikari *et al.* 2006). The urban poorest of the poor are often engaged in some forms of waste management through reclaiming, recycling and reusing (Satterthwaite 2003). New forms of waste collection could have the potential to include the urban poorest of the poor in climate change mitigation with an important synergy with poverty alleviation.

Waste handling and disposal is eligible for funding under one of the 15 different scopes (No. 13) in the CDM (where 'scope' refers to an issue such as energy provision, transport or agriculture) and is the second largest scope with almost 20 per cent of the registered projects. However, none of the 298 projects listed in the CDM database addresses waste collection explicitly in a way that would be conducive to poverty alleviation for the poorest of the poor. Our screening of the registered projects shows that they almost exclusively deal with various engineering solutions on how to extract gases from landfills and animal manure. A possible synergy of adaptation and mitigation could be to include the whole chain of waste collection, transport, treatment and gas recovery and involve the poorest of the poor in the collection phase under organized, fair and healthy conditions. Yet this is not to say that the poorest of the poor should be engaged only in waste management or that this represents their best long-term occupation. For waste management in general there is also need for technological solutions.

18.4.4 Avoided deforestation

A new potential climate change mitigation mechanism based on preventing emissions from deforestation was first discussed in the eleventh conference of the parties to the climate convention in 2005 and subsequently adopted at the thirteenth conference in 2007. The basic idea is that industrialized countries pay developing countries to reduce deforestation within their boundaries. It is easy to agree with the fundamental principles and objectives of the reducing emissions from deforestation and forest degradation in developing countries (REDD), but in practice there are several important concerns (Griffiths 2007). One concern relates to the risk that the REDD policy may be co-opted by interests that would harm poor people. Another concern relates to the risk that only inhabitants in regions with a spectacular biodiversity would benefit. Given the fact that the poorest of the poor often occupy land that is degraded, this implies that the biodiversity of such areas is neither now nor in the near future a priority for REDD.

18.4.5 Stand-alone adaptation policies

Migration is a common response to natural and human-induced disasters (Hugo 1996). In certain cases there might in fact be no other option than to relocate people affected by severe climate impacts, either through individual effort or through government programmes. This is particularly important in cases where large areas are expected to become uninhabitable for long periods to come, such as low-lying coastal regions (for example Bangladesh) or river estuaries (for example Mekong, Ganges–Brahmaputra or the Nile). It should be noted though that international migration is very restricted and regulated (Jerneck and Olsson 2008) and in risky situations stand-alone adaptation in the form of a refugee status would be an important adaptation (on climate refugees, see in more detail Biermann and Boas, this volume, Chapter 16).

18.4.6 Conclusions

We have analysed how policies on adaptation to climate change, including synergies with mitigation, can be identified and formulated from the perspective of the poorest of the poor. The poorest of the poor is an analytical category of people found across the world and sharing the common denominator of being systematically, but not necessarily intentionally, marginalized – by nature, society and politics. In urban areas they live and work in polluted and unhealthy areas where they depend on degraded facilities and hazardous income opportunities. Here the poorest of the poor have no absolute advantages, only comparative advantages. These can be described as options so unattractive that nobody but the poorest of the poor will grasp them for income generation and survival. What is learned from such examples is that in the short term the poorest will benefit more than others in the form of income from those particular options because others do not take them. But many of these options identified by the poor are insecure and very risky owing to uncertainties, pollutants and unhealthy environments. This is the fallacy of the easy-to-grasp but not-so-good opportunity.

In regard to nature the poorest of the poor live in and depend on degraded ecosystems where they experience food insecurity. We introduce the concept of the food imperative to describe a condition faced by food-insecure people. The food imperative implies the constant superordinated goal of putting food on the table thus preventing longer-term planning and risk-taking. This is also the reason for their inability to grasp opportunities and avoid marginalization. If not targeted specifically, the poorest of the poor will not be reached by policies because of this marginalization. Opportunities offered in the social system are often better suited for the not so poor and this is how the poorest of the poor are often incapacitated and fall outside of the system boundaries.

Lessons from the development field also show that policies intended for poverty eradication, like the green revolution in India and many micro-finance schemes, may have an unintended consequence of marginalizing certain groups thereby strengthening the social stratification of the poor and contributing to the reproduction of 'the poorest of the poor'. Policy-making for adaptation must seek to avoid such marginalization. We see only two policy mechanisms that currently have the potential of targeting poor people in poor countries; CDM and the forthcoming REDD. We show, however, that none of these mechanisms is particularly likely to provide benefits for the poorest of the poor. This is related both to the very nature of the problem of reaching the poorest of the poor but also due to the political construction of the policies. This calls for further action.

Regarding new norms and institutions, we therefore need to rethink development from a sustainability perspective rather than mainstreaming climate change and adaptation into the narrower paradigm of development. Yet mainstreaming may be the only option for the short term, provided that the short term does not become the enemy of the long term through path dependence or lock-in situations.

Regarding policy-making at the national and local level, there is therefore a need to specifically target policies so that the poorest of the poor are able to reap the benefits in spite of their marginalization. At the global level, there is an urgent need to assess climate change policies from the perspective of the poorest of the poor in order to avoid damaging policies, such as the current policies on biofuels. This links the ethics of global policy to food security and our notion of the food imperative of the poorest of the poor.

References

Adhikari, B. K., S. Barrington and J. Martinez 2006. 'Predicted growth of world urban food waste and methane production', *Waste Management and Research* **24**: 421–433.

Agrawala, S. and M. van Aalst 2008. 'Adapting development cooperation to adapt to climate change', *Climate Policy* **8**: 183–193.

Angel, M. T. and D. P. Rock 2005. *Industrial Transformation in the Developing World*. Oxford, UK: Oxford University Press.

Barnett, T. P., J. C. Adam and D. P. Lettenmaier 2005. 'Potential impacts of a warming climate on water availability in snow-dominated regions', *Nature* **438**: 303–309.

Biermann, F. 2007. '"Earth system governance" as a crosscutting theme of global change research', *Global Environmental Change* **17**: 326–337.

Cameron, R. E. and L. Neal 1996. *A Concise Economic History of the World: From Paleolithic Times to the Present*. New York: Oxford University Press.

Commission on Legal Empowerment of the Poor 2008. *Making the Law Work for Everyone*, Vol. 1. New York: Commission on Legal Empowerment of the Poor.

Collier, P. 2007. *The Bottom Billion: Why the Poorest Countries Are Failing and What Can Be Done about It*. Oxford, UK: Oxford University Press.

de Soto, H. 2000. *The Mystery of Capital: Why Capitalism Triumphs in the West and Fails Everywhere Else*. New York: Basic Books.

Delali, B. K. D., E. T. F. Witkowski and C. M. Shackleton 2004. 'The fuelwood crisis in southern Africa: relating fuelwood use to livelihoods in a rural village', *GeoJournal* **60**: 123–133.

Dunford, C. 2000. 'The holy grail of microfinance: "helping the poor" and "sustainable"?', *Small Enterprise Development* **11**: 40–44.

Eversole, R. 2000. 'Beyond microcredit: the trickle-up program', *Small Enterprise Development* **11**: 45–58.

FAO 2008. *Hunger on the Rise: Soaring Prices add 75 Million People to Global Hunger Rolls*. Rome: FAO.

Füssel, H. M. 2007. 'Adaptation planning for climate change: concepts, assessment approaches and key lessons', *Sustainability Science* **2**: 265–275.

Global Commission on International Migration 2005. *Migration in an Interconnected World: New Directions for Action*. Geneva: Global Commission on International Migration.

Griffiths, T. 2007. *Seeing 'RED'? Avoided Deforestation and the Rights of Indigenous Peoples and Local Communities*. Moreton-in-Marsh, UK: Forest Peoples Programme.

Grothmann, T. and A. Patt 2005. 'Adaptive capacity and human cognition: the process of individual adaptation to climate change', *Global Environmental Change* **15**: 199–213.

Gurenko, E. N. 2006. 'Introduction and executive summary', *Climate Policy* **6**: 600–606.

Heller, P. and M. Mani 2002. 'Adapting to climate change', *Finance and Development* **39**: 29–31.

Hugo, G. 1996. 'Environmental concerns and international migration', *International Migration Review* **30**: 105–131.

Hulme, D. 2000. 'Is microdebt good for poor people? A note on the dark side of microfinance', *Small Enterprise Development* **11**: 26–28.

IFAD (International Fund for Agricultural Development) 2001. *The Rural Poverty Report*. Rome: International Fund for Agricultural Development.

IPCC (Intergovernmental Panel on Climate Change) 2007. *Climate Change 2007: Impacts, Adaptation and Vulnerability*. Cambridge, UK: Cambridge University Press.

Jackson, J. B. C. *et al.* 2001. 'Historical overfishing and the recent collapse of coastal ecosystems', *Science* **293**: 629–637.

Jerneck, A. 2010. 'Globalization, growth and gender: poor workers and vendors in urban Vietnam', in H. Rydstrøm (ed.), *Gendered Inequalities in Asia: Configuring, Contesting and Recognizing Women and Men*. Copenhagen: NIAS/Curzon Press, pp. 99–123.

Jerneck, A. and L. Olsson 2008. 'Adaptation and the poor: development, resilience, transition', *Climate Policy* **8**: 170–182.

Klasen, S. 2004. 'In search of the Holy Grail: How to achieve pro-poor growth?', paper presented at *Annual World Bank Conference on Development Economics*, 2003.

Klein, R. J. T., E. L. Schipper and S. Dessai 2003. *Integrating Mitigation and Adaptation into Climate and Development Policy: Three Research Questions*, Working Paper No. 40. Norwich, UK: Tyndall Centre for Climate Change Research.

Laska, S. 2008. 'What if Hurricane Ivan had not missed New Orleans?', *Sociological Inquiry* **78**: 174–178.

Linnerooth-Bayer, J. and R. Mechler 2006. 'Insurance for assisting adaptation to climate change in developing countries: a proposed strategy', *Climate Policy* **6**: 621–636.

Lister, R. 2004. *Poverty*. Cambridge, UK: Polity Press.

Lobell, D. B. *et al.* 2008. 'Prioritizing climate change adaptation needs for food security in 2030', *Science* **319**: 607–610.

Martinez-Alier, J. 2002. *The Environmentalism of the Poor: A Study of Ecological Conflicts and Valuation*. Cheltenham, UK: Edward Elgar.

Martinussen, J. 1997. *Society, State and Market: A Guide to Competing Theories of Development*. London: Zed Books.

Mikkelsen, B. 2005. *Methods for Development Work and Research: A New Guide for Practitioners*. London: Sage Publications.

Mitchell, D. 2008. 'A note on rising food prices', cited in: *Soaring Food Prices: Facts, Perspectives, Impacts and Actions Required*. Rome: FAO.

Möhner, A. and R. J. T. Klein 2007. *The Global Environment Facility: Funding for Adaptation or Adapting to Funds?* Climate and Energy Working Papers. Stockholm: Stockholm Environment Institute.

Morton, J. F. 2007. 'Climate change and food security special feature: the impact of climate change on smallholder and subsistence agriculture', *Proceedings of the National Academy of Sciences of the USA* **104**: 19 680–19 685.

Müller, B. 2006. *Adaptation Funding and the World Bank Invesment framework Initiative*. Mexico City: Gleneagles Dialogue Government Working Groups.

Müller, B. 2007. *Trust and the Future of Adaptation Funding*. Oxford, UK: Oxford University.

Myrdal, G. and P. Streeten 1958. *Values in Social Theory*. London: Routledge and Kegan Paul.

Navajas, S. *et al.* 2000. 'Microcredit and the poorest of the poor: theory and evidence from Bolivia', *World Development* **28**: 333–346.

Olsen, K. H. 2007. 'The clean development mechanism's contribution to sustainable development: a review of the literature', *Climatic Change* **84**: 59–73.

Olsson, L. and J. Ardö 2002. 'Soil carbon sequestration in degraded semiarid agro-ecosystems: perils and potentials', *Ambio* **31**: 471–477.

Paavola, J. and W. Adger 2006. 'Fair adaptation to climate change', *Ecological Economics* **56**: 594–609.

Pande, R. 2007. 'Gender, poverty and globalization in India', *Development* **50**: 134–140.

Pauly, D. 2005. 'Rebuilding fisheries will add to Asia's problems', *Nature* **433**: 457.

Pielke, R., G. Prins, S. Rayner and D. Sarewitz 2007. 'Climate change 2007: lifting the taboo on adaptation', *Nature* **445**: 597–598.

Pinstrup-Andersen, P. and A. Herforth 2008. 'Food security: achieving the potential', *Environment* **50**: 48–61.

Razavi, S. 1999. 'Gendered poverty and well-being: introduction', *Development and Change* **30**: 409–433.

Reed, B. (ed.) 2006. *Unnatural Disaster: The Nation on Hurricane Katrina*. New York: Nation Books.

RIVM 2008. *Lessons from Global Environmental Assessments*. Bilthoven, Netherlands: Environmental Assessment Agency.

Rogaly, B. 1996. 'Micro-finance evangelism, "destitute women", and the hard selling of a new anti-poverty formula', *Development in Practice* **6**: 100–112.

Rutherford, S. 2000. 'Raising the curtain on the "microfinancial services era"', *Small Enterprise Development* **11**: 13–25.

Sachs, J. D. 2005. *The End of Poverty: How We Can Make It Happen in Our Lifetime*. London: Penguin Books.

Sachs, J. D. 2008. *Common Wealth: Economics for a Crowded Planet*. London: Penguin Books.

Sanchez, P. A. 2000. 'Linking climate change research with food security and poverty reduction in the tropics', *Agriculture, Ecosystems and Environment* **82**: 371–383.

Satterthwaite, D. 2003. 'The links between poverty and the environment in urban areas of Africa, Asia and Latin America', *Annals of the American Academy of Political and Social Science* **590**: 73–92.

Schlesinger, W. H. 1999. 'Carbon sequestration in soils', *Science* **284**: 2095.

Schmidhuber, J. and F. N. Tubiello 2007. 'Climate change and food security special feature: global food security under climate change', *Proceedings of the National Academy of Sciences of the USA* **104**: 19 703–19 708.

Sen, A. 1999. *Development as Freedom*. Oxford, UK: Oxford University Press.

Sengupta, A., K. P. Kannan and G. Raveendran 2008. 'India's common people: who are they, how many are they and how do they live?', *Economic and Political Weekly* **43**: 49–64.

Silayan, A. H. 2005. *Equitable Distribution of CDM Projects among Developing Countries*. Hamburg: Institut für Wirtschaftsforschung.

Smit, B. and M. W. Skinner 2002. 'Adaptation options in agriculture to climate change: a typology', *Mitigation and Adaptation Strategies for Global Change* **7**: 85–114.

Smit, B. and J. Wandel 2006. 'Adaptation, adaptive capacity and vulnerability', *Global Environmental Change* **16**: 282–292.

Stern, N. 2006. *The Economics of Climate Change: The Stern Review*. Cambridge, UK: Cambridge University Press.

Sutter, C. and J. C. Parreno 2007. 'Does the current Clean Development Mechanism (CDM) deliver its sustainable development claim? An analysis of officially registered CDM projects', *Climatic Change* **84**: 75–90.

Sweetman, C. 2005. 'Editorial', *Gender and Development* **13**: 2–8.

Tompkins, E. L. and W. N. Adger 2004. 'Does adaptive management of natural resources enhance resilience to climate change?', *Ecology and Society* **9**: article 10.

Tschakert, P. and L. Olsson 2005. 'Post-2012 climate action in the broad framework of sustainable development policies: the role of the EU', *Climate Policy* **5**: 329–348.

UNEP 2007. *Global Environmental Outlook 4*. Nairobi: United Nations Environment Programme.

Wallerstein, I. 2007. 'The ecology and the economy: what is rational?', In A. Hornborg, R. J. McNeill and J. Martinez-Alier (eds.), *Rethinking Environmental History: World-System History and Global Environmental Change*. New York: Altamira Press, pp. 379–390.

World Bank 2000. *Attacking Poverty*, World Development Report. Washington, DC: World Bank.

World Bank 2007. *Agriculture for Development*, World Development Report. Washington, DC: World Bank.

19

Conclusions: options for effective climate governance beyond 2012

FRANK BIERMANN, PHILIPP PATTBERG AND FARIBORZ ZELLI

19.1 Introduction

The diversity of the contributions to this volume illustrates the complexity of the challenge. Climate change is a governance problem that needs to be analysed, and addressed, at multiple levels, in multiple sectors, and with a view to multiple actors. In searching for policy options that go beyond current negotiations, the contributions thus addressed issues as diverse as international carbon markets, overlaps between the climate convention and world trade law, the role of non-state actors in technological change, climate refugees, or the vulnerability of the poorest of the poor. The chapters approached these issues from a variety of methodological approaches, showing that the governance challenge of global climate change can be framed very differently.

In light of this complexity, this book did not seek to present a silver bullet for future climate governance. An all-inclusive and perfectly coherent account of policy options would be neither feasible nor would it be desirable given the diversity of interests, perspectives and issues. As Einstein reportedly advised, it is important to simplify a problem to the extent possible – but not more. Instead of applying a structural straitjacket, this book thus offers a broad array of policy options organized under the three research themes of architecture, agency and adaptation. We summarize these options now in Sections 19.2–19.4. Finally, in Section 19.5 we map the policy options according to their political dimensions and institutional settings, illustrating their differences but also the opportunities for joint negotiation and implementation.

19.2 Architecture

A core element of the quest for long-term stable and effective climate governance is the overall institutional architecture. We define the term 'global governance

Global Climate Governance Beyond 2012: Architecture, Agency and Adaptation, eds. F. Biermann, P. Pattberg and F. Zelli. Published by Cambridge University Press.

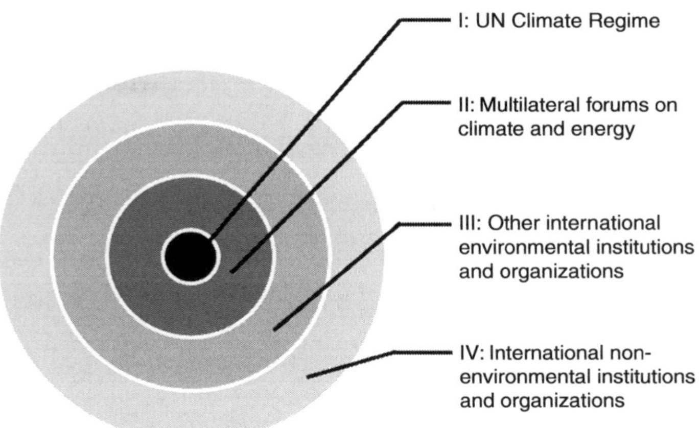

Figure 19.1 Spheres of institutional fragmentation in global climate governance. *I: UN climate regime*: includes for example the Ad Hoc Working Group on Long-Term Co-operative Action under the Convention and the Ad Hoc Working Group on Further Commitments for Annex I Parties under the Kyoto Protocol. *II: Multilateral forums on climate and energy*: includes for example the Asia–Pacific Partnership on Clean Development and Climate; the Methane to Markets Partnership; the Carbon Sequestration Leadership Forum; the International Carbon Action Partnership; the International Partnership for the Hydrogen Economy; and the Major Economies Process on Energy Security and Climate Change. *III: Other international environmental institutions and organizations*: includes for example the World Meteorological Organization; the Convention for the Protection of the Marine Environment of the North-East Atlantic; the Ramsar Convention on Wetlands of International Importance Especially as Waterfowl Habitat; the Convention on Biological Diversity; and the United Nations Convention to Combat Desertification; the World Trade Organization; the International Civil Aviation Organization; the International Maritin.

architecture' as the overarching system of public and private institutions – that is, principles, norms, regulations, decision-making procedures and organizations – that are valid or active in a given issue area of world politics. Architecture can thus be described as the *meta-level* of governance (Biermann *et al.*, this volume, Chapter 2).

In policy and academic debates, there is increasing concern for widespread fragmentation of global governance architectures. Global climate governance, in particular, is marked by a plethora of institutions that are not always effectively related to the overarching climate convention (see also for example Haas *et al.* 2004; Kanie 2008). Regarding intergovernmental institutions, there are four different spheres of fragmentation in international climate politics, which can be arranged concentrically from 'purely' climate-specific institutions towards regimes and organizations with universal or cross-cutting portfolios (see Figure 19.1 for an overview). If one considers in addition private and public–private initiatives, the global climate architecture appears even more fragmented (see Chapters 9–13, this volume, on agency beyond the state).

Fragmentation may have advantages (Zelli *et al.*, this volume, Chapter 3). One benefit of institutional fragmentation is that it may permit laggards to get to the negotiation table. For instance, the current internal fragmentation or duplication in the UN climate regime – with various parallel tracks for negotiating a future regime – allows for the direct involvement of countries that have not ratified the Kyoto Protocol to participate in discussions about a successor agreement. Notably, the United States participated in the Convention Dialogue in 2006 and 2007 and afterwards in the Ad Hoc Working Group on Long-Term Co-operative Action under the Convention. Similarly, a fragmented governance architecture may provide more venues for including non-state and sub-state actors. For instance, major businesses are involved in multilateral technology initiatives such as the International Partnership for the Hydrogen Economy. Another advantage of fragmentation is the potential for a meaningful division of labour among institutions. Instead of overburdening the UN climate regime, other institutions can take over certain functions. Fragmentation might also allow for deeper or faster agreements by circumventing deadlocks in larger forums. For instance, the 2007 meeting of the Group of Eight was the first multilateral arena where major developed country emitters made (soft) commitments to reduce greenhouse gas emissions by at least 50 per cent by 2050. This agreement also helped to reinvigorate debates in other institutions, by providing a major impetus on the road to the conference of the parties 2007 in Bali.

Yet there are also many, and possibly more severe costs involved with heavy fragmentation of governance architecture (Zelli *et al.*, this volume, Chapter 3). First, fragmentation of governance architectures gives room for many initiatives that serve only particular interests. The bulk of multilateral partnerships on climate and energy do not include least-developed countries or small island states. They hence largely focus on the interests of the participating industrialized or newly industrialized countries, while sidelining preferences of poorer countries. Notably, adaptation has marginal roles in the Asia–Pacific Partnership and in the first session of the United States-initiated Major Economies meeting. Moreover, fragmentation might increase coordination gaps among institutions. For instance, at present coordination on adaptation is poor between the climate convention and other institutions, for example the United Nations Food and Agriculture Organization or the desertification convention. Regulatory uncertainty is another severe downside of fragmentation, especially where clear price signals and investment security are important. For example, the variety of unlinked emission trading schemes yields a patchwork of different conditions for the generation and transfer of emission credits and permits (Flachsland *et al.*, this volume, Chapter 5). Scholars have also pointed to the imminent danger of 'chill effects' (Eckersley 2004). In light of the strong dispute settlement system under the World Trade Organization, parties might have been reluctant to include further trade-restrictive measures in the UN climate regime, let alone strengthening the regime's own dispute settlement system. Finally, institutional diversity implies the risk of 'forum shopping' (Raustiala and Victor 2004: 280). The Asia–Pacific Partnership for instance has provided a forum for the United States (and initially Australia) to circumvent the UN climate regime. In the same vein, the success of such initiatives might reduce compliance incentives for parties of the Kyoto Protocol (van Asselt 2007).

How can the environmental effectiveness of different scenarios of institutional fragmentation be quantified in both the short and the long term? This has been analysed in detail by Hof *et al.* (this volume, Chapter 4), building on earlier projections made with the FAIR meta-model for different levels of institutional cooperation among countries (Boeters *et al.* 2007) as well as on a review of other quantitative studies about the costs and environmental effectiveness of different universal and fragmented regimes. One of their chief conclusions is that it is more cost-effective to reduce emissions in a universal regime than in a fragmented regime. Even with high participation, fragmentation implies that emission are not reduced where it is cheapest, since emission trading is usually impossible between regions that participate in different agreements. However, despite the higher overall costs, a fragmented regime consisting of multiple agreements could be more feasible to attain, as it limits incentives for free riding.

In sum, in light of the findings from both qualitative and quantitative research, a strongly integrated climate architecture appears to be the most effective solution. However, in current climate governance as well as in many other areas of world politics, such integrated architectures are not always realistic. The second-best solution may thus be a well coordinated 'web of institutions' (IPCC 2007b: 791) that ensures an enhanced division of labour not only among climate-related institutions, but also with institutions from different issue areas, including the world trade regime.

Building on these overall findings, several contributions to this volume study specific institutional overlaps around the UN climate regime. One chapter analyses internal fragmentation within core climate institutions with regard to emissions trading and prospects for a global carbon market (Flachsland *et al.*, this volume, Chapter 5). Article 17 of the Kyoto Protocol as specified in the later Marrakech Accords establishes a top–down approach, that is, the implementation of emissions trading through multilateral negotiations. On the other hand, there are so-called bottom–up approaches associated with decentralized decision-making of individual nations or sub-national entities that implement emissions trading systems unilaterally, bilaterally or plurilaterally. Members to the International Carbon Action Partnership – including the EU Commission and several EU Member States, Australia, New

Zealand and some US states – emphasize the implementation and linking of such bottom–up schemes. This would imply a stepwise implementation of a global carbon market, compared to the instantaneous implementation of a Kyoto-type trading system.

Using the REMIND model, Flachsland *et al.* (this volume, Chapter 5) analysed the economic costs of delaying the implementation of a comprehensive global trading system. They found that when a global carbon market is implemented by 2020 instead of 2010, global mitigation costs would increase from 1.3 to 2.8 per cent of the global discounted gross domestic product. If the global carbon market is initiated later (that is, 2025 and after), the model predicts that it becomes impossible to limit global temperature increases to 2 °C.

While a global top–down trading approach under a universal architecture is the best solution to control global emissions but may not be realistic in the short term, the second-best option, similar to the conclusions by Hof *et al.* in Chapter 4, would again be a web of institutions. For emissions trading, such a web implies combining elements of different carbon market architectures. For instance, governments could agree on a system where a group of countries that want to adopt binding economy-wide caps continues the intergovernmental cap-and-trade system implemented by the Kyoto Protocol after 2012. By linking their domestic trading systems within this government-level framework, they can devolve trading to companies, which will enhance the efficiency of the international carbon market. This architecture could be designed as an open system that enables other countries to join later with some or all sectors of their economy. This approach could be environmentally and economically more effective than pure bottom–up approaches and less prone to political stalemates and high transaction costs than the top–down approach (Flachsland *et al.*, this volume, Chapter 5).

Another case study analysed fragmentation between the UN climate regime and a non-environmental institution, namely the world trade regime. There are various overlapping policies in both regimes (Biermann and Brohm 2005; van Asselt and Biermann 2007; Zelli 2007), including trade in emission allowances, unilateral policies and measures to level the playing field (for example border tax adjustments, subsidies and technical standards), as well as the transfer of climate-friendly goods, services and technologies. Zelli and van Asselt (this volume, Chapter 6) conducted a theory-guided policy analysis of these overlaps, along with a major international stakeholder workshop jointly organized with the Economics and Trade Branch of the UN Environment Programme in Geneva. One policy option that emerged is to better integrate scientific expertise, for example in the Committee on Trade and Environment of the World Trade Organization, the major forum where environment–trade overlaps are discussed. Another option to involve expertise is the introduction of science-based sustainability criteria for the removal of trade barriers for climate-friendly goods and services. A third policy recommendation is to broaden coordination across institutions to overcome negotiation deadlocks in this committee. Such a dialogue could cut across ministries instead of continuing separate ministerial gatherings. Moreover, at the governmental level, strategic issue linkages could lead to package deals. One option would be to link positions on farm subsidies, trade barriers for environmental goods and services and trade barriers for biofuels. Concessions on biofuels or environmental goods and services might help reinvigorate the larger debate on farm subsides.

The last two chapters in the first part of this volume addressed the global climate architecture as a whole, emphasizing the North–South dimension. Winkler (this volume, Chapter 7) explored options for long-term cooperation based on the principle of equity. He focused on two possible scenarios for a future architecture. First, he discussed a multi-stage package where countries progress from one level of participation and commitment to another. For this package, equity implies that transitions between stages are based on income levels, population size, historical responsibility and the potential to mitigate. As a second option, Winkler considered an 'ambitious transitional' package. This package has a stronger bottom–up character than the first one, but nonetheless requires more urgent action by all parties, especially in terms of quantifiable commitments. Such architecture not only implies stricter mitigation targets for industrialized countries, but also incentives for enhanced mitigation activities by developing countries. Here, the equity principle requires a differentiated approach, allowing developing countries to take quantifiable actions based on their respective national circumstances. Winkler concludes that both packages should not be seen as alternatives, but as different stages in the evolution of the climate regime over the next years and decades.

A similar perspective is taken by Shrivastava and Goel (this volume, Chapter 8). They emphasize the relevance of technological capability and financial support for developing countries and the need for support from industrialized countries. They suggest a two-tier architecture of global climate governance with two distinct but integrated components: a set of institutions, policies and programmes at the national level to identify the direction of technological development within the country; and a network of global institutions, financial mechanisms and technological programmes to support the institutions, policies and programmes in developing countries. In their view, these institutional arrangements at the global level would give a strong signal to developing countries and may alleviate their concerns in taking a more active part in global efforts to address climate change.

19.3 Agency

A number of scholars have voiced concerns about the problem-solving capacity of the state and the international state system. Increasingly, scholars and practitioners alike acknowledge that solutions to the challenges of global change do not exclusively originate from governments and international organizations but are co-produced by a host of actors beyond the state, whose authority is contested and whose legitimacy is questionable. On this account, climate governance is no longer the domain of states and intergovernmental cooperation alone. Instead, scholars observe a growing relevance of non-state actors, such as industry and environmentalist groups, as well as public actors other than central governments, such as cities, local communities or international bureaucracies (Benecke *et al.* 2008; Kolk *et al.* 2008; Kern and Bulkeley 2009; Okereke *et al.* 2009). Increasingly, such actors assume a role in rule-setting institutions that regulate certain sectors, or in market-based mechanisms, such as emissions trading. This emergence of 'transnational' and often 'privatized' climate governance required, first, a detailed conceptualization of this new phenomenon (Stripple and Pattberg, this volume, Chapter 9), which drew on political science and international relations studies of the public/private divide and different spheres of authority (for example Börzel and Risse 2005).

The starting point has been that an 'increasingly pertinent feature of the global public order in and beyond environmental protection and sustainability is the dynamic mixing of the public and the private, with state-based public power being exercised by state institutions alongside and along with the exercise of private power by market and civil society institutions and other actors committed to the public interest and public weal' (Thynne 2008: 329). Especially in climate governance, numerous actors form institutions to address the problem of climate change without being forced, persuaded or funded by states and other public agencies. This transnational institutionalization of climate governance is in line with what Ruggie (2004) has called the reconstitution of a global public domain. As a domain, it does not replace states but 'embed[s] systems of governance in broader global frameworks of social capacity and agency that did not previously exist' (Ruggie 2004: 519). The original claim about 'agency beyond the state' concerns the role and relevance of different actors. The power of individual and collective actors to change the course of events lies increasingly in sites beyond the state and its international organizations (Stripple and Pattberg, this volume, Chapter 9). Based on this conceptualization of the emergent transnational climate governance arena and agency beyond the state in climate governance, Stripple and Pattberg developed a typology that distinguishes different climate governance approaches. These range from governance through markets – including the Clean Development Mechanism and voluntary offsets – to networked governance, which includes public non-state actors such as cities along with transnational corporations and non-governmental organizations.

Subsequently, more detailed research focused on particular elements of the emergent transnational climate governance. First, Pattberg (this volume,

Chapter 10) analysed *networked forms of global climate governance*. Public–private partnerships – that is, networks of different societal actors, including governments, international agencies, corporations, research institutions and civil society organizations – are cornerstones of current global environmental governance, both in discursive and material terms. Within the United Nations, partnerships have been endorsed through the establishment of the Global Compact, a voluntary partnership between corporations and the United Nations, as well as through the 'partnerships for sustainable development' (also known as 'type-2' outcomes) concluded at the 2002 World Summit on Sustainable Development in Johannesburg. Both the 'partnerships for sustainable development' and the Global Compact have been criticized for privatizing parts of the policy response to global change (Biermann *et al.* 2007; Rieth *et al.* 2007). Pattberg (this volume, Chapter 10) analysed public–private partnerships in global climate governance based on three criteria: problem-solving capacity; participation and inclusiveness; and synergies or dysfunctional linkages with international climate governance.

As for problem-solving capacity, several obstacles prevent the realization of the full potential of partnerships. In particular, the geographical bias towards global partnerships indicates that partnerships reflect pre-existing interest structures and therefore seldom deliver benefits that may not have been realized in more traditional multilateral or bilateral implementation arrangements. Regarding increased participation through public–private partnerships, the analysis highlights the overrepresentation of governments in climate partnerships as compared to the total sample of all partnerships for sustainable development registered with the United Nations. Climate partnerships are also largely dominated by states, in terms of both leadership and membership. This finding is in line with the expectation that politically contested areas such as climate politics remain overall under the control of governments. Finally, it appears that a stronger link with the UN climate regime may benefit both the 'partnerships for sustainable development' – by giving them guidance and a clear goal – and the climate regime, by assisting its implementation.

Second, Stripple and Lövbrand (this volume, Chapter 11) analysed the processes that drive the current transformation of current carbon markets. Instead of asking who governs carbon markets, they rather explore by which procedures carbon markets are rendered thinkable and operational in the first place. To this end, Stripple and Lövbrand analysed baseline-and-credit markets in particular, where a complex measurement of counterfactuals (current emissions vis-à-vis a business-as-usual scenario) enables reductions of carbon dioxide-equivalents to be assigned market value and transformed into various offset currencies. Through the detailed analysis of the global supply chain of two concrete carbon offset contracts, they scrutinized the role of a wide range of actors beyond the state, including investors, developers, managers, auditors, brokers, retailers and buyers as well as individuals.

The analysis suggests that these actors perform a range of governance functions, including enhancing the credibility of offsets, providing information, enabling aggregation, facilitating transactions, influencing regulation and adjudicating conflicts. While their empirical observations signify a shift from hierarchical forms of government to more decentralized forms of regulation, Stripple and Lövbrand (this volume, Chapter 11) did not interpret carbon market governance as a retreat of politics or of the state in favour of the market. Instead, they understood agency beyond the state as a distinct form of political organization that governs social behaviour 'at a distance'.

Third, den Elzen *et al.* (this volume, Chapter 12) modelled mitigation efforts from economic sectors in industrialized and developing countries. They drew on the 'Triptych approach', a method for allocating future greenhouse gas emission reductions among countries under an international mitigation regime that may follow the Kyoto Protocol and be based on technological criteria at sector level. Targets are defined for industry (manufacturing and construction), domestic (including carbon dioxide emissions from the residential, commercial, agriculture and inland transport sectors), power production, fossil fuel production, non-carbon dioxide emissions in agriculture, and waste. Defining targets for separate sectors allows linking real-world emission reduction strategies and makes it possible to take diverse national circumstances of countries better into account. The major advantage of this sectoral approach is that it puts internationally competitive industries on the same level playing field. However, one of the major challenges is establishing reliable, uniform sectoral emissions registrations for all countries, as currently reliable sectoral emissions data for many (especially developing) countries are lacking.

Finally, Alfsen, Eskeland and Linnerud (this volume, Chapter 13) have analysed the role of non-state actors with regard to research and development and technological change. While standard economic theory recommends that governments set a price on emissions, they argued that market imperfections and dynamic inconsistencies may require that in addition governments support far-reaching technological change by means of publicly funded research, development and demonstration. In fact, public funding of research and development and carbon pricing policies are, at least in theory, mutually supportive and should not be seen as alternatives. An international agreement on research and development funding and cap-and-trade systems are mutually supportive precisely because research and development reduces future abatement costs and thus makes it feasible for politicians to agree on tighter caps. With regard to the policy instruments used, Alfsen, Eskeland and Linnerud contend that in the near future, a mix of different policies will coexist, including standards and labelling, instruments that reward not effort but results (for example prizes for a given solution), public procurement as well as research contracting that involves research institutes and industry.

19.4 Adaptation

It becomes increasingly clear that despite all mitigation efforts, some degree of global warming cannot be prevented, and impacts from climate change will become a reality of the twenty-first century. This poses the question of optimal adaptation governance. While a number of research programmes have addressed adaptation governance at local and national levels (including in the ADAM Project: see Hinkel *et al.* 2010; Mechler *et al.* 2010), the chapters gathered here ventured into a largely unexplored research terrain: *global* adaptation governance. How can we build over the course of the next decades systems of global governance that will cope with the global impacts of climate change that require adaptation? What institutions are in need of redesign and strengthening? To what extent, and in what areas, do we need to create new institutions and governance mechanisms from scratch?

As Biermann and Boas (this volume, Chapter 14) illustrate, global adaptation governance will affect most areas of world politics, including many core institutions and organizations. The need to adapt to climate change will influence for example the structure of global food regimes, global health governance, global trade flows and the world economic system as well as many other sectors from tourism to transportation or even international security.

Yet how can the damages of climate change, as well as the possible costs of adaptation, be assessed and, if possible, quantified? Hof *et al.* (this volume, Chapter 15) report on the most recent quantitative research on adaptation costs that underscores the urgency for international action. They combined the FAIR meta-model and the AD-RICE model (de Bruin *et al.* 2009) to analyse the mitigation costs, adaptation costs and residual damages of climate change on a global as well as regional scale. For a 'contraction and convergence' emission allocation regime (with per capita emissions converging in 2050, a climate sensitivity of 3.0 °C and the United Kingdom Green Book discounting method), the projected global adaptation costs are of the same order of magnitude as the recent adaptation cost estimates of the World Bank (2006) and the secretariat of the climate convention (UNFCCC 2007). They show that although the share of adaptation costs in the total climate change costs is relatively small, adaptation plays a major role by reducing potential damages. The extra costs if no adaptation measures are taken (defined as the increase in residual damages minus the decrease of adaptation costs) are projected to amount to USD 30 billion globally in 2010 and increase sharply to USD 3.4 trillion in 2100. Investment in adaptation is therefore very effective: residual damages are on average reduced by about five dollars for every dollar invested in adaptation. Furthermore, adaptation and mitigation cannot be regarded as substitutes, but rather complement each other. Adaptation can effectively reduce climate change damages in the shorter run, but is much less effective in the end since it does not reduce climate change itself. Mitigation is very effective in reducing climate change damages in the long run. Implementing both adaptation and mitigation gives the best results according to the FAIR meta-model.

Building on these insights, this project analysed three challenges for future global adaptation governance: climate-change-induced migration (Biermann and Boas, this volume, Chapter 16); climate-change-induced food insecurity (Massey 2008); and the need for coordinated adaptation funding (Klein and Persson 2008). In addition, two specific analyses focused on the perspectives of developing countries as a group of nations (Ayers, Alam and Huq, this volume, Chapter 17) and the interests of the poorest of the poor (Jerneck and Olsson, this volume, Chapter 18).

As for migration, it is likely that climate change will fundamentally affect the lives of millions of people who may be forced over the next decades to leave their villages and cities to seek refuge in other areas. Biermann and Boas (this volume, Chapter 16) defined these people as 'climate refugees': as people who have to leave their habitats, immediately or in the near future, because of sudden or gradual alterations in their natural environment related to at least one of three impacts of climate change: sea-level rise, extreme weather events, and drought and water scarcity. The exact numbers of such future climate refugees are unknown and vary from assessment to assessment depending on underlying methods, scenarios, time-frames and assumptions, and Biermann and Boas (this volume, Chapter 16) concur that estimation methods and assumptions are complex and controversial. Yet despite these remaining uncertainties, a meta-analysis of all available studies indicated that the climate-change-induced refugee crisis is most likely to surpass all known refugee crises in terms of the number of people affected (Biermann and Boas, this volume, Chapter 16).

Yet the current refugee protection regime of the United Nations is poorly prepared, and does not cover climate refugees in its mandate. At a meeting in the Maldives in 2006, delegates proposed therefore an amendment to the 1951 Geneva Convention Relating to the Status of Refugees that would extend the mandate of the UN refugee regime to cover also climate refugees. But such an amendment, as argued by Biermann and Boas (this volume, Chapter 16), leads into the wrong direction. They argue therefore for a separate regime: a legally binding agreement on the recognition, protection and resettlement of climate refugees under the climate convention. This could be a separate protocol under the convention ('climate refugee protocol'), but also integral part of a larger legal instrument, such as a protocol on adaptation, or even a single undertaking that regulates all future measures on climate governance (Biermann and Boas, this volume, Chapter 16). Importantly, the protection of climate refugees must be seen as a global problem and a global responsibility. In most cases, climate refugees will be poor, and their own responsibility for the past accumulation of greenhouse gases will be small. By a large measure, the rich industrialized countries have caused most emissions in past and present, and it is thus these countries that have most moral, if not legal, responsibility for the victims of global warming. Industrialized countries should hence do their share in financing, supporting and facilitating the protection and the voluntary resettlement of climate refugees.

A second case study focused on a related challenge – food security (Massey 2008). A changing climate will significantly affect many communities that are faced today with hunger and malnutrition. Key impacts on agriculture are a depletion of groundwater, reduced precipitation and changes – primarily a shortening – of the growing season, all of which may reduce yields. For example, the IPCC Fourth Assessment Report suggests that a 2–3 °C range of warming by 2020 could decrease agricultural yields in Africa by as much as 50 per cent (IPCC 2007a: 447–448). Therefore, some form of adaptation must occur to ensure greater food security in the most vulnerable regions. Our research indicates that there needs to be a mechanism that allows for adaptation at the local level to help farmers and communities and at the same time ensures that there is a well-functioning institutional system at the global level that supports the financing and implementation of adaptive measures, including improved farming techniques and technologies.

One potential means of adaptation to meet this challenge could be improved access of farmers in developing countries to state-of-the-art research on farming technologies. So far, developing countries are at a competitive disadvantage as a result of funding for agricultural research in general, including the protection offered to more adaptive crop seeds due to international intellectual property rights. Developed countries as well as the private sector may thus have a special role in aiding the farming sector in developing countries to adapt. This support could come in the form of an adaptation levy to fund agricultural research in developing countries as well as a renegotiation of international intellectual property rights in the domain of agriculture. The overall institutional context could be strengthened through a legally binding agreement on adaptation and food security under the climate convention (Massey 2008). This could be a single agreement – such as a protocol to the climate convention – but also be integrated (possibly with the agreement on climate refugees outlined above) into a larger legal instrument, such as an adaptation protocol to the climate convention. In addition, as discussed earlier under the 'architecture' domain, discussions on farm subsidies and transfer of technologies could be coupled with adaptation-related concerns, for example through sustainability criteria for trade barrier removals.

Adaptation is clearly a key priority for most developing countries, many of which have contributed only marginally to the build-up of greenhouse gases in the atmosphere but which will be especially affected by climatic change. Ayers, Alam and Huq (this volume, Chapter 17) thus examined the current discourses and negotiations on adaptation to climate change from the perspective of developing countries. Their analysis also took into account debates on a major workshop on Southern perspectives that the ADAM Project organized in 2008 in New Delhi, India. Ayers, Alam and Huq concluded that although significant progress has been made on empowering the adaptation agenda within the climate governance architecture, this resulted in a framing of adaptation that is inappropriate for addressing the many developing country concerns. First, they argue, adaptation remains under existing frameworks an undervalued policy option relative to mitigation. Second, they see the type of adaptation favoured by the climate convention as not conducive to building the broader resilience that is necessary to reduce the vulnerability of developing countries. Third, they view the adaptation discourse under the climate convention as largely technical and not open to alternative types of expertise that are locally generated and non-technical. In sum, Alam, Ayers and Huq suggest that it is both necessary and possible to refine the adaptation agenda under the climate convention. According to them, more deliberative policy-making processes must be created for adaptation that are better able to engage with vulnerable communities and citizens to create bottom–up, locally meaningful adaptation strategies. This would require a reframing of the adaptation discourse that is more open to non-technical expertise generated from indigenous and locally based knowledge.

In addition to a comprehensive analysis of the perspectives of the developing countries, this research programme also explored the special situation of the poorest people in these countries (Jerneck and Olsson, this volume, Chapter 18). In the context of the poorest of the poor, mitigation is not a priority because their contribution to the global emission of greenhouse gases is miniuscule and their capacity to reduce emissions is low. This makes adaptation their main priority. Today, there are 923 million hungry people worldwide, who are in general also extremely vulnerable to climate change impacts. The already large number of poor people is expected to increase further and remain large for a long time while people exposed to climate change are expected to become even more vulnerable due to increasing incidence of extreme climate events. In relation to the poorest of the poor, adaptation to climate change should thus be seen as a process of profound social change away from livelihoods threatened at their roots by climate change.

Several policy options were considered to increase the adaptive capacities of the poorest of the poor. These include mainstreaming climate change into development assistance; identifying synergies with other mechanisms, such as climate change mitigation, biodiversity or desertification; as well as a number of stand-alone adaptation policies, such as special support for climate refugees (Jerneck and Olsson, this volume, Chapter 18). Regarding new norms and institutions, the study argued for rethinking development from a sustainability perspective rather than mainstreaming climate change and adaptation into the narrower paradigm of development, even though mainstreaming may be the only option for the medium term.

The integrated assessment modelling of adaptation costs and our studies on climate refugees, food insecurity, the perspectives of developing countries and the needs of the poorest among the poor signal the need for an enhanced and targeted set of funding mechanisms for adaptation. It is thus not only important to better endow existing funds and to add new funds, but to coordinate the various financial mechanisms in order to reach a meaningful division of labour. We therefore also studied adaptation funding, including a participatory appraisal exercise with stakeholders and experts from developing and developed countries in Brussels (Klein and Persson 2008).

19.5 Synthesis

This chapter has summarized a three-year research effort on policy options for stable, long-term climate governance, carried out by seven research institutions in Europe and India. The research reported in this volume focused on three areas of rapid political development as well as increasing concern: the research problem of increasing fragmentation of the overall architecture of global climate governance; the research problem of increasing privatization and marketization of global climate governance; and the research problem of developing new mechanisms for global adaptation governance. All themes are interlinked. For instance, most options discussed under agency and adaptation include elements of a future climate architecture, for example reform of the Clean Development Mechanism, or protocols on climate refugees and food security. Options discussed under the 'architecture' theme involve non-state actors (for example the linking of emissions trading schemes) or may be relevant for adaptation to climate change (for example technology transfer).

This concluding section highlights connections between the various policy options. To this end, Table 19.1 restructures the options in terms of the international institutional environment where they could be pursued: under the UN climate regime, in other international organizations and forums, or in cross-institutional collaboration. Moreover, the table distinguishes options depending on their political and legal dimension: either they suggest new political 'hardware', that is, new norms, treaties or institutions, or they propose specific policies, measures or standards. These two dimensions take into account two crucial aspects to be considered when feeding recommendations into the negotiation process: *where?* (institutional setting) and *what*? (nature of proposal, level of ambition). These criteria are more suitable to structure policy-relevant findings, while the three themes have helped structuring and guiding research.

Table 19.1 *Overview of options for global climate governance beyond 2012*

	UN climate regime	Other international institutions and forums	Cross-institutional collaboration between UN climate regime and others
Norms and institutions	Ambitious comprehensive successor agreement to the Kyoto Protocol with differentiated commitments of countries Strengthened international institutions on adaptation, including: • a legally binding agreement on the recognition, protection and resettlement of climate refugees under the climate convention • an agreement on adaptation and food security under the climate convention • a climate refugee protection and resettlement fund	Cross-ministerial dialogue among environment, trade and development ministries Opening World Trade Organization Committee on Trade and Environment for regular scientific inputs on climate–trade overlaps Public funds to stimulate private research and development Multilateral agreements on research and development of climate-friendly technologies	Open EU emissions trading scheme and link emissions trading schemes bottom–up and top–down International body of experts on technological needs and adaptive capacities Network of technology research and development institutes Fund for technology research, development and diffusion
Policies, measures and standards	Differentiation among Clean Development Mechanism target countries, project types and technologies Sectoral Clean Development Mechanism pilot phase with discounted sectoral credits Sectoral mitigation targets Science-based sustainability standards for Clean Development Mechanism projects	Science-based sustainability criteria for removal of trade barriers for climate-friendly goods and services Issue-linking and package deals on related discussions in the World Trade Organization Doha Round (for example farm subsidies, transfer of environmental goods and services, biofuels) Deliberative adaptation policy-making processes	Focused national, regional and local policies targeting the poorest of the poor – incentivized by international framework

Based on these two dimensions, Table 19.1 highlights the commonalities among policy options that have been analysed under the three research themes. The columns show to what extent some options can be pursued in the same institutional arena and might hence be linked in a comprehensive negotiation approach (for example protocols on climate refugees and food security). Most suggestions fall under the UN umbrella or in the middle column that at least involves the UN regime. This is in line with our general finding that in spite of some benefits of institutional fragmentation, it is pivotal to strengthen the UN regime as the chief institution to address global climate change.

All policies, measures and standards listed in Table 19.1 relate to different institutional settings (inside and outside the UN system), with some sharing features such as sustainability criteria based on scientific advice for both CDM and trade barrier removals. There is an obvious linking potential here, since a scientific body as the IPCC could for instance provide broad expertise to develop criteria across different topics. The distinction between institutional and policy-based options also points to the variant political feasibility of options. Other things being equal, one can expect that agreement on new policies is easier to achieve than on new institutional instruments, for example, an open emissions trading scheme or a food security protocol.

One could also combine the dimensions according to technical or material commonalities, in the attempt to advance options in parallel in negotiations. Consider, for example, issues of funding (climate refugee fund, public research and development funds); scientific advice (for sustainability criteria for CDM and technology transfer and for the World Trade Organization Committee on Trade and Environment); trade (linkage of emissions trading schemes, issue-linking in the Doha Round on world trade); technology (research and development funding, CDM reform proposals, technology transfer); and sectoral approaches (sectoral CDM, sectoral mitigation targets, sector-based emissions trading schemes as part of an open trading system).

In the final analysis, and in light of the complexity of climate negotiations and the multitude of actors involved, it will be important, however, not to 'over-integrate' options before communicating them in the policy process. 'Optimal' yet highly complex and demanding combinations might overburden negotiations. The potential for concrete combinations of options in the governance process will depend on political bargaining as well as on ad hoc opportunities of daily politics. Future climate policy does not only need well-designed strategies for long-term effective, equitable and efficient governance architectures, but also a high degree of flexibility in actual operationalization and implementation. For better or worse, climate governance, as most areas of policy-making, will always combine long-term vision with short-term incrementalism.

References

Asselt, H. van 2007. 'From UN-ity to diversity? The UNFCCC, the Asia-Pacific Partnership and the future of international law on climate change', *Carbon and Climate Law Review* **1**: 17–28.

Asselt, H. van and F. Biermann 2007. 'European emissions trading and the international competitiveness of energy-intensive industries: a legal and political evaluation of possible supporting measures', *Energy Policy* **35**: 497–506.

Benecke, G., L. Friberg, M. Lederer and M. Schröder 2008. *From Public–Private Partnership to Market: The Clean Development Mechanism (CDM) as a New Form of Governance in Climate Protection*, SFB Governance Working Paper Series 10. Berlin: Sonderfor\schungsbereich (SFB).

Biermann, F. and R. Brohm 2005. 'Implementing the Kyoto Protocol without the United States: the strategic role of energy tax adjustments at the border', *Climate Policy* **4**: 289–302.

Biermann, F., P. Pattberg, S. Chan and A. Mert 2007. *Partnerships for Sustainable Development: An Appraisal Framework*, Global Governance Working Paper No. 31. Amsterdam: The Global Governance Project.

Boeters, S., M. G. J. den Elzen, T. Manders, P. Veenendaal and G. Verweij 2007. *Post-2012 Climate Change Scenarios*, MNP Report No. 51001004007. Bithoven, The Netherlands: Environmental Assessment Agency (MNP).

Börzel, T. A. and T. Risse 2005. 'Public–private partnerships: effective and legitimate tools of international governance', in E. Grande and L. W. Pauly (eds.), *Reconstructing Political Authority: Complex Sovereignty and the Foundations of Global Governance*. Toronto: University of Toronto Press, pp. 195–216.

de Bruin, K. C., R. B. Dellink and S. Agrawala 2009. *Economic Aspects of Adaptation to Climate Change: Integrated Assessment Modelling of Adaptation Costs and Benefits*. Paris: OECD.

Eckersley, R. 2004. 'The big chill: the WTO and multilateral environmental agreements', *Global Environmental Politics* **4**: 24–40.

Haas, P. M, N. Kanie and C. N. Murphy 2004. 'Conclusion: institutional design and institutional reform for sustainable development', in N. Kanie and P. M. Haas (eds.), *Emerging Forces in Environmental Governance*. Tokyo: United Nations University Press, pp. 263–281.

Hinkel, J., S. Bisaro, T. E. Downing, M. E. Hofmann, K. Lonsdale, D. Mcevoy and J. D. Tàbara 2010 (in press). 'Learning to adapt: narratives of decision makers adapting to climate change', in M. Hulme and H. Neufeldt (eds.), *Making Climate Change Work for Us: European Perspectives on Adaptation and Mitigation Strategies*. Cambridge, UK: Cambridge University Press.

IPCC 2007a. *Climate Change 2007: Impacts, Adaptation and Vulnerability. Contribution of Working Group II to the Fourth Assessment Report of the Intergovernmental Panel on Climate Change*. Cambridge, UK: Cambridge University Press.

IPCC 2007b. *Climate Change 2007: Mitigation of Climate Change. Contribution of Working Group III to the Fourth Assessment Report of the Intergovernmental Panel on Climate Change*. Cambridge, UK: Cambridge University Press.

Kanie, N. 2008. 'Towards diffused climate change governance: a possible path to proceed after 2012', in V. I. Grover (ed.), *Global Warming and Climate Change: Ten Years after Kyoto and Still Counting*. Enfield, NH: Science Publishers, pp. 977–992.

Kern, K. and H. Bulkeley 2009. 'Cities, Europeanization and multi-level governance: governing climate change through transnational municipal networks', *Journal of Common Market Studies* **47**: 309–332.

Klein, R. J. T. and A. Persson 2008. *Financing Adaptation to Climate Change: Issues and Priorities*, ECP Report No. 8. Brussels: Centre for European Policy Studies.
Kolk, A., D. Levy and J. Pinske 2008. 'Corporate responses in an emerging climate regime: the institutionalization and commensuration of carbon disclosure', *European Accounting Review* **17**: 719–745.
Massey, E. 2008. 'Global governance and adaptation to climate change for food security', in F. Zelli (ed.), *Integrated Analysis of Different Possible Portfolios of Policy Options for a Post-2012 Architecture*, ADAM Project report No. D-P3a.2b. Norwich, UK: Tyndall Centre for Climate Change Research, pp. 143–153.
Mechler, R., A. Aaheim, S. Hochrainer, Z. Kundzewicz, J. Linnerooth-Bayer and N. Lugeri 2010. 'A risk management approach to assessing adaptation to extreme weather events in Europe', in M. Hulme and H. Neufeldt (eds.), *Making Climate Change Work for Us: European Perspectives on Adaptation and Mitigation Strategies*. Cambridge, UK: Cambridge University Press, pp.
Okereke, C., H. Bulkeley and H. Schroeder 2009. 'Conceptualizing climate governance beyond the international regime', *Global Environmental Politics* **9**: 58–78.
Raustiala, K. and D. G. Victor 2004. 'The regime complex for plant genetic resources', *International Organization* **58**: 277–309.
Rieth, L., M. Zimmer, R. Hamann and J. Hanks 2007. 'The UN Global Compact in sub-Sahara Africa: decentralization and effectiveness', *Journal of Corporate Citizenship* **7**: 99–112.
Ruggie, J. G. 2004. 'Reconstituting the global public domain: issues, actors, practices', *European Journal of International Relations* **10**: 499–541.
Thynne, I. 2008. 'Climate change, governance and environmental services: institutional perspectives, issues and challenges', *Public Administration and Development* **28**: 327–339.
UNFCCC 2007. *Climate Change: Impacts, Vulnerabilities and Adaptation in Developing Countries*. Bonn: Climate Change Secretariat. http://unfccc.int/files/essential_background/background_publications_htmlpdf/application/txt/pub_07_impacts.pdf.
World Bank 2006. *Clean Energy and Development: Towards an Investment Framework*. Washington, DC: World Bank.
Zelli, F. 2007. 'The World Trade Organization: free trade and its environmental impacts', in K. V. Thai, D. Rahm and J. D. Coggburn (eds.), *Handbook of Globalization and the Environment*. London: Taylor and Francis, pp. 177–216.

Index

Printed in Great Britain
by Amazon.co.uk, Ltd.,
Marston Gate.